Advanced Dairy Science and Technology

Edited by
Trevor J. Britz
University of Stellenbosch
South Africa

Richard K. Robinson
Consultant in Food Science and Technology
Reading
UK

Blackwell Publishing editorial offices:
Blackwell Publishing Ltd, 9600 Garsington Road, Oxford OX4 2DQ, UK
Tel: +44 (0)1865 776868
Blackwell Publishing Professional, 2121 State Avenue, Ames, Iowa 50014-8300, USA
Tel: +1 515 292 0140
Blackwell Publishing Asia Pty Ltd, 550 Swanston Street, Carlton, Victoria 3053, Australia
Tel: +61 (0)3 8359 1011

First published 2008 by Blackwell Publishing Ltd

2 2008

Library of Congress Cataloging-in-Publication Data

Advanced dairy science and technology / edited by Trevor J. Britz and Richard K. Robinson. – 1. ed.
p. cm.
Includes bibliographical references and index.
ISBN: 978-1-4051-3618-1 (hardback : alk. paper)
1. Dairy processing. 2. Dairying. I. Britz, T. J. II. Robinson, R. K. (Richard Kenneth)

SF250.5.A38 2008
637–dc22

2007022840

A catalogue record for this title is available from the British Library

Set in Times New Roman 10/13 pt
by Newgen Imaging Systems (P) Ltd, Chennai, India
Printed and bound in Singapore
by Utopia Press Pte Ltd

The publisher's policy is to use permanent paper from mills that operate a sustainable forestry policy, and which has been manufactured from pulp processed using acid-free and elementary chlorine-free practices. Furthermore, the publisher ensures that the text paper and cover board used have met acceptable environmental accreditation standards.

For further information on Blackwell Publishing, visit our website:
www.blackwellpublishing.com

Contents

Contributors

Dr Lucia E. C. M. Anelich, Department of Biotechnology and Food Technology, Tswane University of Technology, Private Bag X680, Pretoria 0001, South Africa.

Dr Apostolos S. Angelidis, Laboratory of Milk Hygiene and Technology, School of Veterinary Medicine, Aristotle University of Thessaloniki, 541 24 Thessaloniki, Greece.

Dr Thomas Bintsis, 25 Kapadokias, 55134 Thessaloniki, Greece.

Professor Trevor J. Britz, Department of Food Science, Stellenbosch University, Private Bag X1, Matieland 7602, South Africa.

Dr Elna M. Buys, Department of Food Science, University of Pretoria, Pretoria 0001, South Africa.

Dr Peter de Jong, NIZO Food Research, PO Box 20, 6710 BA Ede, The Netherlands.

Mr Evaggelos Doxanakis, Lotus Business Consulting and Training Services Ltd, 68–70, Aeolou Street, Athens 105 59, Greece.

Dr Athanasios Goulas, School of Food Biosciences, University of Reading, Reading RG6 6AP, UK.

Dr Alistair S. Grandison, School of Food Biosciences, University of Reading, Reading RG6 6AP, UK.

Dr Peter J. Jooste, Department of Biotechnology and Food Technology, Tswane University of Technology, Private Bag X680, Pretoria 0001, South Africa.

Mr Asterios Kefalas, Lotus Consulting and Training Services SA, Aiolou 68-70, Athens, 105 59, Greece.

Dr Corné Lamprecht, Department of Food Science, Stellenbosch University, Private Bag X1, Matieland 7602, South Africa.

Dr J. Ferdie Mostert, ARC-Animal Production Institute, Private Bag X2, Irene 0062, South Africa.

Dr Lefki Psoni, School of Food Technology and Nutrition, Technological Education Institution (TEI), Thessaloniki, Greece.

Dr Gunnar O. Sigge, Department of Food Science, Stellenbosch University, Private Bag X1, Matieland 7602, South Africa.

Preface

The dairy industry is the largest sector of the food-supply chain, in that not only does it supply retailers with numerous ready-to-eat products like liquid milk, butter and cheese, but also it provides ingredients, e.g. milk powders and condensed milk, to a variety of food processors. It is also an industry that relies for success on a unique blend of 'craft and science', with the balance changing in relation to the product. A dairy supplying retail milk, for example, will heat-treat the milk within limits set down by microbiologists to ensure that the milk is free from vegetative pathogens, but while a cheese plant may well use pasteurized milk as the base material, the flavour of the finished cheese is largely dependent on the intuitive skills of the cheesemaker.

Nevertheless, every company dealing with milk and milk products has to be thinking about the future. What new products could find a niche in the market place? How can production costs be held steady or even reduced? Answering such questions will often depend to a large extent on scientific expertise, for it may well be the laboratory that is charged with devising new formulations and testing that they pose no risk to the consumer, or finding an original approach to refining a novel ingredient or reducing the cost of effluent treatment. As such tasks are not product-specific, the aim of this book is to examine some of the scientific and technical options that many dairy companies will need to consider, and maybe embrace, over the next decade. Certainly all aspects of food safety are being prioritized by retailers and regulatory authorities alike, as are approaches for manufacturing products that are free from microbiological, chemical or physical faults. Reducing or holding costs is another feature that motivates manufacturers, and many technological advances have arisen in response to this pressure.

If this book contributes to the discussion of these problems, then its publication will have served a useful purpose, and with this thought in mind, the Editors acknowledge with gratitude the expert contributions of the authors. The support of Blackwell Publishing has been much appreciated as well.

Trevor J. Britz and Richard K. Robinson

Chapter 1

Thermal Processing of Milk

Peter de Jong

1.1 Introduction

1.1.1 Background

Chemical reactor engineering is not counted as one of the classical disciplines in dairy science. Nevertheless, in the dairy industry, physical, chemical and biochemical phenomena are the basis for the strong relation between the product quality and the process operation and design. This is particularly valid for heat treatment, where a number of heat-induced transformations of milk components determine the functional properties of the final product, e.g. biological safety, shelf-life, flavour, taste and texture. Fresh milk, cheese, milk powder and fermentation products such as yoghurt all require a different heat treatment, i.e. a specific temperature–time history.

The heat treatment or pasteurization of milk derives its principles from the work of Louis Pasteur (1822–1895). In 1864, he developed a method to prevent abnormal fermentation in wine by destroying the organisms responsible by heating to 60°C. Since 1880, milk for babies has been heated to reduce the risk of infection by heating milk in a continuous flow to temperatures of about 60–75°C. A hundred years later, the International Dairy Federation gives the following definition of pasteurization: 'Pasteurization is a process applied to a product with the objective of minimizing possible health hazards arising from pathogenic microorganisms associated with milk by heat treatment which is consistent with minimal chemical, physical and organoleptic changes in the product'. Today, the basic heat treatments in the dairy industry can be summarized by five types: thermization (e.g. 20 s, 65°C) for inactivation of psychotropic microorganisms, low pasteurization (e.g. 20 s, 74°C) for inactivation of pathogenic microorganisms, high pasteurization (e.g. 20 s, 85°C) for inactivation of all microorganisms but not spores and finally sterilization and UHT (ultra-high-temperature) treatment (e.g. 30 min, 110°C; 5 s, 140°C) to destroy spores. The effect of the heat treatment on the final product quality depends on the combination of temperature and time applied; this determines the equipment selection.

Not only is product quality affected by the heat-induced reactions, but fouling of equipment by deposit formation on walls is governed by specific reactions of milk components. These typical undesired reactions reduce the heat transfer coefficient, increase the pressure drop over heat treatment equipment, and increase product losses, resulting in higher operating costs.

As a result of the complexity of the milk reaction system, the design and operation of equipment for thermal treatment of milk have been based mainly on simplifying assumptions and empirical experience. However, now that kinetic data of relevant transformations are becoming available, a reaction engineering approach appears to be applicable to the optimal design and operation of dairy heat-treatment equipment (De Jong and Van der Linden, 1992; De Jong, 1996).

1.1.2 Outline

In this chapter, the impact of heating on the product properties will be reviewed. A classification of the heat-induced (bio)-chemical reactions in milk is given and how the effect of the temperature-time history can be quantified. Next the current types of heating processes with their specific aims are classified. Also, new, more advanced heating systems are discussed.

The main limitations and quality aspects of heating systems, for example, due to fouling and adherence and growth of bacteria in the equipment, are discussed in the following section. Mathematical models are given to handle these limitations in the design of heating processes.

The last section covers the methodology of optimization of heat treatments, in both the process design phase and in-line during operation of the heating process. The chapter ends with some conclusions and future trends.

1.2 Heat-induced changes of milk

1.2.1 Heat-induced reactions in milk – bulk reactions

During the heat treatment, milk behaves like a complex reaction system. A large number of chemical, physical and biochemical reactions take place. Some of these transformations are important because they change those characteristics of milk that are easily recognized by a potential consumer. Others may change the nutritional value and increase the biological safety of the milk, or may be of use as an indicator to assess the severity of the process applied.

The heat-induced reactions in the bulk of the milk can be subdivided into five groups: destruction of microorganisms, inactivation of enzymes, denaturation of proteins, loss of nutrients and formation of new components. Most of the reactions can be described by a single irreversible reaction step

$$A \rightarrow B$$

with the rates of disappearance and formation given by a standard reaction-rate equation

$$r_A = -kC_A^n, \quad r_B = -r_A \tag{1.1}$$

where r is the reaction rate (g l^{-1} s^{-1}), k is the reaction rate constant ($g^{(1-n)}$ $l^{(n-1)}$ s^{-1}), and n is the reaction order. The way in which the reaction rate constant k is affected by temperature is important in determining the extent of the overall conversion after heat treatment. Although, in the dairy industry, several relations are used, particularly with the destruction of microorganisms (Burton, 1988; Holdsworth, 1992; Kessler, 1996), the most appropriate and pragmatic description of the temperature dependence is given by the Arrhenius relationship (Hallström *et al.*, 1988).

$$k = k_0 \exp\left(\frac{-E_a}{RT}\right) \tag{1.2}$$

where k_0 is the so-called pre-exponential factor ($g^{(1-n)}$ $l^{(n-1)}$ s^{-1}), E_a the activation energy (J mol^{-1}), R the gas constant (8314 J mol^{-1} K^{-1}) and T the absolute temperature in K. The kinetic parameters k_0 and E_a are dependent on the reaction involved and the composition of the product. Table 1.1 gives an overview of the available kinetic constants of the heat-induced reaction, partly calculated from experimental data reported in the literature.

Table 1.1 Overview of kinetic constants for reactions in milk.

Component	Product	Concentration (gl^{-1})	Temperature range (°C)	ln k_0	E_a (kJ-mol^{-1})	n	Reference
Destruction of microorganisms							
Escsherichia coli	Milk	-	62–82	132.22	378	1	Evans *et al.* (1970)
	Cream (40% fat)	-	52–80	132.42	375	1	Read *et al.* (1961)
Micrococcus luteus	Milk	-	60–90	112.71	330	1	Peri *et al.* ((1988)
Mycobacterium tuberculosis	Milk	-	60–90	173.88	498	1	Klijn *et al.* (2001)
Bacillus cereus Vc1 spores	Milk	-	90–105	56.47	179.4	1	Stadhouders *et al.* (1980)[a]
Bacillus cereus Vc1 spores	Milk	-	95–110	91.92	294.5	1	Stadhouders *et al.* (1980)[a]
Bacillus coagulans spores	Milk	-	116–123	151.29	509	1	Hermier *et al.* (1975)
Bacillus stearothermophilus spores	Milk	-	100–140	101.15	345.4	1	Peri *et al.* (1988)
Clostridium botulinum spores	Milk (1.5% fat)	-	104–113	107.50	351	1	Denny *et al.* (1980)
Inactivation of enzymes							
Catalase	Milk	-[b]	60–80	180.72	529	1	Walstra and Jennes (1984)[a]
Chymosin	Milk	-	50–70	134.10	372	1	Luf (1989)[a]
Phosphatase	Milk	-	60–82	95.17	275	1	De Jong and Van den Berg (1993)
Lipase	Milk	-	60–90	53.70	160	1	Walstra and Jennes (1984)[a]
Lipase *Pseudomonas*	Milk	-	150	22.60	91	1	Adams and Brawley (1981)
Lipase *Pseudomonas fluorescens*	Milk	-	150	21.16	83	1	Fox and Stepaniak (1983)
Peroxidase	Milk	-	70–90	225.26	663	1	Walstra and Jennes (1984)[a]
Protease	Milk	-	70–150	15.19	64	1	Peri *et al.* (1988)
Protease *Pseudomonas fluorescens*	Milk	-	70–130	24.64	101	1	Driessen (1983)
Xanthine oxidase	Milk	-	60–80	127.29	380	1	Walstra and Jennes (1984)[a]

Denaturation of proteins							
Bovine serum albumin	Skim milk	0.4	60–82	23.68	80	1	De Jong and Van den Berg (1993)
			82–150	13.18	49	1	De Jong and Van den Berg (1993) and Hiller and Lyster (1979)
Immunoglobulin	Milk	0.7	60–76	90.38	275	1	Luf (1989)[a]
			76–82	54.21	170	1	De Jong and Van den Berg (1993)
α-Lactalbumin	Skim milk	1.2	70–85	84.92	269	1	Dannenberg and Kessler (1988)[a,b]
			85–150	16.95	69	1	Dannenberg and Kessler (1988)[a,b]
β-Lactoglobulin	Skim milk	3.2	70–90	89.43	280	1.5	Dannenberg and Kessler (1988)[a,b]
			90–150	12.66	48	1.5	Dannenberg and Kessler (1988)[a,b]
	Milk	3.2	75–85	120.64	374	1.8	De Wit and Klarenbeek (1988)
Lysine (amino acid)	Milk	2.9	130–160	8.77	109	2	Kessler and Fink (1986)
Loss of nutrients							
Thiamin (vitamin B_1)	Milk	0.0004	120–150	29.78	100.8	0	Kessler and Fink (1986)
Formation of components (i.e. Maillard type)						0	
Pigment brown	Milk	0[c]	50–160	29.09	116	0	Kessler and Fink (1986)
Furosin	Milk	0	120–150	16.01	81.6	0	Peri *et al.* (1988)
5-Hydroxymethylfurfural (HMF)	Milk	0	75–130	30.31	135.1	0	Peri *et al.* (1988)
			130–160	30.72	139	0	Kessler and Fink (1986)
Lactulose	Milk	0	60–145	30.52	120.2	0	Peri *et al.* (1988)
Lysinoanaline	Milk	0	110–130	18.39	101.4	0	Peri *et al.* (1988)

[a] Calculated from published experimental data.
[b] Enzyme activity in raw milk is defined as 100%.
[c] 1 = no change in colour (perception threshold); 2 = light ivory; 4 = saffron-yellow; 10 = brown-yellow.

1.2.1.1 Destruction of microorganisms

According to the high activation energies (>300 kJ mol^{-1}) of most of the reactions, the required destruction of microorganisms is strongly dependent on temperature. A slight increase in temperature causes a relatively much higher reaction-rate constant, i.e. destruction rate. Since destruction is assumed to be a first-order reaction, the degree of destruction is generally expressed as the number of decimal reductions (log N_0/N) of active microorganisms, independent of the initial concentration.

1.2.1.2 Inactivation of enzymes

In general, the inactivation of enzymes is a desired reaction because of their negative effect on the flavour and shelf-life of the dairy product. However, some enzymes stimulate the development of flavours. An example is xanthine oxidase, which has a positive effect on the quality of cheese during ripening (Galesloot and Hassing, 1983). The degree of inactivation is expressed as a percentage of the initial enzyme activity.

1.2.1.3 Denaturation of proteins

The most reactive proteins in milk are the whey proteins: serum albumin, immunoglobulin, α-lactalbumin and β-lactoglobulin. In particular, the denaturation of β-lactoglobulin is strongly related to many functional properties of dairy products (De Wit and Klarenbeek, 1988; Rynne *et al.*, 2004). For example, the texture of yoghurt is improved (Dannenberg and Kessler, 1988b), and the solubility of milk powder is decreased with the denaturation degree of β-lactoglobulin (De Wit and Klarenbeek, 1988). Although the denaturation is recognized as a complex reaction with many reaction steps (Singh, 1993), it can still be quantitatively described with one overall reaction. This explains the broken orders in Table 1.1.

A special case of protein denaturation is the flocculation of casein micelles as a result of dephosphorization, partial hydrolysis and several cross-linking reactions at high temperatures (Walstra and Jennes, 1984). To date, there have been no quantitative models available with reasonable accuracy.

1.2.1.4 Loss of nutrients

The major nutrient which is lost due to heating is thiamine (vitamin B_1). For example, in the case of UHT milk, the loss of thiamine should be less than 3% (Kessler, 1996).

1.2.1.5 Formation of new components

In general, the formation of new components is undesirable and is governed by the Maillard reaction. The Maillard reaction is a complex series of steps (Adrian, 1974; Brands, 2002) in which reactions occur between lactose and proteins. Various pathways lead to the production of brown pigments which are responsible for the brown colour of heated milk. In some cases, e.g. for coffee creamers, this colour formation is a desired phenomenon. Hydroxymethylfurfural (HMF) appears in the early stages of the Maillard reaction, and it has been suggested as a measure of the severity of the heat treatment to which milk has been subjected (Burton, 1988).

1.2.2 Heat-induced reactions in milk – surface reactions

The transformations of some milk components in the bulk have an additional undesirable effect: deposition of milk components on heat-treatment equipment walls. The exact mechanism of deposition is presently unknown, and there are no mathematical models with sufficient physical background which can be applied for process optimization. However, the correlation between protein denaturation and fouling of heat exchangers has been confirmed by several investigators (Lalande *et al.*, 1985; Dannenberg, 1986; Fryer, 1989; De Jong *et al.*, 1992). It appears that particularly the denaturation of β-lactoglobulin plays an important role in the fouling process of heat exchangers.

Besides proteins and minerals, a number of microorganisms are able to adhere to the surface of the heating equipment. Well-known examples are *Streptococcus thermophilus* (Bouman *et al.*, 1982) and *Bacillus cereus* spores (Te Giffel *et al.*, 1997). These microorganisms multiply on surfaces during operating and may cause reduced shelf-life of the product.

In Section 1.4, the surface reactions and their role in heating processes are described in more detail.

1.2.3 Reaction engineering approach

In principle, the heat treatment should be designed in such a way that all the components with a positive effect on the product quality should not be damaged (e.g. vitamins, proteins). On the other hand, components with a negative impact (e.g. spoilage bacteria and enzymes, lactulose) should be at minimal concentration level. Of course, this will be a trade-off. Moreover, the quality of a specific milk product depends on the conversion of the so-called key-components. For example, the texture of yoghurt is related to the degree of denaturation of the protein β-lactoglobulin, and the consumer desires a fresh product without pathogens: objectives that make different demands of the heat treatment. This implies that every milk product should have its dedicated heat treatment with an optimal result.

Since much of the kinetics of the key components has been quantified (see also Table 1.1) the heating process can be considered as a chemical reactor in which the conversion of these components has to be optimized. In other words, the selectivity of the reactor should be optimal. This is explained in Fig. 1.1.

In general, the reaction rate of the key components increases with temperature. Since the temperature dependence differs for each reaction, it is possible to find the optimal temperature profile in the equipment with the aid of numerical optimization routines. This means a high conversion of the desired reactions and a low conversion for the undesirable reactions. For example, when in Fig. 1.1 component A represents a pathogenic microorganism and component B a nutrient, a short heat treatment at high temperature will result in a better final product. The UHT procedure is based

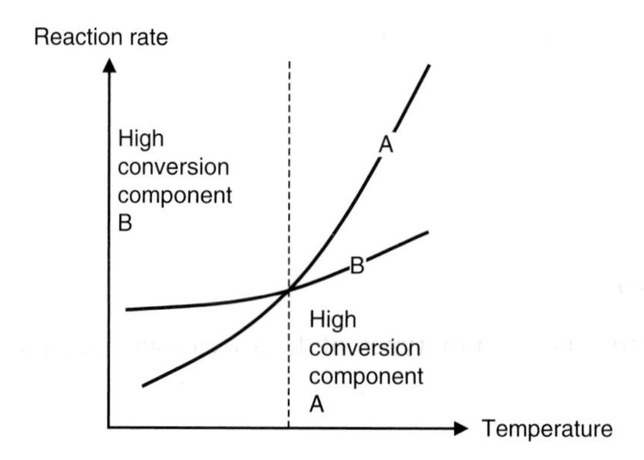

Fig. 1.1 Principle of selectivity with heat-induced reactions.

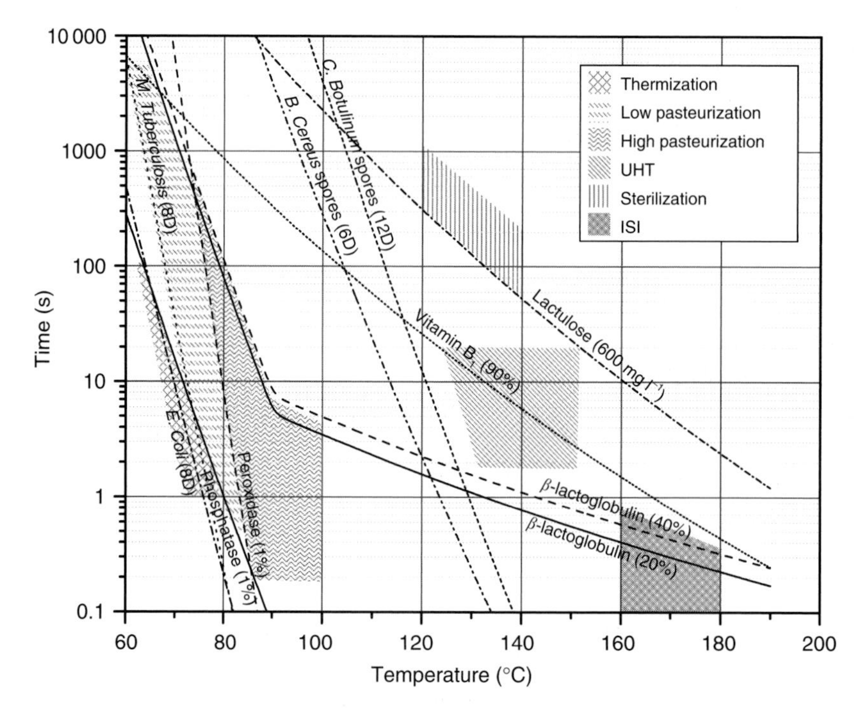

Fig. 1.2 Temperature–time dependency of several heat-induced reactions based on model calculations and applied heating processes in the dairy industry. Heating and cooling are not taken into account (D = decimal reduction).

on this principle. By integration of a quantitative fouling model in the optimization procedure, it becomes possible to obtain the desired conversion of the key components and simultaneously to reduce energy, cleaning, depreciation, and emission costs.

Figure 1.2 demonstrates the effect of the combination of temperature and time on the conversion of milk components and microorganisms. It is obvious that the conversion of some components, such as bacterial spores, is highly dependent on the heating temperature. Others, such as proteins, are relatively more dependent on the heating time. Considering that the heat process itself has a large impact on the temperature–time history of the product, it is clear that process design cannot be done without careful evaluation of the conversion of the specific key components for product quality.

1.3 Processes

1.3.1 Equipment

In the dairy industry, two major types of heat-treatment equipment can be distinguished according to the heat-exchange system which is applied. These are the so-called direct and indirect heating systems with steam and hot water, respectively, as the heating medium. Which type of heating system is selected mainly depends on the desired heating rate; the direct system is used for a high heating rate (10–100 K s^{-1}), the indirect system for lower rates (0.01–10 K s^{-1}).

With both direct and indirect systems, a part of the heat transfer is achieved in the form of heat recovery, i.e. heat is extracted from the product to cool it from the heating temperature and is transferred to the incoming product. Heat recovery results in considerable energy savings and is therefore an important factor in operating costs (Burton, 1988).

1.3.1.1 Continuous indirect heating systems

The indirect heating systems can be subdivided according to the shape chosen for the heat-transfer surface. The heat exchanger can consist of an assembly of plates or a tubular heat-exchange system.

Figure 1.3a shows schematically the operation principle of a plate heat exchanger. The specific surface texture of the plates increases the degree of turbulence and thus stimulates the heat transfer. The plates are assembled in packs and clamped in a frame, each adjacent pair of plates forming a flow channel with the two media flowing in

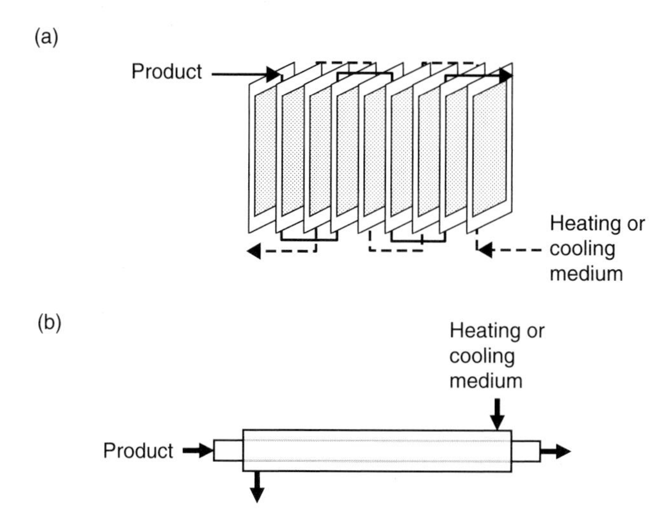

Fig. 1.3 Operation principle of indirect heating systems: (a) plate heat exchanger; (b) tubular heat exchanger.

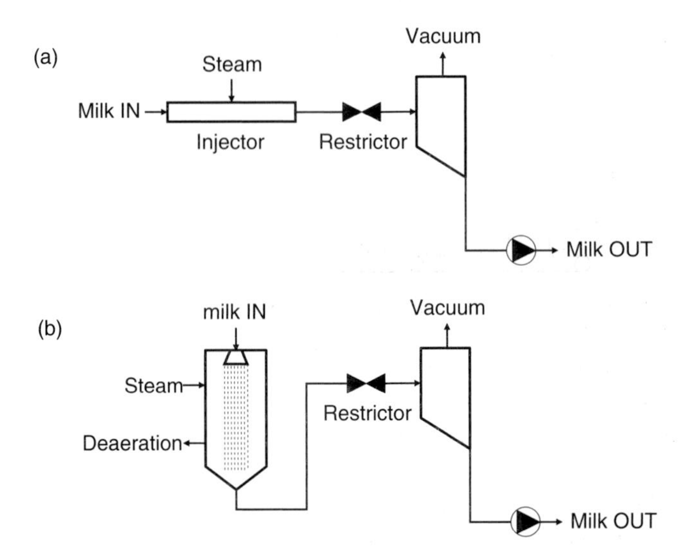

Fig. 1.4 Operation principle of indirect heating systems: (a) steam injection; (b) steam infusion.

alternate channels. Different channel groupings can be chosen to give the desired pressure-drop characteristics and flow pattern.

The tubular heat exchangers (Fig. 1.3b) are mainly of two types: concentric tube exchangers and shell-and-tube heat exchangers. Tubular systems are more robust than plate heat exchangers, but the specific heat-exchange area (area per m^3) is smaller than that of plate exchangers, and mainly natural turbulence as a result of a high Reynolds number is used to improve the heat-transfer coefficient.

1.3.1.2 Continuous direct heating systems

With direct heating systems, the heating is performed by mixing the product with steam under pressure. The steam condenses, transferring its latent heat of vaporization to the product and giving a much more rapid heating rate than is available with indirect systems. After the required time at the heating temperature, the product is expanded through a restrictor into an expansion cooling vessel to realize a similar rapid rate of cooling.

There are two types of system, depending on the method used for mixing the product with steam. Figure 1.4 gives a schematic representation of the two systems. In one type (Fig. 1.4a), steam, at a pressure higher than that of the product, is injected into the product stream through a suitable nozzle and condenses to give the required temperature. This system is called the injection or steam-into-product type. Alternatively (Fig. 1.4b), a vessel is pressurized with steam. The product is sprayed into the top of the vessel, and while it falls, steam condenses on the product. This system is called the infusion or product-into-steam type.

1.3.1.3 (Semi) Batch heating systems

Traditionally, products such as evaporated milk are sterilized in glass bottles, tins or, more recently, plastic bottles (e.g. HDPE). Heating is then performed in a batch system.

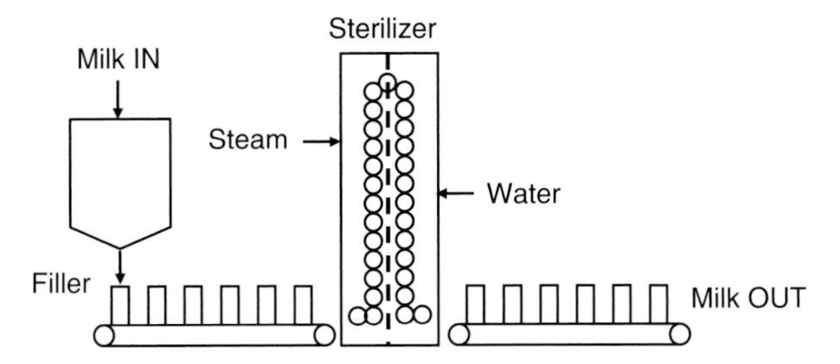

Fig. 1.5 Operation principle of a batch heating system (sterilization tower).

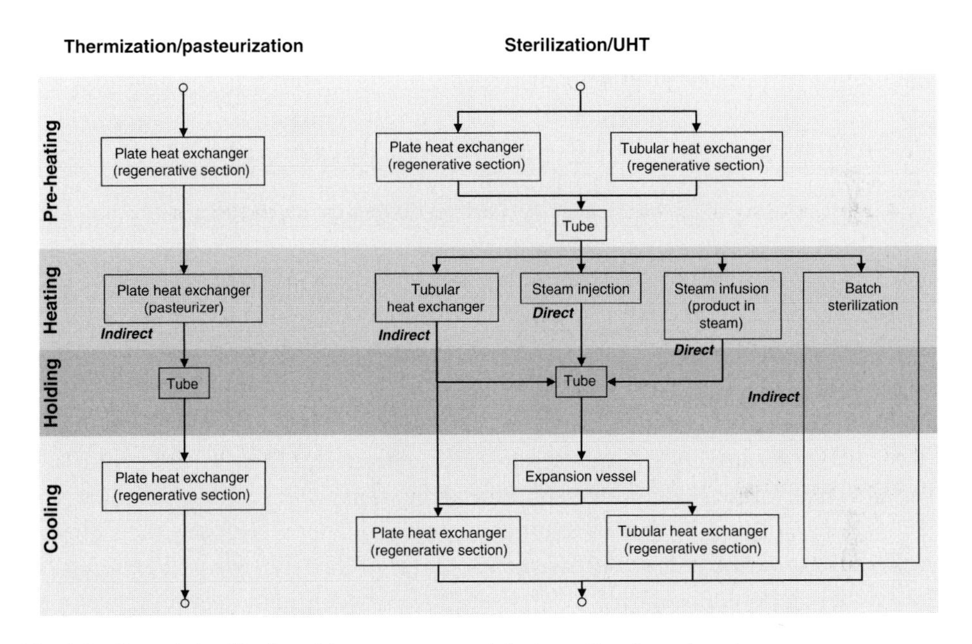

Fig. 1.6 General classification of heat treatment and the type of equipment used.

Figure 1.5 shows schematically the operation principle of such a batch in-container sterilization process. After preheating, the product is packaged. The second heating stage takes place in a sterilizer tower more than 10 m high. The tower is divided in several heating and cooling sections through which the filled bottles are transported.

1.3.2 Classification of heating processes

As concluded at the end of section 1.2, each milk product should have a dedicated heating process. In practice, this means that, for a given heat treatment, a combination of different types of heating equipment is used. Figure 1.6 shows a classification for several heating processes and the different types of equipment used. In general, plate heat exchangers are used for non-fouling heating conditions, depending on the type of product. For viscous products and conditions causing severe fouling problems, tubular heat exchangers are used. When high heating temperatures are necessary but the heat load should be minimal, direct heating systems are applied. For liquid milk products with a high added value, such as infant formulae, batch sterilization is used.

For an overview of the applied temperature–time combinations for the different heating processes, see Fig. 1.2.

1.3.2.1 Thermization

Thermization aims at reducing the number of microorganisms by a factor of 10^3 or 10^4 (log 3 or log 4). Surviving microorganisms will be heat-stressed and become more vulnerable to subsequent microbial control measures (Codex Alimentarius, 2003; FAO, 2005). Typical temperature–time combinations are 68°C for 10 s or 65°C for 20 s. There are no general verification methods. In some cases, it is desired that the alkaline phosphatase test gives a positive result.

1.3.2.2 Pasteurization

Pasteurization is a heat treatment aimed at reducing the number of harmful microorganisms, if present, to a level at which they do not constitute a significant health hazard. Pasteurization can be carried out either as a batch operation ('batch pasteurization') with the product heated and held in an enclosed tank or as a continuous operation ('High Temperature Short Time, HTST pasteurization') with the product heated in a heat exchanger and then held in a holding tube for the required time (FAO, 2005).

Pasteurization conditions are designed to effectively destroy the organisms *Mycobacterium tuberculosis* (Rowe *et al.*, 2000; Klijn *et al.*, 2001) and *Coxiella burnettii*. As *C. burnettii* is the most heat-resistant non-sporulating pathogen likely to be present in milk, pasteurization is designed to achieve at least a 5 log reduction of *C. burnettii* in whole milk (Codex Alimentarius, 2003; FAO, 2005). According to validations carried out on whole milk, the minimum pasteurization conditions are those having bactericidal effects equivalent to heating every particle of the milk to 72°C for 15 s (continuous-flow pasteurization) or 63°C for 30 min (batch pasteurization).

To confirm sufficient pasteurization, the residual alkaline phosphatase levels in heat-treated milk have to be below 10 μg *p*-nitrophenol equivalent ml^{-1} (FAO, 2005). Low-pasteurized milk should also be lactoperoxidase-positive. For highly pasteurized milk, the lactoperoxidase test has to be negative. Lactulose in pasteurized milk should be below the detection limit and may not exceed 50 mg l^{-1} in highly pasteurized milk (Mortier *et al.*, 2000). In the near future, it will also be required that the concentration of non-denatured β-lactoglobulin should be more than 2600 mg l^{-1} for pasteurized milk and 2000 mg l^{-1} for highly pasteurized milk (Mortier *et al.*, 2000).

1.3.2.3 UHT

Thermal processes necessary to obtain commercially sterile products are designed to result in preferably 12 log reductions of *Clostridium botulinum*, and in the absence of viable microorganisms and their spores capable of growing in the treated product when kept in a closed container at normal non-refrigerated conditions at which the food is likely to be held during manufacturing, distribution and storage (Codex Alimentarius, 2003).

UHT treatment is normally in the range of 135–150°C in combination with holding times necessary to achieve commercial sterility. The products should be microbially stable at room temperature, either measured after storage until end of shelf-life or incubated at 55°C for 7 days (or at 30°C for 15 days). On the upper limit of UHT milk, the lactulose content has to be below 600 mg l^{-1} and the non-denatured concentration of β-lactoglobulin should be higher than 50 mg l^{-1} (Mortier *et al.*, 2000).

1.3.2.4 Sterilization

Sterilized milk usually is produced in two steps consisting of a continuous heating step at 120–140°C for several seconds and an in-bottle sterilization at 110–120°C over 10–20 minutes. European norms for sterilized milk will be a lactulose concentration above 600 mg l^{-1} and non-denatured concentration of β-lactoglobulin below 50 mg l^{-1} (Mortier *et al.*, 2000).

1.3.3 Advanced processes

1.3.3.1 ESL

There is no legal, internationally accepted definition of extended shelf-life (ESL) milk. Generally, ESL milk can be defined as a product with a shelf-life longer than pasteurized milk under equivalent chilled distribution and storage conditions. ESL technology can be applied to all chilled liquid foods. Examples include white and flavoured milk, fermented products, cream, dairy desserts, soy drinks, and iced tea and coffee (Te Giffel *et al.*, 2005).

Consolidation in the dairy industry has resulted in fewer plants and wider distribution. Therefore, a few extra days of shelf-life are a significant benefit to dairy companies and retailers, allowing them to extend their distribution chain. In addition, extending shelf-life opens new opportunities for innovative, value added, dairy products and marketing milk products to consumers demanding safety and high, consistent quality.

ESL is a complete systems approach along the entire processing chain combining processing, packaging systems and distribution. The ESL technology is the result of optimization of the temperature–time history with respect to the key components for chilled liquid milk. Moreover, some additional technologies are introduced to lower the initial count of bacterial spores. However, the key to ESL is hygiene. Using complementary or different processing solutions to pasteurization, reduction of the risk of recontamination by pathogenic and spoilage bacteria from the production environment and advanced filling equipment, it is possible to improve product quality and extend shelf-life. Modified heat treatment and microfiltration combined with heat treatment are the two major processing technologies applied for shelf-life extension of chilled dairy products. Producers can also use bactofugation to reduce the initial levels of microorganisms and gain a few days of additional shelf-life. This is, for example, applied in the

Netherlands to guarantee extended shelf-life of various types of fresh drinking milk. The typical shelf-life of liquid ESL products is 15–25 days stored at 7°C.

1.3.3.2 ISI

A new preservation method is based on an innovative steam injection (ISI) approach. The patented IS technology is based on existing UHT treatment, but very short heating is combined with very high temperatures: less than 0.1 s at 150–200°C (Huijs *et al.*, 2004). The heating is directly followed by flash cooling in a vacuum vessel. The process costs of the ISI technology are some 10 percent higher than the prevailing UHT process, making the ISI technology substantially less expensive than alternative technologies such as microfiltration, normally used for ESL products. The setup is shown in Fig. 1.7.

Using the ISI, a significant inactivation of heat-resistant spores can be achieved while preserving the functionality of important ingredients. Table 1.2 shows the typical time–temperature combinations used in the dairy industry, and the effect of these processes on denaturation of whey proteins, *Bacillus stearothermophilus* and *Bacillus sporothermodurans* spores in milk. The denaturation of whey proteins is a measure of product damage caused by the stress of heating.

For sustained stability at room temperature, the enzymes that cause spoilage must also be inactivated. The extremely heat-resistant enzyme plasmin, instigator of a bitter taste, plays the leading role here in milk products. Plasmin does not become inactivated in the pasteurization process, but that is unimportant to pasteurized milk. This is first because pasteurized milk is stored under refrigeration, and the enzymatic activity of

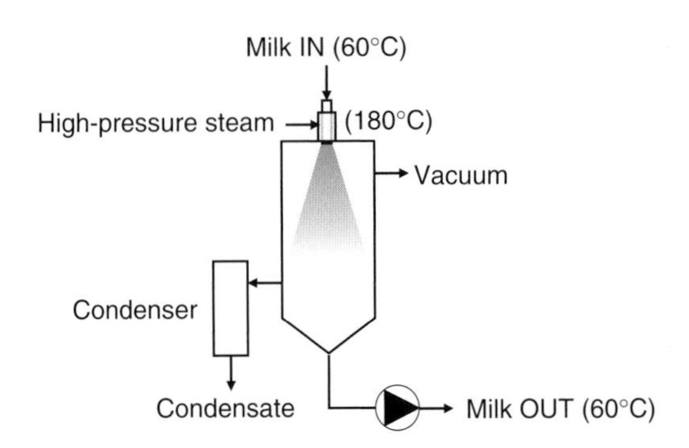

Fig. 1.7 Schematic overview of the ISI heater.

Table 1.2 Effects of different heating technologies on whey proteins and bacterial spores in milk.

Technology	Time (s)	Temperature (°C)	Denaturation of whey proteins (%)	Log reduction *B. stearothermophilus* spores	Log reduction *B. thermodurans* spores
Pasteurization	15–1	72–80	4–8	0	0
Indirect UHT	20–5	130–145	75–90	>6	0–3
Direct UHT	6–2	142–150	60–85	>6	0–4
ISI	0.1	160–180	<25	>6	3–4

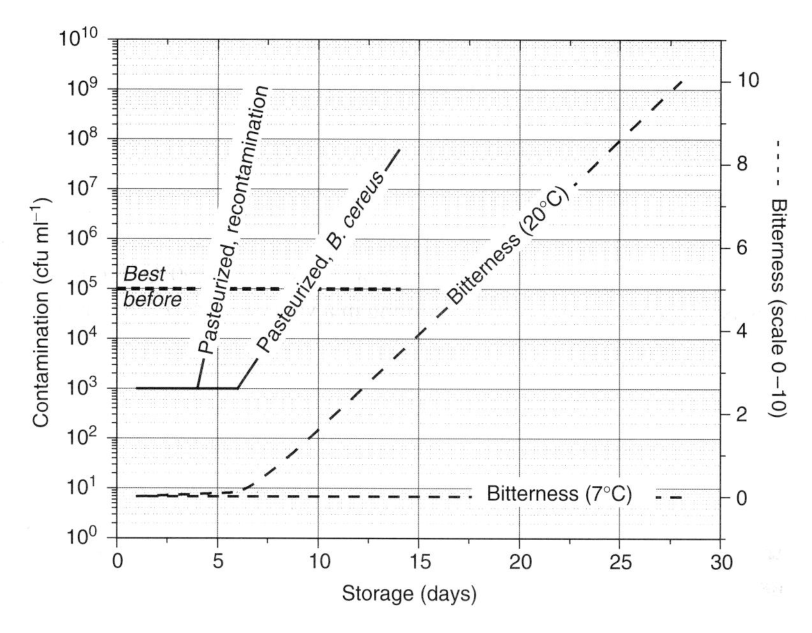

Fig. 1.8 Limitation of shelf life of pasteurized milk by growth of *B. cereus* compared with the formation of bitterness in ISI–ESL milk caused by enzymatic activity of plasmin during storage, cooled (7°C) and at room temperature.

plasmin is minimal at low temperature. Second, the germination of *B. cereus* spores in pasteurized milk is limiting for the shelf-life, long before the effects of plasmin become observable. To obtain an ISI milk that is stable for months at room temperature, an additional preheating step is necessary for inactivation of plasmin. As shown in Fig. 1.8, if this preheating is performed, the ISI milk has, under chilled conditions, a substantially longer shelf-life of up to 60 days.

1.4 Operational considerations and limitations

1.4.1 Flow characteristics

In order to meet the desired product specifications, the residence time distribution of the heated product should be as narrow as possible. Since laminar flow results in more variation in residence time, the design and operation of heating processes have to be directed on obtaining turbulent flow. The flow conditions existing in the equipment are taken into account by calculating the Reynolds number of the product.

$$\mathrm{Re} = \frac{\rho v D}{\eta} \tag{1.3}$$

where ρ is the density (kg m^{-3}), v the velocity (m s^{-1}), D the hydrodynamic diameter (m) and η the dynamic viscosity (Pa s^{-1}). Turbulent flow is assumed when the Reynolds number exceeds 4000 (Knox, 2003). In cases where turbulent flow is not possible (e.g. with viscous products), the effective heating time has to be corrected.

From a microbial point of view, for Re<2300 the effective time is reduced by 50%, and for Re<4000 the effective time is reduced by 25%.

1.4.2 Protein and mineral fouling

The operating costs of milk processing are mainly related to the heat-induced formation of deposits. The operation time of equipment at temperatures above 80°C is determined largely by the deposition of proteins and minerals. In the dairy industry, every product is heated at least once, and so heat treatment is by far the major unit operation. Taking into account that approximately 80% of the operation costs are related to fouling, it is clear that the control of fouling is an important condition in heating a variety of milk products (De Jong, 1996).

1.4.2.1 Types of fouling

An important limitation of heating milk products is the deposition of proteins and mineral on the equipment walls (Fig. 1.9). Undesirable side effects of such deposits are: decreased heat transfer coefficients, increased pressure drops, product losses, and increased cleaning costs and environmental load.

There are two distinct types of deposits, A and B (Burton, 1988), depending on the actual limiting reactions of the fouling mechanism. The first is a relatively soft, bulky material that is formed at temperatures between 75°C and 115°C (Lalande *et al.*, 1985). Owing to the high protein content (50–70%, w/w), this type of fouling is known as protein fouling. The second type of deposit is formed at higher temperatures, that is, above 110°C. It is hard and has a granular structure with high mineral content (up to 80%, w/w), and is therefore known as mineral fouling (Lalande *et al.*, 1985). In the case of whey, the content of total solids, as well as temperature, determines the dominant type of fouling (Schraml and Kessler, 1996). At solids concentrations up to 25%, the deposit layer consists mainly of protein. At higher concentrations, relatively

Fig. 1.9 Fouled plate of a heat exchanger after processing milk.

more calcium phosphates and calcium citrates become insoluble, which increases mineral precipitation during the formation of the protein layer.

Because of the bulk of the first type of deposit, it is protein fouling that restricts the area of the flow passages and the heat transfer in the heat exchanger, and is therefore of major importance for process optimization. However, with higher concentrations of minerals, the structure of the deposit layer becomes more compact. This influences the ease of removal of deposits during the cleaning procedures (De Jong, 1997).

1.4.2.2 Fouling mechanism

The exact mechanism and underlying reactions between milk components that describe the fouling phenomena are unknown. It appears that the absolute fouling level is not only related to the process design and conditions but also affected by more intrinsic factors such as the age of the milk and its composition due to seasonal influences. However, the correlation between protein denaturation in milk and fouling of heat exchangers has been confirmed by many investigators (Lalande *et al.*, 1985; Dannenberg, 1986; Fryer, 1989; De Jong *et al.*, 1992). Experimental results have shown that β-lactoglobulin plays a dominant role in the fouling process of heat exchangers (Tissier *et al.*, 1984; De Jong *et al.*, 1992, 1993; Delplace and Leuliet, 1995). It appears that the denaturation of β-lactoglobulin and the formation of deposits occur simultaneously as the milk flows through the heat exchangers. Further research has shown that up to temperatures of ~115°C, the fouling rate is related to the concentration of unfolded β-lactoglobulin, an intermediate of the denaturation reaction (Fig. 1.10). The active form of β-lactoglobulin is able to aggregate with other proteins or is adsorbed at the deposit layer. Fouling due to deposition of proteins and minerals can therefore be described by an adsorption reaction (De Jong *et al.*, 1992):

$$J_{x,t} = k'' C_{x,t}^{1.2} \tag{1.4}$$

where $J_{x,t}$ is the local flux (kg m^{-2} s^{-1}) of food components to the wall, k'' the adsorption rate constant of proteins (m$^{1.6}$ kg$^{-0.2}$ s^{-1}) and $C_{x,t}$ the local bulk concentration of

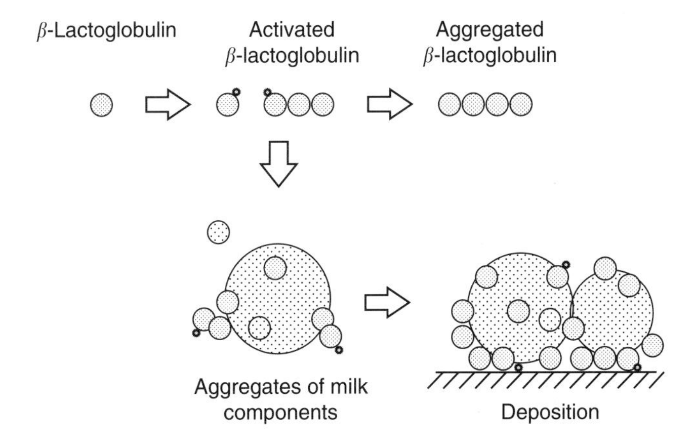

Fig. 1.10. Mechanism of protein and mineral fouling (schematic representation).

the key component related to fouling (e.g. whey proteins). The amount of fouling is obtained by integrating the flux over operating time and surface area of the equipment walls. More generally, the local fouling process is considered to be a heterogeneous adsorption reaction of milk constituents at the surface with mass transfer and reaction in series transport in the boundary layer and the reaction process at the heat transfer surface. In other words, aggregated proteins are formed in the bulk phase, transported to the wall and finally adsorbed at the wall of the heat treatment equipment. This implies that it is not the temperature difference across the wall but the absolute wall temperature that affects the local fouling rate. Hence, in principle, it is possible to get fouling in coolers (0, 0, 0), although the wall temperature is lower than the bulk temperature. The parameters of the fouling model have been quantified on the basis of experiments with skim milk and whole milk. The model has been used successfully to predict the effects of changing process conditions on the total fouling rate of equipment (De Jong, 1996; Benning *et al.*, 2003; De Jong *et al.*, 1993). Although the exact fouling mechanism is undoubtedly more complex, protein denaturation is considered to be the key mechanism that explains most of the phenomena. Some of the factors that affect the fouling rate of milk are discussed below.

1.4.2.3 Factors affecting fouling

- *Calcium* – As well as affecting the heat stability of milk, calcium plays a role in the formation of deposits during milk processing. This is not only because the solubility of calcium phosphate decreases on heating, but also because calcium influences the denaturation of whey proteins and the precipitation of the proteins, that is, casein micelles (Delsing and Hiddink, 1983). Experiments have shown that either increasing or decreasing the calcium concentration in the milk leads to a lower heat stability and to more fouling in comparison with that in normal milk (Jeurnink and De Kruif, 1995). In addition, there is a shift in the protein composition of the deposit from serum proteins to caseins. Obviously, the increased instability of the casein micelles, with denatured β-lactoglobulin at their surface, causes an increase in protein fouling.
- *pH* – Decreasing the pH of milk, for example from 6.8 to 6.4, causes a strong increase in fouling, mainly owing to the additional deposition of caseins (Skudder *et al.*, 1986; Patil and Reuter, 1988). In the case of whole milk, the deposition of both protein and fat is increased, but the deposition of minerals is reduced. It appears that the increased deposit formation of pH-reduced milk is mainly due to the reduced stability of protein to heat (Skudder *et al.*, 1986).
- *Air* – The solubility of air in milk decreases on heating. If the local pressure is too low (below the saturation pressure), air bubbles can arise (Burton, 1988). The formation of air bubbles may also be enhanced by mechanical forces that are induced by valves, expansion vessels or free-falling streams. It has been suggested that air in milk encourages fouling only if it forms bubbles on the heating surface, which then act as nuclei for deposit formation (Thom, 1975; Jeurnink, 1995). In addition, the composition of the deposit is influenced by the evaporation of the boundary layer of the air bubbles, which are mainly associated with caseins, resulting in the fouling layer having a higher casein content.

- *Age of milk* – Aged milk causes more fouling in a heat exchanger than does fresh milk. Experiments with skim milk (Jeurnink, 1991; De Jong *et al.*, 1993) have shown that the action of proteolytic enzymes, produced by psychrotropic bacteria, is responsible for the increase in deposition. For example, when raw milk is stored at 5°C for 6 d before processing, the degree of fouling may increase fourfold.
- *Seasonal influences* – The run time of an ultra-high-temperature (UHT) plant, which is determined by the degree of deposition, varies throughout the year. Especially at heating temperatures >120°C, the composition of the deposit layer is influenced by season (Grandison, 1988). Reasons for this might include the switch of the cows' diet from forage to grain feed. It is likely that this seasonal variation in run time is related not only to compositional variations of the raw milk but also to its changing heat stability throughout the year (Lyster, 1965).
- *Induction phase* – Before fouling of the heat-transfer surface occurs, there may be an induction period during which only a very thin layer of deposits is formed (Fryer, 1989). This layer has a negligible resistance to heat transfer and, by enhancing the surface roughness, actually increases the heat-transfer coefficient (Delsing and Hiddink, 1983). In tubes, the duration of the induction phase depends mainly on temperature, velocity and surface conditions (Belmar-Beiny *et al.*, 1993) and lasts for ~1–60 min (0,0). In general, the duration of the induction phase is loosely related to run time. There is no induction phase in plate heat exchangers (Paterson and Fryer, 1988). It is concluded that this is due to their geometry, in that they contain low-shear areas in which deposition starts immediately and builds up rapidly.
- *Coating* – Several studies of fouling have found that the nature of the surface becomes unimportant once the first layers have been adsorbed (Britten *et al.*, 1988; Yoon and Lund, 1989). One such study showed that different coatings on the heating surface did not affect the amount of deposit formed but did affect the strength of adhesion (Britten *et al.*, 1988). To date, the use of coatings has not been shown to influence the length of the operating time of heat exchangers significantly. Thus, it appears that coatings may enhance the cleaning rate but not reduce the fouling rate.

Summarizing, some important measures to minimize the amount of fouling in heat treatment equipment are:

- optimization of the temperature–time profile;
- lowering the surface temperature of the equipment wall of the heat exchangers by changing water flow rates and/or configuration of the plates or tubes;
- increasing the velocity of the product (to more than 2 m s^{-1}).

1.4.3 Adherence and growth of microorganisms

Adherence of microorganisms and bacterial spores on heat-exchanger surfaces in cheese and liquid milk factories is an important source of bacterial contamination of dairy products. Typically, this type of contamination occurs in heating processes below 80°C (Fig. 1.11). Growth of microorganisms in final products leads to spoilage and defects, e.g. excessive openness in cheese or taste deviations of milk. The operating time of pasteurizers is limited due to the growth of thermoresistant streptococci. The

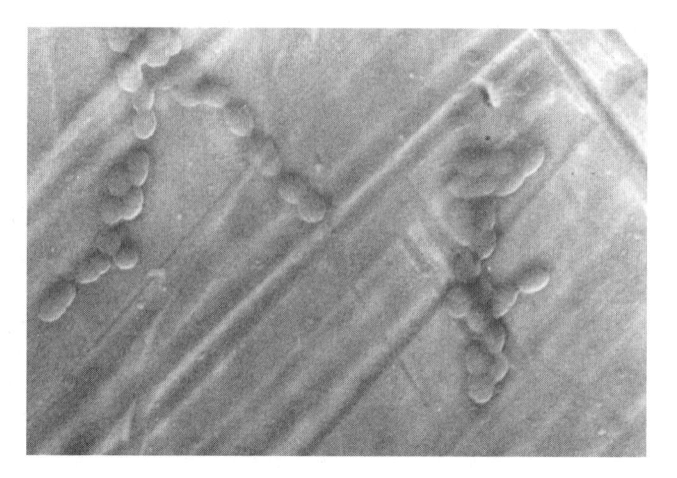

Fig. 1.11 *Strep. thermophilus* adhering to stainless steel of a pasteurizer after 8 h of production (pasteurized-milk side, local temperature 40°C).

streptococci isolated from milk and cheese were identified as mainly *Streptococcus thermophilus* (Bouman *et al.*, 1982; Te Giffel *et al.*, 2001). The increase in the levels of bacteria in the product during process operation is partly the result of growth in the product, but the release of bacteria that have grown on the walls of the equipment also plays a significant role (Bouman *et al.*, 1982; Langeveld *et al.*, 1995; Flint *et al.*, 2002; Te Giffel *et al.*, 2005). The microbial adhesion to surfaces is governed by a complex interplay of Van der Waals forces, electrostatic interactions, hydrodynamic conditions, cell–cell interactions and possible anti-adhesive biosurfactant production by the bacteria (Rosenberg, 1986; Mozes *et al.*, 1987; Zottola and Sasahara, 1994; Austin and Bergeron, 1995).

Since the continuous production of safe products is becoming increasingly important, predictive models for product contamination would greatly benefit the food industry, especially if they made it possible to optimize the production process in relation to the desired product quality. Computer models have been used for the prediction of product spoilage, for example, during storage and in quantitative risk assessments (Van Gerwen and Zwietering, 1998). An example of a industrial validated predictive model for contamination of heat treatment equipment is given below (De Jong *et al.*, 2002a).

The mechanism of contamination as used by the model is shown in Fig. 1.12. Bacteria adhere to the surface and multiply. A number of bacterial cells are released to the milk flow. Two mass balances form the basis of the model: one balance for the wall on which bacteria adhere and one for the liquid. The bacterial growth as a function of the operating time t at position x on the wall in a tubular plug flow reactor is defined by the transfer equation as

Change in wall coverage with time = Produced at the surface − Released + Adhered

$$\frac{dn_w}{dt} = \mu_T n_w (1-\beta) + k_a c \tag{1.5}$$

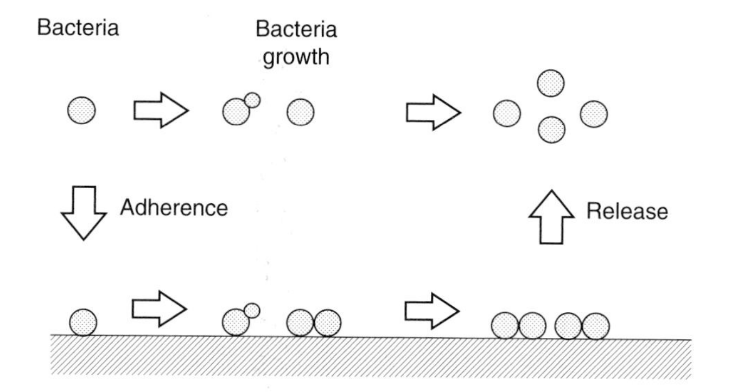

Fig. 1.12 Mechanism of growth of microorganisms in heating equipment.

where n_w is the wall coverage in cfu m^{-2}, μ_T is the bacterial growth rate at temperature T (s^{-1}), β is the fraction of generated bacteria which is released into the bulk, and k_a is the adhesion constant (m s^{-1}).

Furthermore, the release of bacteria from the wall is related to the wall coverage according to the following equation:

$$\beta = 1 - A \cdot e^{-k_r n_w} \tag{1.6}$$

where A is equal to $1-\beta$ at $n_w = 0$, and k_r is a release constant. This equation implies the assumption that, at complete wall coverage, all the bacteria grown at the bacteria–liquid interface are released to the bulk.

The local bulk concentration c (cfu m^{-3}) at operating time t follows from the component (bacteria) balance:

Change in bulk concentration with position
= Released − Adhered + Produced in the bulk − Destroyed

$$\frac{\mathrm{d}c}{\mathrm{d}x} = \frac{\pi d}{\phi}\left(\beta \mu_T n_w - k_a c\right) + \frac{\pi d^2}{4\phi}\left(\mu_T - k_d\right)c \tag{1.7}$$

where ϕ is the product flow in (m^3 s^{-1}), k_d the destruction constant (s^{-1}) and d the hydraulic diameter of the equipment (m). Since the wall temperature is a function of the position in the equipment, the differential Equations (1.5) and (1.7) have to be solved numerically in parallel. In the case of *Strep. thermophilus*, the contamination can be predicted using the following values of the model parameters: $k_a = 4.14 \times 10^{-8}$ m s^{-1}, $A = 0.82$ and $k_r = 6.1 \times 10^{-12}$ m^2 cfu^{-1} (De Jong *et al.*, 2002a).

Figure 1.13 shows the effect of adherence of microorganisms to the wall of the equipment on product contamination (industrial conditions). The predicted concentration of thermophilic bacteria in whey after pre-heating the milk before it enters an evaporator is shown as a function of operating time. In the case where no adherence or growth occurs (the lower line, consisting of short dashes in Fig. 1.13), the outlet concentration will be nearly two $\log_{10}$ cycles lower than the concentration in the raw material as a result of pasteurization. If the growth of bacteria in the whey (product) phase is taken into account, the upper line (consisting of longer dashes in Fig. 1.13)

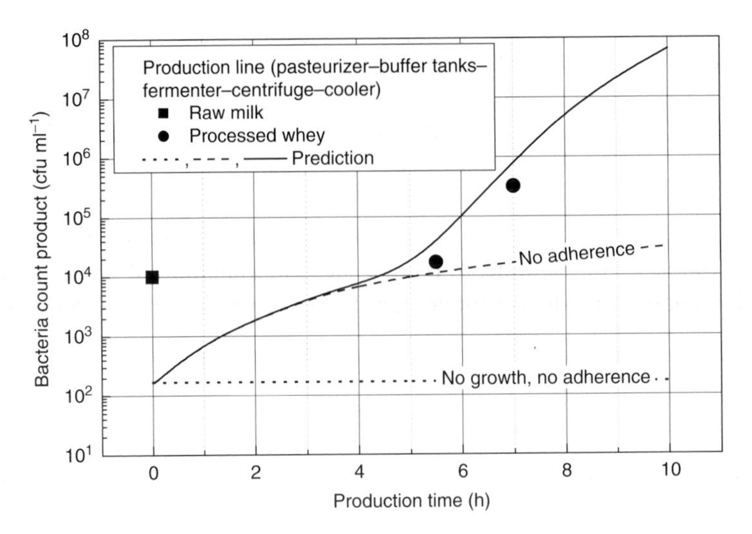

Fig. 1.13 Concentration of *Strep. thermophilus* in the product as a function of the operating time of downstream processed whey: effect of the local adherence and growth in processing equipment.

will give the actual situation. Figure 1.13, however, shows clearly that the adherence of bacteria is a major factor in describing the actual increase in the bacterial load with operating time.

Important design factors determining the bacterial concentration arising from adherence are the temperature profile and the surface/volume ratio of the installation, and the local shear stress in the equipment. Predictive models may be applied with good results to minimize bacterial contamination (De Jong *et al.*, 2002a).

1.5 Optimization

1.5.1 Introduction

Since the production of safe products is becoming increasingly important, predictive models for product contamination greatly benefit the food industry, especially if it is possible to optimize the process operation in relation to the desired product quality and safety.

In general, three types of predictive models are necessary for optimization and improvement of food heat treatments:

(1) Process models that describe the production chain in terms of model reactors. In general, process models are based on energy and mass balances of the liquid phase and not on the food components or contaminants. For example, a plate heat exchanger can be described by at least four plug flow reactors in series: upstream regenerative section, heater, holding tube and downstream regenerative section. All the plug flow reactors must have the same volume and specific surface area as the equipment itself (De Jong, 1996). The main output of such models is a temperature–time history of the food product. In cases where water is removed

(e.g. evaporating, drying), the local water content is important, since an additional concentration change of food components and contaminants is introduced.

(2) Kinetic models that predict the transformation of food components and contaminants related to the food properties recognized by the consumer. These models include, for example, the denaturation and aggregation of proteins, the inactivation of enzymes, bacteria and spore inactivation, contamination and the formation of reaction products (pigments, (off-)flavours). In some cases, the models are quite complex. For example, to predict the contamination of bacteria in the production chain, a predictive model for the concentration of microorganisms as a result of growth, adherence, release and inactivation is needed.

(3) Predictive kinetic models for estimation of the operating costs related to process operation. In many processes, the operating costs are governed by microbial and physical fouling. In cases where it is possible to predict the amount of protein and mineral deposits and the number of adhered and growing bacteria, it is relatively simple to estimate the operating costs.

In order to simulate a heat treatment in the food production chain with respect to food properties and operating costs, the model types II and III are integrated with the process model (type I). All three types of kinetic models have been developed and validated for industrial application. In this section, a general procedure for optimization of the heat treatment in the food production chain is described.

The operating costs of many food production chains primarily depend on microbial and physical fouling of the equipment (Sections 1.4.2 and 1.4.3). In general, process operating times at relatively low temperatures (<70°C) are due to adherence and growth of bacteria. The operating time of equipment at temperatures above 80°C is determined largely by the deposition of protein and minerals. The amount of fouling can be related to the costs due to cleaning, changeover (rinsing losses), depreciation, energy, operator, pollution and product losses (De Jong, 1996).

1.5.2 Approach

For the application of models in the food industry, an approach is needed that integrates the three types of models (process, product and costs model). Another need is kinetic data. The increasing availability of predictive kinetic models and necessary kinetic data has stimulated a reaction engineering approach to obtain optimal product quality (0). The functional properties of the product and the operating costs of the equipment are largely determined by conversion of so-called key components in the raw materials processed. The main control factors for product and process optimization are the temperature–time relationship and the configuration of the processing equipment. In order to determine the optimal values of the control variables, a general objective function is used:

$$F(u,x) = \alpha c_{\text{quality}}(u,x) + \beta c_{\text{operation}}(u) \tag{1.8}$$

where u is a vector of process control variables (e.g. temperature, flow), and x is a vector of desired product properties related to food quality and safety. The value of c_{quality} depends on the outcomes of the predictive models for contamination and transformation of food components. $c_{operation}$ is related to the operating costs. The optimal configuration and operation of a production chain are achieved by minimization of the objective function. To avoid trivial and undesired solutions, the weight factors α and β are introduced. These weight factors give the relative importance of each term of the objective function. For example, too high a value of β may result in a very clean and cheap production process but an inferior product quality.

Figure 1.14 shows a general approach for process and product development by use of predictive kinetic models. In order to optimize a production chain, first the available raw materials and ingredients, the desired product properties and a general process description should be known. Based on the desired product properties, the desired conversion of key components is determined. Embedding the predictive models for product properties (II) and for physical and microbial fouling (III) into the process

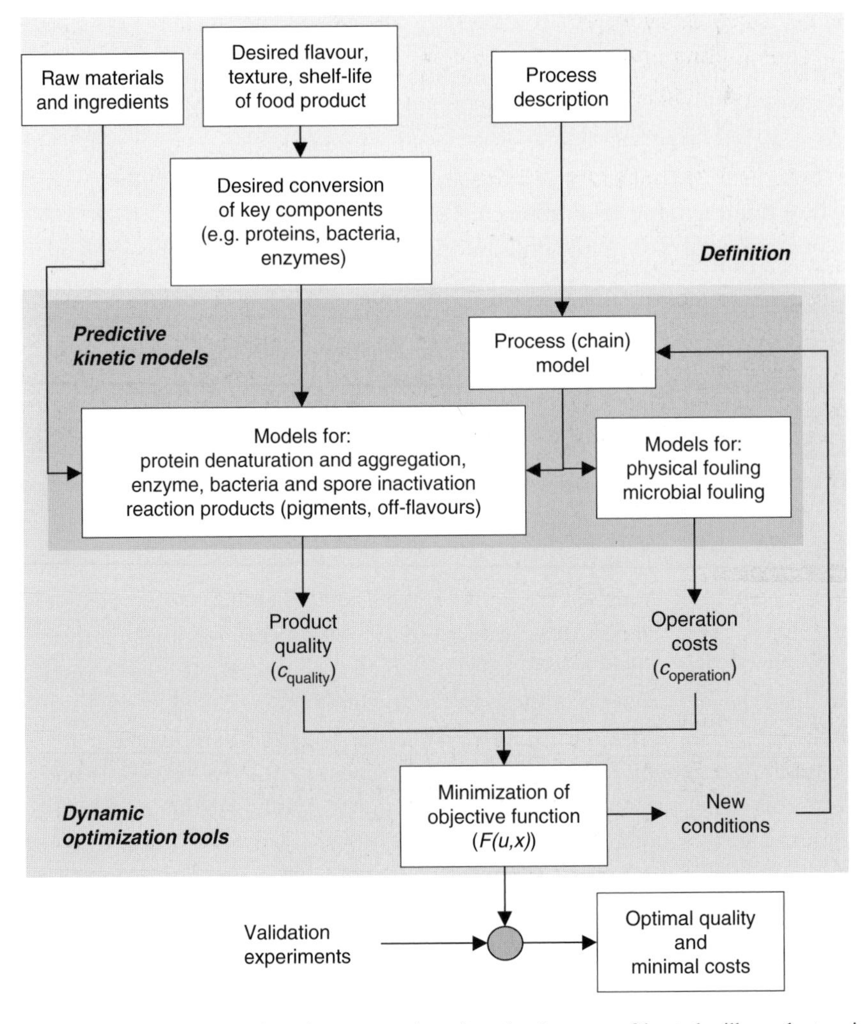

Fig. 1.14 Schematic representation of process and product development of heated milk products using predictive kinetic models.

model (I), the values of c_{quality} and $c_{\text{operation}}$ can be calculated. Next, the evaluation of objective function results in improved conditions (i.e. control factors) for the production chain and the evaluation of the predictive models is repeated. This process goes on until the minimum of the objective function is obtained, i.e. the optimal conditions are found. Before the optimization results are applied, some validation experiments can be performed.

Some examples of recent industrial applications that accelerated the process and product development are:

- improvement of the performance (extended operating time) of a cheese milk pasteurizer;
- examination of two evaporator designs with respect to bacterial growth;
- determination of critical points in the downstream processing of whey;
- extended operating time (200%) of a production chain for baby food.

For the Dutch dairy industry, it has been calculated that in terms of energy, the reduction of fouling and contamination by predictive models has already a savings potential of 90 million m^3 of gas (De Jong and Van der Horst, 1997).

1.5.3 Case study: pasteurization

To illustrate the application of the described procedure for optimizing food production chains, the following case study has been performed. A heating process with a capacity of 40 tonnes of skim milk per hour consists of a regenerative section, a heating section, two holder sections and a cooler. In Fig. 1.15, the scheme of the process is shown with some preliminary temperatures and residence times. In order to have a process model, the equipment is transformed to a cascade of model reactors. Details of the characteristic dimensions are given in the literature (De Jong *et al.*, 2002b). The objective is to develop a process that meets the specifications as given in Table 1.3. The objective function is defined as:

$$F(u,x) = \sum_{i=1}^{3} \alpha_i \left(\frac{x_{i,\text{des}} - x_i(\text{u})}{x_{i,\text{des}}} \right)^2 + F_{\text{cos}t} \tag{1.9}$$

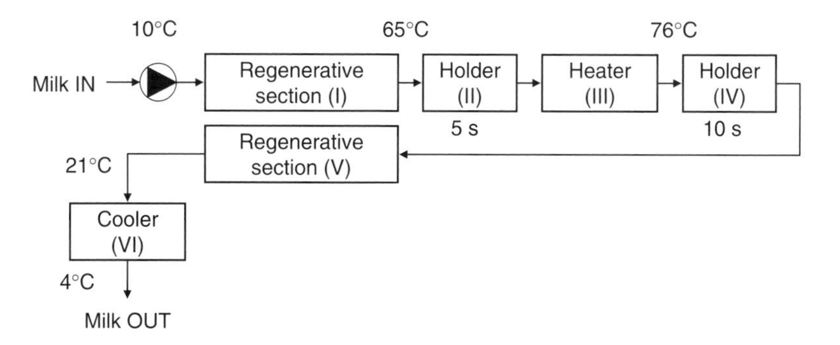

Fig. 1.15 Scheme of the pre-design of the pasteurization process.

Table 1.3 Product and process specifications of pasteurization (process variables).

Process variable	Desired value ($x_{i,\mathrm{des}}$)	Weight factor (α_i)
β-Lactoglobulin denaturation	72–80	4–8
Decimal reduction *Strep thermophilus*	130–145	75–90
Production costs	142–150	60–85

where

$$F_{\cos t} = \frac{c_{\mathrm{operation}} t_{\mathrm{operation}} + c_{\mathrm{solids}} t_{\mathrm{production}} \iint\limits_{x,t} J_{x,t} \mathrm{d}t\,\mathrm{d}x}{t_{\mathrm{production}} \phi} \tag{1.10}$$

where the integral term is the total amount of deposits after 1 h of production, and ϕ is the capacity of the process in tonnes per hour and:

$$t_{\mathrm{production}} = \frac{t_{\mathrm{operation}} t_{\mathrm{run}}}{t_{\mathrm{run}} + t_{\mathrm{cleaning}}} \tag{1.11}$$

and:

$$t_{\mathrm{run}} = t \text{ if } (C_{\mathrm{Strep.thermophilus}} > 0.0001 \text{ cfu ml}^{-1}) \tag{1.12}$$

Values of several constants are given in the literature (De Jong *et al.*, 2002b). The weight factor α_i is introduced to avoid trivial and undesired solutions; for example, a low-cost process resulting in an inferior product quality. The chosen values of the weight factors are determined by the relative importance of the different product properties. However, since the relationship between the weight factor values and the optimization results are not clear beforehand, the determination of the weight factor value is an iterative process in consultation with industrial experts.

In this case, the control variables (u) are limited to two, the heating temperature and the residence time at this temperature in the second holder section. With two control variables, surface plots can present the results of the computer model simulations. Figure 1.16 shows the results of the model evaluations.

According to Equation 1.12, in this case it is assumed that the maximum run time is limited by contamination with *Strep. thermophilus* and not limited by the deposition of proteins and minerals on the wall surface. At a temperature lower than 79°C and run times shorter than 30 h, the deposition layer does not result in insufficient heat transfer. Related to that, the objective function accounts for the increasing amount of product losses. According to Fig. 1.16d, the operating costs per tonne of heated product decrease with temperature and residence time. This is due to the increased operating time (Fig. 1.16e) resulting in a extended annual production time (Fig. 1.16b). However, at higher temperatures and longer residence times, the amount of denatured proteins exceed the desired value of 2.5% resulting in a substantial contribution to

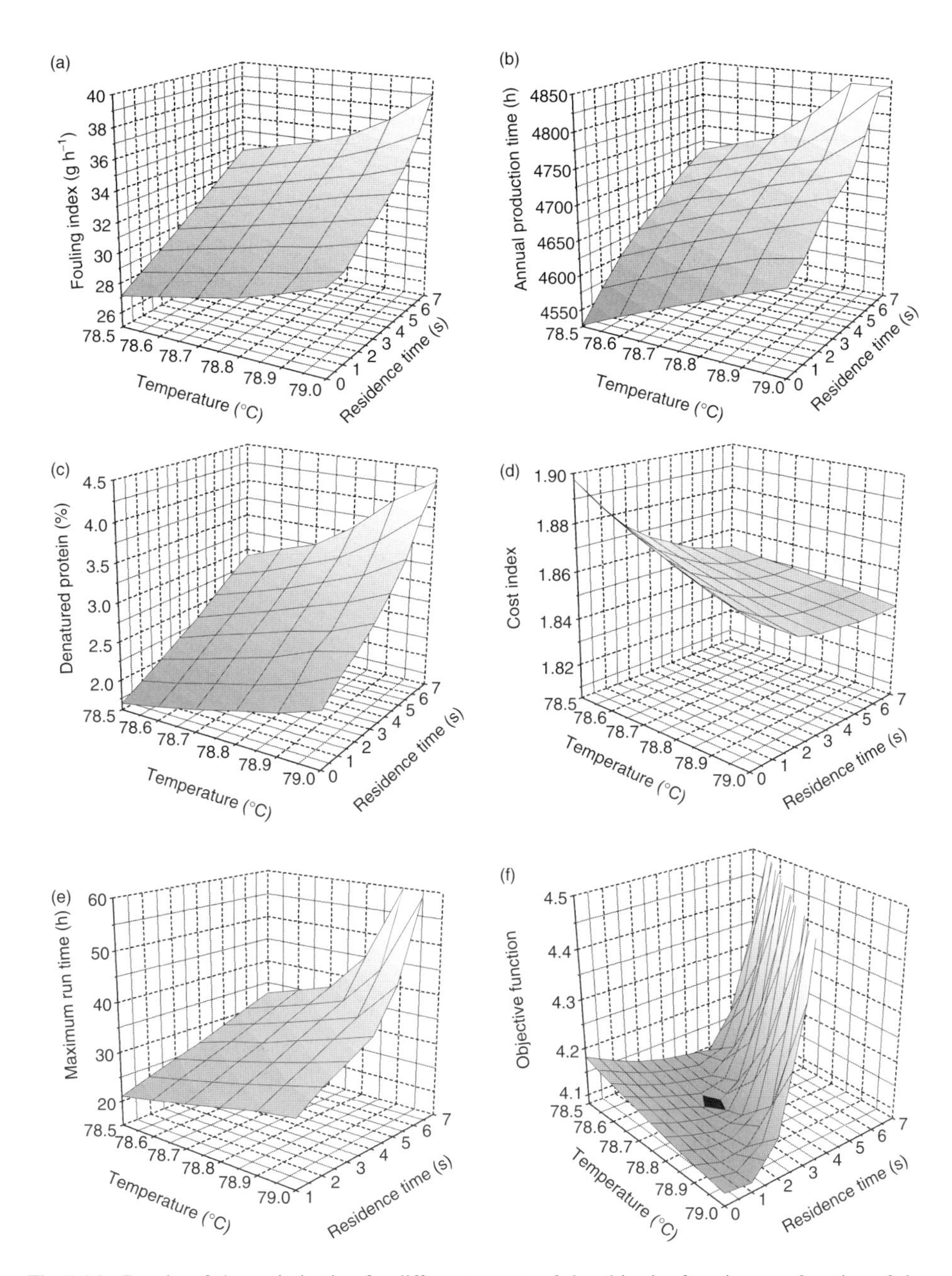

Fig. 1.16 Results of the optimization for different aspects of the objective function as a function of the control variables: (a) fouling index; (b) annual production; (c) protein denaturation; (d) cost index (€ ton^{-1}); (e) maximum run time; (f) objective function evaluation.

the objective function (Fig. 1.16f). Catalase activity was not a key parameter in the temperature and residence time region applied. At a temperature of 78.5–79.0°C and a residence time of 1 s or longer, the activity was <0.1%.

In Table 1.4, the optimal values of the control variables and the related process variables are listed. Compared with the initial preliminary design (10 s, 76°C), the

Table 1.4 Optimization of the pasteurization process.

Variable	Reference	Optimal
Control variable	76.0	78.7
Heating temperature (°C)	10	3
Residence time (s)		
Process variable		
Catalase activity (%)	1.6	0.01
β-Lactoglobulin denaturation (%)	0.84	2.46
Decimal reduction *Sterptococcus thermophilus*	6	6
Production costs (€ ton^{-1})	2.16	1.86

operating costs could be decreased by 14%. At an annual production time of 4700 h, this means an estimated cost saving of €58 000.

1.6 Conclusions and future trends

Heating of milk has been rationalized to a great extent by introducing the chemical engineering approach. In this approach, the milk is described as a fluid with a number of key components and the equipment is described as a number of chemical model reactors. Processes for heating can be designed on the basis of the desired product specifications. After determination of the optimal temperature–time combination, the appropriate heating equipment can be selected and designed.

Although heating is a well-developed and relatively robust preservation technology, there are still a number of challenges for improvement. For example, to improve the nutritional value of heated dairy products, there is a need for heating technologies that realizes hyper-short treatment at high temperature. Developments such as the ISI technology should be encouraged. Also, for a number of products, the (bio)fouling of the heating equipment and its related negative consequences limits the application.

1.6.1 Longer operating times

In order to increase the operating time, further new technologies have to be developed and applied. The following routes can be followed:

- Increasing the induction phase of protein and mineral fouling. There are some indications that the induction is increased when the equipment surface is very smooth (e.g. $Ra < 0.2$ µm by electropolishing of steel, see also Fig. 1.17) and the flow is well developed. Some researchers claim that, with this polished steel, the adherence of bacteria is also reduced substantially.
- Removing the fouling components and treating them separately, i.e. split stream processing. This is, for example, done by removal of the bacteria by membrane filtration. However, the maximum reduction is around 4 log, not sufficient to eliminate adherence and growth in equipment. Reducing the amount of protein fouling

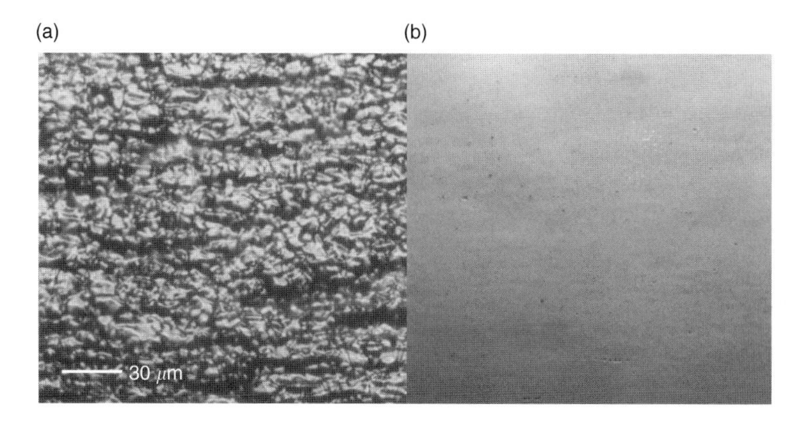

Fig. 1.17 Effect of electro-polishing: (a) normal steel; Ra = 0.4 μm; (b) electro polished steel; Ra = 0.25 μm (Dockweiler).

by removal of β-lactoglobulin before the heating process would also seem to be an effective approach, albeit not easy.

- Dedicated design of critical parts of the heating equipment. For example, higher liquid velocities or surface treatment in the regions where fouling occurs.

1.6.2 Integrating technologies

In recent years many new technologies have been proposed that could replace heating as a preservation technology of milk. However, none of these technologies has succeeded so far. One of the proposed alternatives to heating is the application of pulsed electric fields (PEF). Here, a product is exposed to a number of very short pulses, which cause damage to the cell wall of the bacteria. Moreover, it is stated that the cells that survive this process multiply less quickly than cells that survive a thermal treatment (De Jong and Van Heesch, 1998). The most important disadvantages of PEF are that spores are not, or only marginally, inactivated (Pol *et al.*, 2001) and that the energy costs are relatively high. Various groups have been carrying out research into combining PEF with other technologies in order to increase the effectiveness of the technique (Wouters *et al.*, 2001).

When subjected to high pressure (>400 MPa), the cell walls of bacteria become damaged. Milk treated at 500 MPa for 3 min, for example, can be stored for 10 days at 10°C (Rademacher *et al.*, 1999). When the pressure treatment is applied at room temperature, there is almost no inactivation of spores. In combination with heating, it was observed that spores became inactivated (Meyer *et al.*, 2000).

Individually, these alternative methods are not robust, do not have any significant effect on bacterial spores and are therefore unsuitable for use in the preparation of milk products with a long shelf-life. Furthermore, the enzymes that cause spoilage are also not, or only scarcely, inactivated. The search is continuing for synergies through combinations of new technologies or combinations of new and conventional technologies. This is a direction that will still require much research and development work before use will be possible in the dairy industry.

1.6.3 Model-based control of heating processes

In Section 1.5, the optimization of heat treatment was discussed. This static, off-line optimization helps to find the optimal set-points in case of the average composition of the raw milk and constant processing conditions. However, the composition of raw milk and other ingredients is only constant to a limited extent. The same is valid for the processing conditions. For example, fouling results in changing temperature profiles in the equipment which affect the product properties of the heated product. In fact, the optimization should be repeated every minute during optimization.

This can be performed by model-based control. The predictive models and the objective function (Section 1.5) enable the in-line optimization. If predictive models are integrated into process control systems, based on actual process data and the composition of the raw materials, the models can predict the state of the process (e.g. amount of fouling, bio-film thickness and energy usage) and the state of product (degree of contamination, stability, texture). This means that the process can be controlled on product specifications instead of process conditions. By adding cost-related models, the system can continuously optimize the production process with respect to the product quality and the production costs. In Fig. 1.18, the principle of this approach is shown.

In the literature (Smit *et al.*, 2004) an example of a model-based controller is described. This system is based on predictive models, mostly similar to the models described in this chapter. It is operational in a pilot plant and shows a 10% reduction of operating costs of heating processes of milk products.

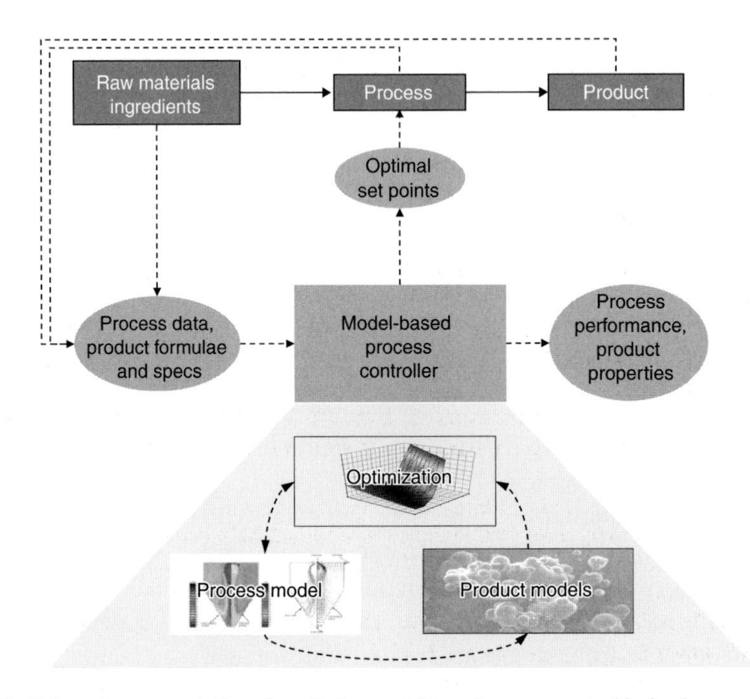

Fig. 1.18 Schematic representation of predictive-model based process control in food processing.

References

Adams, D.M. and Brawley, T.G. (1981) *Journal of Dairy Science* **64**, 1951.

Adrian, J. (1974) *World Review of Nutrition and Dietetics*, Vol. 19, p. 71. Karger, Basel.

Austin, J.W. and Bergeron, G. (1995) Development of bacterial biofilms in dairy processing lines. *Journal of Dairy Research* **62**, 509–519.

Belmar-Beiny, M.T., Gotham, S.M., Paterson, W.R., Fryer, P.J. and Pritchard, A.M. (1993) The effect of Reynolds number and fluid temperature in whey protein fouling. *Journal of Food Engineering* **19**, 119–139.

Benning, R., Petermeijer, H., Delgado, A., Hinrichs, J., Kulozik, U. and Becker, T. (2003) Process design for improved fouling behaviour in dairy heat exchangers using hybrid modelling. *Food and Bioproducts Processing* **81**, 266–274.

Bouman, S., Lund, D.B., Driessen, F.M. and Schmidt, D.G. (1982) Growth of thermoresistant streptococci and deposition of milk constituents on plates of heat exchangers during long operating times. *Journal of Food Protection* **45**, 806–812.

Brands, C. (2002) *Kinetic Modelling of the Maillard Reaction Between Proteins and Sugars.* Wageningen University.

Britten, M., Green, M.L., Boulet, M. and Paquin, P. (1988) Deposit formation on heated surfaces: effect of interface energetics. *Journal of Dairy Research* **55**, 551–562.

Burton, H. (1988) *Ultra-High-Temperature Processing of Milk and Milk Products.* Elsevier Applied Science, London.

Codex Alimentarius (2003). *Report of the Thirty-Fourth Session of the Codex Committee on Food Hygiene*, Orlando, FL. Appendix III. Proposed draft of Code of Hygienic Practice for Milk and Milk Products.

Dannenberg, F. and Kessler, H.G. (1988b) Effect of denaturation of β-lactoglobulin on texture properties of set-style non-fat yoghurt II. Firmness and flow properties. *Milchwissenschaft* **43,** 700–704.

Dannenberg, F. and Kessler, H.G. (1988a) Reaction kinetics and functional aspects of whey proteins in milk. *Journal of Food Science* **53**, 258–263.

Dannenberg, F. (1986) Zur Reaktionskinetik der Molkenproteindenaturierung und deren technologischer Bedeutung. Thesis, Technical University Munich.

De Jong, P., Bouman, S. and Van der Linden, H.J.L.J. (1992) Fouling of heat treatment equipment in relation to the denaturation of β-lactoglobulin. *Journal of the Society of Dairy Technology* **45**, 3–8.

De Jong, P (1997)., Impact and control of fouling in milk processing. *Trends in Food Science and Technology* **8**, 401–405.

De Jong, P. (1996) *Modelling and Optimization of Thermal Treatments in the Dairy Industry.* Delft University.

De Jong, P., Te Giffel, M.C. and Kiezebrink, E.A. (2002a) Prediction of the adherence, growth and release of microorganisms in production chains. *International Journal of Food Microbiology* **74**, 13–25.

De Jong, P., Te Giffel, M.C., Straatsma, J. and Vissers, M.M.M. (2002b) Reduction of fouling and contamination by predictive kinetic models. *International Dairy Journal* **12**, 285–292.

De Jong, P. and Van den Berg, G. (1993) Unpublished results. NIZO Ede.

De Jong, P. and Van der Horst, H.C. (1997) *Extended Production and Shorter Cleaning Times, a Review of Research Projects Financed by the Dutch Ministry of Economics Affairs.* Novem, Sittard, Netherlands.

De Jong, P. and Van der Linden, H.J.L.J. (1992) Design and operation of reactors in the dairy industry. *Chemical Engineering Science* **47**, 3761–3768.

De Jong, P. and Van Heesch, E.J.M. (1998) Review: effect of pulsed electric fields on the quality of food products. *Milchwissenschaft* **53**, 4–8.

De Jong, P., Waalewijn, R. and Van der Linden, H.J.L.J. (1993) Validity of a kinetic fouling model for heat treatment of whole milk. *Lait* **73**, 293–302.

De Wit, J.N. and Klarenbeek, G. (1988) Technological and functional aspects of milk proteins. In: *Milk Proteins in Human Nutrition, Proceedings of the International Symposium*. Steinkopff, Kiel, Germany.

Delplace, F. and Leuliet, J.C. (1995) Modelling fouling of a plate heat exchanger with different flow arrangements by whey protein solutions. *Food Bioprod. Process* **73**, 112–120.

Delsing, B.M.A. and Hiddink, J. (1983) Fouling of heat transfer surfaces by dairy liquids. *Netherlands Milk and Dairy Journal* **49**, 139–148.

Denny, C.B., Shafer, B. and Ito, K. (1980) Inactivation of bacterial spores in products and on container surfaces. In: *Proceedings of the International Conference on UHT Processing*, Raleigh, NC.

Driessen, F.M. (1983) Lipases and proteinases in milk. Thesis, Agricultural University Wageningen.

Evans, D.A., Hankinson, D.G. and Litsky, W. (1970) Heat resistance of certain pathogenic bacteria in milk using a commercial plate heat exchanger. *Journal of Dairy Science* **53**, 1659–1665.

FAO (2005). http://www.fao.org/docrep/meeting/008/j2308e/j2308e02.

Flint, S., Brooks, J., Bremer, P., Walker, K. and Hausmann, E. (2002) The resistance to heat of thermo-resistant streptococci attached to stainless steel in the presence of milk. *Journal of Industrial Microbiology and Biotechnology* **28**, 134–136.

Fox, P.F. and Stepaniak, L. (1983) Isolation and some properties of extracellular heat-stable lipases from pseudomonas-fluorescens-strain AFT-36. *Journal of Dairy Research* **50**, 77–89.

Fryer, P.J. (1989) The uses of fouling models in the design of food process plant. *Journal of the Society of Dairy Technology* **42**, 23–29.

Galesloot, T.E. and Hassing, F. (1983) Effect of nitrate and chlorate and mixtures of these salts on the growth of coliform bacteria: results of model experiments related to gas defects in cheese. *Netherlands Milk and Dairy Journal* **37**, 1–10.

Grandison, A.S. (1988) UHT-processing of milk: seasonal variation in deposit formation in heat exchangers. *Journal of the Society of Dairy Technology* **41**, 43–49.

Hallström, B., Skjöldebrand, C. and Trägårdh, C. (1988) *Heat Transfer and Food Products*. Elsevier Applied Science, London.

Hermier, J., Begue, P. and Cerf, O. (1975) Relationship between temperature and sterilizing efficiency of heat treatments of equal duration. Experimental testing with suspensions of spores in milk heated in an ultra-high-temperature sterilizer. *Journal of Dairy Research* **42**, 437–444.

Hillier, R.M. and Lyster, R.L.J. (1979) Whey protein denaturation in heated milk and cheese whey. *Journal of Dairy Research* **46**, 95–102.

Holdsworth, S.D. (1992) *Aseptic Processing and Packaging of Food Products*. Elsevier Applied Science, London.

Huijs, G., Van Asselt, A.J., Verdurmen, R.E.M. and De Jong, P. (2004) High speed milk, a new way of treating milk. *Dairy Industries International* November, pp. 30–32.

Jeurnink, T.J.M. and De Kruif, C.G. (1995) Calcium concentration in milk in relation to heat stability and fouling. *Netherlands Milk and Dairy Journal* **49**, 139–148.

Jeurnink, T.J.M. (1991) Effect of proteolysis in milk on fouling in heat exchangers. *Netherlands Milk and Dairy Journal* **45**, 23–32.

Jeurnink, T.J.M. (1995) Fouling of heat exchangers by fresh and reconstituted milk and the influence of air bubbles. *Milchwissenschaft* **50**, 189–193.

Jeurnink, T.J.M., Walstra, P. and De Kruif, C.G. (1996) Mechanisms of fouling in dairy processing. *Netherlands Milk and Dairy Journal* **50**, 407–426.

Kessler, H.G. and Fink, R. (1986) Changes in heated and stored milk with interpretation by reaction kinetics. *Journal of Food Science* **51**, 1105–1111.

Kessler, H.G. (1996) *Lebensmittel- und Bioverfahrenstechnik, Molkereitechnologie*. Verlag A. Kessler, Freising, Germany.
Klijn, N., Herrewegh, A. and de Jong, P. (2001) Heat inactivation data for *Mycobacterium avium* subspp. *paratuberculosis*: implications for interpretation. *Journal of Applied Microbiology* **91**, 697–704.
Knox, T. (2003) *Nzfsa Dairy and Plants Standard. Dairy Heat Treatments, Version 121.1.*
Lalande, M., Tissier, J.P. and Corrieu, G. (1985) Fouling of heat transfer surfaces related to β-lactoglobulin denaturation during heat processing of milk. *Biotechnol. Progress* **1**, 131–139.
Langeveld, L.P.M., van Montfort-Quasig, R.M.G.E., Weerkamp, A.H., Waalewijn, R. and Wever, J.S. (1995) Adherence, growth and release of bacteria in a tube heat exchanger for milk. *Netherlands Milk and Dairy Journal* **49**, 207–220.
Luf, W. (1989) Erhitzungsbedingte chemische Veränderungen in Milch und deren Nachweis. *Milchwirtschaftliche Berichte* **101**, 217–223.
Lyster, R.L.J. (1965) The composition of milk deposits in a UHT plant. *Journal of Dairy Research* **32**, 203–208.
Meyer, R.S., Cooper, K.L., Knorr, D. and Lelieveld, H.L.M. (2000) High pressure sterilization of foods. *Food Technology* **54**, 67–72.
Mortier, L., Braekman, A., Cartuyvels, D., Van Renterghem, R. and De Block, J. (2000) Intrinsic indicators for monitoring heat damage of consumption milk. *Biotechnologie Agronomie Société et Environnement* **4**, 221–225.
Mozes, N., Marchal, F., Hermesse, M., Haecht, J.L. van, Reuliaux, L., Léonard, A.J. and Rouxchet, P.G. (1987) Immobilization of microorganisms by adhesion; interplay of electrostatic and non-electrostatic interactions. *Biotechnology and Bioengineering* **30**, 439–450.
Paterson, W.R. and Fryer, P.J. (1988) A reaction engineering theory for the fouling of surfaces. *Chemical Engineering Science* **43**, 1714–1717.
Patil, G.R. and Reuter, H. (1988) Deposit formation in UHT plants. III. Effect of pH of milk in directly and indirectly heated plants. *Milchwissenschaft* **43**, 360–362.
Peri, C., Pagliarini, E. and Pierucci, S. (1988) A study of optimizing heat treatment of milk I. Pasteurization. *Milchwissenschaft* **43**, 636–639.
Pol, I.E. Van Arendonk, W.G.C., Mastwijk, H.C., Krommer, J., Smid, E.J. and Moezelaar, R. (2001) Sensitivities of germinating spores and carvacrol-adapted vegetative cells and spores of *Bacillus cereus* to nisin and pulsed-electric-field treatment. *Applied and Environmental Microbiology* **67**, 1693–1699.
Rademacher, B., Hinrichs, J., Mayr, R. and Kessler, H.G. (1999), Reaction kinetics of ultra-high pressure treatment of milk. In: Ludwig, H. (ed.) *Advances in High Pressure Bioscience and Biotechnology*, pp. 449–452. Springer-Verlag, Berlin.
Read, R.B., Schwartz, C. and Lisky, W. (1961) Studies on the thermal destruction of *Escherichia coli* in milk and milk products. *Applied Microbiology* **9**, 415–418.
Rosenberg, E. (1986) Microbial surfactants. *CRC Critical Reviews in Biotechnology* **3**, 109–132.
Rowe, M.T., Grant, I.R., Dundee, L. and Ball, H.J. (2000) Heat resistance of *Mycobacterium avium* subspp. *paratuberculosis* in milk. *Irish Journal of Agricultural and Food Research* **39**, 203–208.
Rynne, N.M., Beresford, T.P., Kelly, A.L. and Guinee, T.P. (2004) Effect of milk pasteurization temperature and *in situ* whey protein denaturation on the composition, texture and heat-induced functionality of half-fat Cheddar cheese. *International Dairy Journal* **14**, 989–1001.
Schraml, J.E. and Kessler, H.G. (1996) Effects of concentration on fouling of whey. *Milchwissenschaft* **51**, 151–154.
Schreier, P.J.R., Green, C., Pritchard, A.M. and Fryer, P.J. (1996) Heat exchanger fouling by whey protein solutions. In: Fryer, P.J., Hasting, A.P.M. and Jeurnink, T.J.M. (eds), *Fouling and Cleaning in Food Processing*, pp. 9–17. ECSC-EC-EAEC, Brussels.

Singh, H. (1993), Heat-induced interactions of proteins in milk. In: *Protein and Fat Globule Modifications by Heat Treatment and Homogenization*, International Dairy Federation Special Issue No. 9303, pp. 191–204.

Skudder, P.J., Brooker, B.E., Bonsey, A.D. and Alvarez-Guerrero, N.R. (1986) Effect of pH on the formation of deposit from milk on heated surfaces during ultra-high-temperature processing. *Journal of Dairy Research* **53**, 75–87.

Smit, F., Straatsma, J., Vissers, M.M.M., Verschueren, M. and de Jong, P. (2004) NIZO Premia and Premic: Off-line and in-line product and process control tools for the food industry. In: de Jong, P. and Verschueren, M. (eds), *Proceedings FOODSIM 2004*, pp. 100–102.

Stadhouders, J., Hup, G. and Langeveld, L.P.M. (1980) Some observations on the germination, heat resistance and outgrowth of fast-germinating and slow-germinating spores of *Bacillus cereus* in pasteurized milk. *Netherlands Milk and Dairy Journal* **34**, 215–228.

Te Giffel, M.C., Beumer, R.R., Langeveld, L.P.M. and Rombouts, F.M. (1997) The role of heat exchangers in the contamination of milk with *Bacillus cereus* in dairy processing plants. *International Journal of Dairy Technology* **50**, 43–47.

Te Giffel, M.C., Meeuwisse, J. and De Jong, P. (2001) Control of milk processing based on rapid detection of microorganisms. *Food Control* **12**, 305–309.

Te Giffel, M.C., Van Asselt, A.J. and De Jong, P. (2005) Shelf-life extension: technological opportunities for dairy products. In: *Proceedings of the IDF-Conference*, Vancouver.

Thom, R. (1975) Über die Bildung von Milchansatz in Plattenerhitzern. *Milchwissenschaft* **30**, 84–89.

Tissier, J.P., Lalande, M. and Corrieu, G. (1984) A study of milk deposit on heat exchange surface during UHT treatment. In: *Engineering and Food*, McKenna, B.M. (ed.), pp. 49–58. Elsevier Applied Science, London.

Van Gerwen, S.J.C. and Zwietering, M.H. (1998) Growth and inactivation models to be used in quantitative risk assessments. *Journal of Food Protection* **61**, 1541–1549.

Walstra, P. and Jennes, R. (1984) *Dairy Chemistry and Physics*. John Wiley, New York.

Wouters, P.C., Álvarez, I. and Raso, J. (2001) Critical factors determining inactivation kinetics by pulsed electric field food processing. *Trends in Food Science and Technology* **12**, 112–121.

Yoon, J. and Lund, D.B. (1989) Effect of operating conditions, surface coatings and pretreatment on milk fouling in a plate heat exchanger. In: Kessler, H.G. and Lund, D.B. (eds.), *Fouling and Cleaning in Food Processing*, pp. 59–80. University of Munich, Munich.

Zottola, E.A. and Sasahara, K.C. (1994) Microbial biofilms in the food processing industry – Should they be a concern? *International Journal of Food Microbiology* **23**, 125–148.

Chapter 2

Applications of Membrane Separation

Athanasios Goulas and Alistair S. Grandison

2.1 Introduction

Membrane filtration processes are defined as separations, based primarily on size differences, between two or more components in the liquid phase. Their spectrum ranges from the millimetre (for coarse filters) to the Angstrom (Å) scale (for reverse osmosis and gas-separation membranes) (Lewis, 1996a; Cheryan, 1998; Pellegrino, 2000). Commercial membrane processing has developed over the last 40 years, and is becoming increasingly important in the food industry for concentration and fractionation operations.

The term membrane filtration describes an operation in which the filter is a membrane. A membrane is defined as a structure with lateral dimensions much greater than its thickness through which mass transfer may occur under a variety of driving forces (Pellegrino, 2000). The membrane, in this case, acts as a selective barrier (an interface) that permits passage of certain components and retains others (Cheryan, 1998). Usually, the main criterion for separation is size, although other factors, such as surface charge, or shape of the molecule or particle, may have an effect. The material that passes through the membrane is the permeate stream, which is depleted with respect to the retained components, whereas the fraction that is retained by the membrane is known as the concentrate or, more correctly, the retentate stream, which is enriched in the non-permeable components.

Membrane separation processes have the following significant advantages over competing approaches to concentration or separation used in the food and biotechnology industries:

(1) No phase or state change of the solvent is required – hence, they are frequently more cost-effective.
(2) Operation is usually at relatively low temperatures – therefore, they are suitable for processing of thermolabile materials, reducing changes in flavour or other quality characteristics, and minimizing heat denaturation of enzymes.
(3) Good levels of separation can be achieved without the need for complicated heat-transfer or heat-generating equipment.
(4) Separations cover a wide spectrum of sizes, ranging over several orders of magnitude, from the smallest ions to particles such as fat globules or bacterial cells.

A major limitation of membrane processes is that they cannot concentrate solutes to dryness. The degree of concentration is limited by the extreme osmotic pressures, high viscosities or low mass-transfer rates generated at the increased solute concentrations (Cheryan, 1998).

Membrane separations have a broad range of applications in the fermentation, medical, electronic, food, beverage, chemical and automobile industries. Developments in membrane materials and technology, and reducing relative costs of membrane processing equipment are leading to wider and more innovative applications of the technology in the dairy industry. This chapter deals with the basic theory of membrane processing and applications in the dairy industry, concentrating on some of the more novel approaches.

2.2 Transport theory of membrane separation processes

2.2.1 Classification of processes

Membrane separations that are commonly used in the dairy industry are pressure-driven processes, where hydraulic pressure is used as a means of overcoming and reversing the osmotic pressure and flow, caused by the difference in solute concentrations between two liquid phases (retentate and permeate), and also the frictional forces generated between the liquid phase and membrane pore walls. The difference between applied hydraulic pressure and osmotic pressure is called the trans-membrane pressure. The major pressure-driven membrane separation processes are microfiltration (MF), ultrafiltration (UF) and reverse osmosis (RO). There are no precisely defined boundaries between them, and conceptually they are similar to each other, the distinction between them being their degree of semi-permeability and consequently their separation characteristics (Gutman, 1987). Generally the following definitions apply: 'RO ideally retains all components other than the solvent (usually water)'; 'UF retains only macromolecules or particles larger than 10–200 Å (0.001–0.02 μm)'; 'MF retains suspended particles in the range of 0.1–5 μm' (Cheryan, 1998). Nanofiltration (NF) is a more recent nomenclature for a separation which lies on the boundaries between RO and UF, and hence permits permeation of some monovalent electrolytes and small organic molecules (unlike RO), while retaining all other solute molecules that a UF membrane would generally allow to permeate. Figure 2.1 shows a classification of membrane filtration processes. It should be noted that the distinction between these processes is not exact, and the spectrum can be considered to be continuous.

These differences in the molecular size of retained and permeating solutes which characterize the processes also have a direct effect on their operating pressure range. Microfiltration and UF operate at fairly low pressures (1–15 bar), since the osmotic pressures generated from suspended particles and macromolecules are low. In contrast, the pressures involved in RO and NF are fairly high (in the range of 30–100 bar

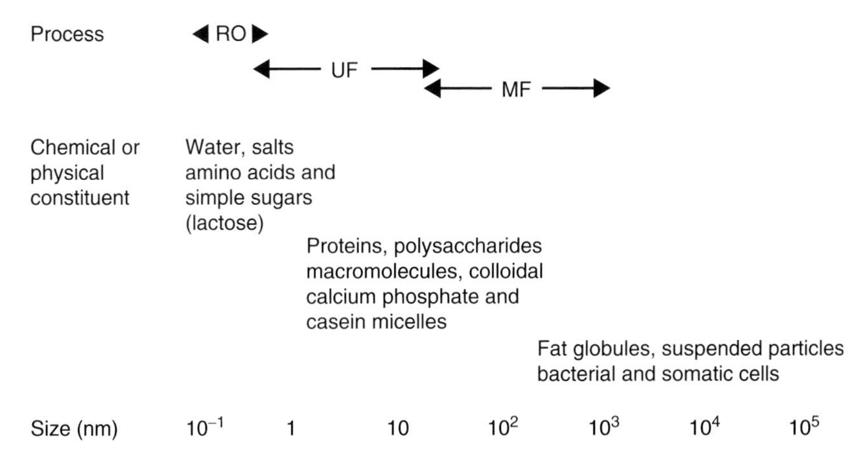

Fig. 2.1 Filtration spectrum and sizes of components.

and 10–30 bar, respectively) due to the high osmotic pressures generated from small solutes, and hence high osmotic pressure differences between retentates and permeates (Cheryan, 1998).

A further membrane separation process is electrodialysis (ED), where membranes are used to provide selectivity, but the principle of separation and the driving forces are different to the pressure-driven processes. Electrodialysis is an electrochemical separation process by which electrically charged (ionic) species are transported from one solution to another through ion-exchange (perm-selective, charged) membranes in a system that forms a combination of dialysis and electrolysis. Separation of solutes takes place according to their electrical charges, and the driving force for the separation is an electrical potential difference generated between two electrodes. Separations and concentrations may take place between salts, acids and bases from aqueous solutions, between monovalent and multivalent ions or multiple charged species, and the separation of ionized from uncharged molecules.

2.2.1.1 Microfiltration and ultrafiltration

Microfiltration and UF membranes are thought to behave like physical sieves; since they are highly porous and, even with asymmetric structures, their surface skin has discernible pores (Gutman, 1987; Lewis, 1996a). Separation is therefore principally on the basis of size, with the solvent of the solution being the component that most readily passes through these membranes. Several models have been developed for describing these processes, taking into account the resistances generated from concentration polarization and fouling (e.g. concentration-polarization model, gel-polarization model, osmotic-pressure model, adsorption and osmotic-pressure model) (Osada and Nakagawa, 1992).

Generally, in UF and MF, solvent and solute mass transfer are controlled by convective transport (Pontalier *et al.*, 1997; Pellegrino, 2000). There is no definite distinction between the two processes, the only difference being the pore size, and even that is not an absolute distinction. Osmotic pressure is very much less important in UF and MF compared with RO, because the membranes are permeable to the smaller molecules and ionic species, which are the main contributors to osmotic pressure.

Permeate flux during UF and MF is often modelled as a purely sieving process in terms of flow through a bundle of capillaries according to the Hagen–Poiseuille equation, where flux per unit area of membrane (J) is as follows:

$$J = \frac{d^2 \Delta P \pi}{32 \mu L} \tag{2.1}$$

where d is the capillary (pore) diameter, μ the dynamic viscosity and L the capillary length (membrane thickness).

In practice, this relationship is complicated by other properties of the membrane (porosity and tortuosity effects) and will change throughout processing due to concentration polarization and fouling (discussed in Section 2.2.2). However, the relationship predicts the strong influence of pore diameter on flux, as well as how increasing viscosity and membrane thickness would lead to reduced flux. It is notable that viscosity often increases during processing

2.2.1.2 Reverse osmosis and nanofiltration

Reverse osmosis is considered to be a more complex separation process since the mechanism by which ionic and small organic molecules are separated from the solvent (usually water) cannot be simply explained by the size difference between the solvent and the solute (Lewis, 1996a). These separations depend on the ionic charge, the ionic composition of the feed and the affinity of the solute for the membrane material (Gutman, 1987). The principle of RO is based on reversing the solvent flow that is caused when two solutions with different concentrations of small molecules (e.g. salts, simple sugars) are separated by a semi-permeable membrane. This pressure, generated due to the concentration difference, is termed osmotic pressure (Π), and the reversal of solvent flow against the concentration gradient by means of applied pressure is called reverse osmosis. A prediction of the osmotic pressure for dilute, non-ionized solutions can be obtained from the van't Hoff equation:

$$\Pi = RT\frac{C}{M} \tag{2.2}$$

where R is the gas constant, T the absolute temperature, C the solute concentration and M the molecular weight of the solute. For ionized solutes, this becomes:

$$\Pi = iRT\frac{C}{M} \tag{2.3}$$

where i is the degree of ionization, e.g. for NaCl $i = 2$, and for $CaCl_2$ $i = 3$. It should be noted, however, that osmotic pressure actually rises in an exponential manner with increasing concentrations, and the van't Hoff equation underestimates the osmotic pressures of more concentrated solutions. In many food fluids, osmotic pressure is derived from the combined contribution of many chemical species, and some examples are given in Table 2.1. The contribution of any species to the total osmotic pressure is inversely proportional to its molecular weight. Therefore, small species such as salts and sugars contribute much more than large molecules such as proteins. This is illustrated by the fact that seawater has a much greater osmotic pressure than a protein (casein) solution

Table 2.1 Osmotic pressures of some fluids (data adapted from Cheryan, 1986 and Lewis, 1996a).

Liquid	Approximate total solids (%)	Osmotic pressure (bar)
Milk	11	6.7
Whey	6	6.7
Orange juice	11	15.3
Apple juice	14	20.0
Sea water[a]	3.5	14.1
Casein solution[a]	3.5	0.03

[a] Calculated from the Van't Hoff equation.

at the same solids concentration. In the same context, the osmotic pressures of whey and milk are identical, while the total solids content of milk is much greater than whey as it contains higher levels of protein and fat, but approximately the same concentrations of dissolved salts and lactose, which actually determine the osmotic pressure.

Several models developed for this process address the mechanism of permeation through the membranes in different ways. The 'solution diffusion model' suggests that there is no bulk flow of solvent or solute through the membranes, but instead, instantaneous dissolution of the two components occurs on the membrane skin, followed by diffusion across the membrane. The 'preferential adsorption capillary flow model' considers that flow occurs through pores, which comprise a certain fraction of the membrane surface, and the component that preferentially will be concentrated in the permeate is that which is adsorbed most strongly on the membrane surface. A third model is based on the formation of hydrogen bonds between the membrane surface and the water of the solution, creating bound water that regulates the dissolution of solutes on the membrane surface and consequently passage through the membrane (Osada and Nakagawa, 1992; Lewis, 1996a).

Another very important principle in understanding the separation mechanism of RO is Donnan exclusion. This exclusion of ions from the membrane is due to the membranes' fixed charges (Gutman, 1987), and the phenomenon takes place as a result of the electro-neutrality conditions occurring in all solutions when more than one ionic solute is present in a mixture with different valences and different sizes. A simple example of the Donnan exclusion phenomenon is that in a solution where monovalent and divalent co-ions are present, both repulsed by the surface charge of the membrane, the monovalent ions will preferably pass through the membrane in order to maintain the electroneutrality of the permeate where their counter-ion passes without difficulty. The majority of the membranes used for RO have a surface charge but charge-free membranes do exist.

In any event, RO requires that the applied pressure exceeds the osmotic pressure of the feed, and the rate of solvent transport across the membrane is proportional to the pressure difference; hence:

$$J_{w} \propto A(\Delta P - \Delta \Pi) \tag{2.4}$$

where J_w is the solvent flux, A the membrane area, ΔP the applied pressure and $\Delta \Pi$ the osmotic pressure difference across the membrane (approximating to osmotic pressure of the solution).

Nanofiltration comprises a mixture of the above separation mechanisms for UF and RO, since it lies between them in terms of size separation, and has characteristics of both processes. Nanofiltration membranes have molecular-weight cutoffs in the range of 200–1000 Da, and reject solutes at least 2 nm in size (Tsuru *et al.*, 2000). It is a pressure-driven membrane separation technique for separating electrolytes with different valences (monovalent or multivalent ions) or ionic species from low-molecular-weight compounds such as simple sugars (Pontalier *et al.*, 1997). Since this operation lies between UF and RO, the mechanism of mass transfer is believed to be a superposition of both convective and diffusive mass transport respectively, in combination with the charge effect when electrolytes are present in the feed solutions (Pontalier *et al.*, 1997; Tsuru *et al.*, 2000). Generally, the solute transport through these membranes depends on the size and nature of the solute particles (neutral or electrolyte) and the characteristics of the membrane (pore size, surface charge). Charge and Donnan exclusion effects are very important when single or binary salt solutions are to be filtered (Timmer *et al.*, 1998), whereas the sieve effect, in both forms, convection (occurring in UF) or diffusion (occurring in RO) governs the mass transport of neutral solutes (e.g. monosaccharides) (Pontalier *et al.*, 1997).

2.2.1.3 Electrodialysis and electro-membrane filtration

Electrodialysis is an electrochemical separation process by which ionic charged species are transported from one solution to another, by transfer through one or more permselective membranes, under the influence of a DC electrical current. Although ED is a membrane separation process, it separates particles primarily according to their electrical charges rather than their sizes.

The operating principle and a typical ED cell are depicted in Fig. 2.2. When an ionized liquid, such as an aqueous salt solution, is pumped through the cells, and an electrical potential is applied between the anode and the cathode, the positively charged cations migrate towards the cathode, and the negatively charged anions migrate towards the anode. The cations pass easily through the negatively charged cation-exchange membranes but are retained by the positively charged anion-exchange membranes, while the opposite applies for the anions in the solution. This results in an increase in concentration of ions in alternate compartments, whereas the remainder become simultaneously depleted. The depleted solution is generally referred to as the 'diluate' and the concentrated solution as the 'concentrate'.

The region between the two adjacent membranes containing the diluate solution, and the concentrate solution, between the two contiguous membranes next to the

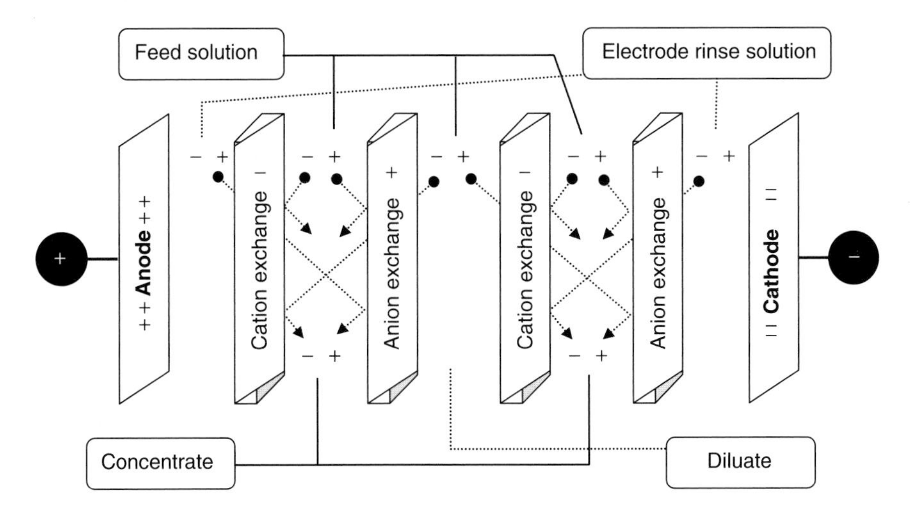

Fig. 2.2 Operating principle of electrodialysis.

diluate chamber together with the two contiguous anion- and cation-exchange membranes make up a 'cell pair'. In commercial practice, ED stacks consist of multiple cell pairs and may have hundreds of ion-exchange membranes. On the other hand, electrolysis cells also exist where two solutions, one for each electrode compartment, are separated by a single membrane. This application is based on electrode redox reactions which are electrolysis-specific properties.

The energy required in an ED process is a combination of the electrical energy required for transfer of the ionic solutes through the membranes, together with the energy necessary to pump the solution through the ED stack. A number of parameters should be taken into account when optimizing the efficiency of such a process, including the applied voltage, the concentrations of ionic species in the solutions (conductivity), solution pH and the water dissociation which occurs when excess current is applied. More information on ED can be found in Strathmann (1992) and Bazinet (2005).

An ion-exchange membrane is composed of a macromolecular support material that carries ionizable groups, similar to ion-exchange resins. Membranes that carry one type of ionizable group, either negative or positive, are called monopolar membranes, and they are permeable to only one type of ion. Bipolar membranes also exist, which consist of a cation-exchange layer, an anion-exchange layer and a hydrophilic transition layer as their junction. The main ionic groups used for the preparation of these membranes are sulfonic acid ($-SO_3^-$), carboxylic acid ($-COO^-$), arsenic acid ($-AsO_3^{2-}$), phosphoric acid ($-PO_3^{2-}$) and alkyl ammonium ($-NR_3^+$, $-NHR_2^+$, $-NH_2R^+$) groups.

A method that combines ED with the sieving principles of separation is electromembrane filtration (EMF). In EMF, conventional filtration membranes are used in combination with ion-exchange membranes, and the driving force is again the electrical potential between two electrodes. The first membranes are responsible for separation according to size, whereas the ion-exchange membranes are used to prevent degradation of the feed and permeate (usually the product) by preventing direct contact with the electrodes. Electro-membrane filtration therefore resembles ED, the main difference being that in the former, conventional MF, UF or NF membranes are used in order to allow the transport (selective isolation) of larger components (e.g. peptides) from the feed to the permeate than is usually achieved by ED (Bargeman *et al.*, 2002). During EMF, transport of solutes through the membranes by convection is undesirable, since it reduces the selectivity of the separations. For that reason, the transmembrane pressures, during EMF, between the feed and permeate compartments are usually kept to a minimum (Bargeman *et al.*, 2002).

2.2.2 Concentration polarization and fouling

In any membrane process, when the feed is switched from water to a solution, there is a marked drop in permeate flux. The flux reduction may be as great as 10 times when switching from water to milk. This phenomenon is caused by concentration polarization, which results from a localized concentration increase in solids as permeate is removed from the feed stream. When a liquid flows over a surface, frictional forces between the liquid and the surface cause a reduction of the flow near the surface, while flow in the middle of the flow channel increases. The line between lower velocity fluid and the region of uniform bulk velocity is called the boundary layer (Cheryan, 1998). If this slow-moving region of fluid is generated on top of a semi-permeable membrane (in a filtration process), it gives rise to the concentration polarization layer. The concentration polarization layer (or just concentration polarization) is essentially a distinct region on the membrane surface where the concentrations of the rejected solutes are higher than in the remainder of the bulk solution, and generally increase as the membrane surface is approached. This increased solute concentration layer occurs due to selective permeation of the solvent and any permeating solutes, compared with rejected solutes (Fig. 2.3) (Lewis, 1996a; Cheryan, 1998).

Concentration polarization complicates the separation characteristics of all the different membrane processes. The polarized concentration layer is in a dynamic equilibrium between the convective transport of each solute to the membrane surface and the back transfer of the accumulated solute into the bulk of the feed (which has a uniform lower concentration) or the permeation of the solute through the membrane. The extent of this layer of accumulated solute is inversely related to the feed flow velocity and results in significant additional resistances to permeate flux and solute permeability through the membranes. These resistances are a consequence first of the reduced trans-membrane pressure caused by the increased osmotic pressure (more pronounced in RO and NF), and second by the formation of secondary dynamic layers of solute on the membrane surface (gel layer, cake layer) causing additional resistance to the

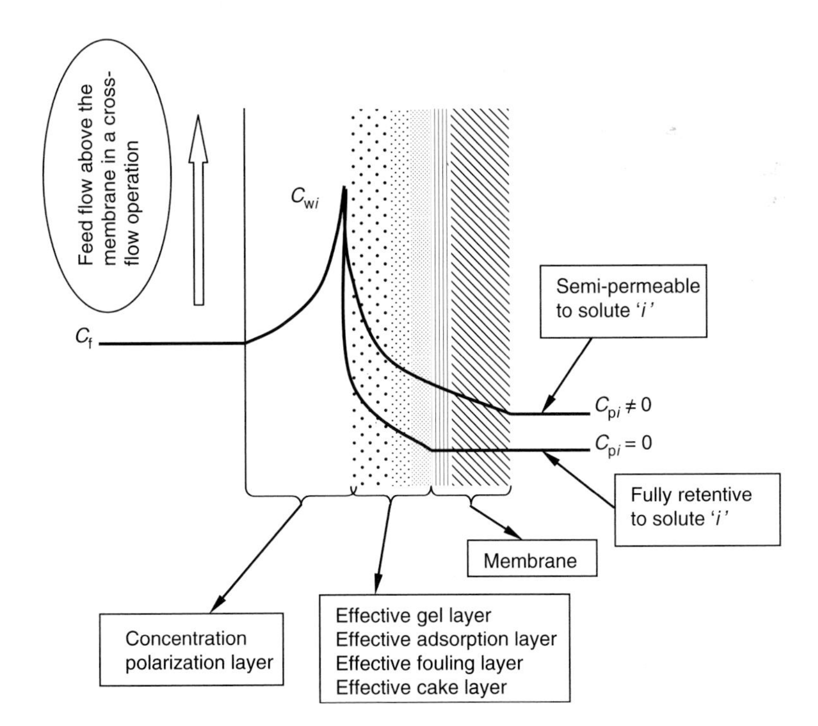

Fig. 2.3 Differential 'slice' of a pressure-driven liquid filtration process also showing concentration polarization and fouling (adapted and modified from Pellegrino, 2002). C_f: solute concentration in the bulk of the feed; C_{wi}: concentration of solute at the interface between the feed and the membranes surface including any surface layer; C_{pi}: concentration of solute in the permeate.

permeation and back-diffusion of the solute (Cheryan, 1998; Lewis, 1996a). As a result of these additional resistances caused by concentration polarization, there is a certain pressure limit (characteristic of each different process) above which there are no further changes in permeate flux and solute rejection as further pressure is applied – this is known as the pressure-independent region. Theoretically, concentration polarization effects should be reversible, the original solvent flux being restored if the solution is substituted with a pure solvent. In practice, however, this is not usually the case because the increased concentrations on the membrane surface give rise to fouling.

Fouling refers to material that becomes attached to the membrane surface and internal porous structure, which leads to an irreversible decline in flux (Gutman, 1987). Fouling affects the separation characteristics of a membrane, and also the composition of the products due to loss of deposited solutes. Fouling is caused by a variety of different interactions between membrane and solutes, and is influenced by their physicochemical nature, including factors such as conformation, charge, zeta potential and hydrophobicity. Process engineering factors, such as cross-flow velocity, pressure and temperature, affect the fouling process. Generally proteins, lipids and salts (especially calcium phosphates) cause the major problems in dairy systems.

There are several models in the literature which describe flux and mass transfer in membrane processes, taking into account the effects of concentration polarization and fouling. A simplified general expression for the solvent and solute flux is given by the following equations (Pellegrino 2000), without taking into account any charge effects:

$$\text{Solvent flux:}\quad J_\mathrm{v} = \frac{\Delta P - \Delta\Pi}{\left[(t_\mathrm{m}/P'_\mathrm{v}) + (t_\mathrm{g}/P'_\mathrm{g})\right]\eta} \tag{2.5}$$

$$\text{Solute flux:}\quad J_i = J_\mathrm{v}(1-\sigma_i)C_{\mathrm{w}i} - \frac{J_\mathrm{v}(1-\sigma_i)(C_{\mathrm{p}i} - C_{\mathrm{w}i})}{\exp\left(J_\mathrm{v}(1-\sigma_i)/P_\mathrm{E}\right)-1} \tag{2.6}$$

where $C_{\mathrm{w}i}$ is the concentration of solute at the interface between the feed and the membrane surface including any surface layer; $C_{\mathrm{p}i}$ is the concentration of the solute in the permeate; J_i is the flux of solute i into the permeate (concentration/membrane area/time); J_v is the solvent flux including solutes (volume/membrane area/time); P'_v is the specific water conductance of a clean membrane ($\mathrm{m}^3\ \mathrm{m}^{-1}$), where $t_\mathrm{m}/P'_\mathrm{v}$ is the compressed membrane hydraulic permeability based on Darcy's relationship; t_m is the thickness of the membrane's permselective layer (m); P'_v is the specific water conductance of the surface layer ($\mathrm{m}^3\ \mathrm{m}^{-1}$), where t_g/P'_g is the specific overall resistance, usually determined experimentally; t_g is the thickness of the surface layer (gel, cake adsorption, fouling); $\Delta\Pi$ is the applied transmembrane pressure (pressure units); $\Delta\pi$ is the actual osmotic pressure based on the solute concentration on the membrane surface (C_w) and the solute concentration in the permeate (C_p) (at a specific point) (pressure units); η is the solution viscosity ($\mathrm{kPa\ s}^{-1}$) (this quantity varies due to concentration polarization and temperature); $P_{\mathrm{m}i}$ is the specific solute's permeability coefficient (m s^{-1}) (this parameter relates the rejection of a solute to diffusive transport); σ_i is the solute reflection coefficient (dimensionless, calculated by measuring the pressure difference at which the volumetric permeation flux is zero for a given osmotic pressure difference across the membrane); this parameter relates the transport of a solute to convective transport. These equations include several components which vary with time, surface and position along a membrane, and hence the predictions are limited. In practice, a complete description requires trial-and-error calculations and prediction through modelling of the concentration at the membrane interface in order to acquire sufficient information (Pellegrino, 2000).

2.2.3 *Physical parameters of membrane processes*

In practical membrane processing, the volumetric flux (J_v) of permeate is measured and expressed as litres per square metre per hour (l m^{-2} h^{-1}) or metres per second (m s^{-1}):

$$J_v = \frac{V_p}{A \times t} \tag{2.7}$$

where V_p is the permeate volume, A the membrane effective area (m^2) and t the time necessary for the removal of V_p litres or cubic metres of permeate.

Solute rejection (R) by a membrane is defined for each main solute or a family of solutes. It is dimensionless and is distinguished as either 'observed' or 'intrinsic' solute rejection. The observed solute rejection is defined as

$$R = 1 - \frac{C_p}{C_f} \tag{2.8}$$

where C_p and C_f are solute concentrations in the permeate and the feed, respectively, measured by sampling. The intrinsic rejection is defined similarly, the only difference being that instead of the feed concentration, it utilizes the real concentration of the solute at the membrane interface (Pellegrino, 2000). Based on these definitions, if a component is completely rejected by a membrane, $C_p = 0$, then $R = 1$. On the other hand, for components which freely permeate the membrane, $C_p = C_f$, then $R = 0$.

Ideally, during RO, all components would have $R = 1$. For UF, large molecules such as proteins would have values of rejection approaching 1, whereas for small components such as dissolved salts or simple sugars, rejections would approach 0. As UF and MF membranes in practice have a pore size distribution, and hence diffuse cutoff points, many species will have rejection values between 0 and 1.

For batch operations, an important parameter is the yield (Y) of a component, which is the fraction of that component from the original feed recovered in the final retentate. It is dimensionless and can be expressed in percentage form as follows:

$$Y = \left(\frac{C_i V_i}{C_f V_f}\right) \times 100 \tag{2.9}$$

where V_f, V_i and C_f, C_i are the volumes and concentrations of the initial and final (retentate) feed, respectively.

The volume concentration ratio or volume concentration factor (VCF) is a useful parameter, especially in processes operating in batch mode, demonstrating the feed concentration achieved:

$$\mathrm{VCF} = \frac{V_\mathrm{f}}{V_\mathrm{i}} \tag{2.10}$$

The VCF is dimensionless and can be used for the calculation of the observed rejection values given for a solute during batch processes:

$$R = \frac{\ln(C_\mathrm{i}/C_\mathrm{f})}{\ln(\mathrm{VCF})} = \frac{\ln[\mathrm{VCF} - (C_\mathrm{p}/C_\mathrm{f})(\mathrm{VCF}-1)]}{\ln(\mathrm{VCF})} \tag{2.11}$$

The observed rejections calculated in this way take into account either the permeate or the retentate concentrations, and the corresponding volume concentration ratio, allowing comparison of rejection values throughout the process (Kulkarni and Funk, 1992).

2.2.4 *Diafiltration*

Diafiltration can be used to improve purification with any of the membrane fractionations. It is the process of adding water to the retentate so that further removal of membrane-permeable solutes can continue along with the added water (Cheryan, 1998). The two main modes of operation are discontinuous diafiltration (DD) and continuous diafiltration (CD). DD refers to the operation where permeable solutes are cleared from the retentate by volume reduction, followed by re-dilution with water and repeated volume reduction in repetitive steps. On the other hand, CD involves the continuous addition of water (at the appropriate pH and temperature) in the feed tank at a rate equal to the rate at which permeate is removed, thus keeping feed volume constant during the filtration (Lewis, 1996b; Cheryan, 1998).

A simple mathematical analysis of a CD process, described by Lewis (1996b), considers that when $R \neq 0$ for a given solute and $C_\mathrm{p} = C(1 - \mathrm{R})$, the concentration of that solute in the feed can be calculated using the following equation:

$$\ln\left(\frac{C_\mathrm{f}}{C_\mathrm{i}}\right) = (1-R)\times\left(\frac{V_\mathrm{c}}{V_\mathrm{f}}\right) \tag{2.12}$$

where V_f and V_c are the volume of the feed and the cumulative permeate volume removed during the course of the process, and C_i and C_f are the concentrations of the final and initial feed, respectively. In the same way, the variation in concentrations of a specific solute in the permeate can be monitored from the equation:

$$\ln\left(\frac{C_\mathrm{f}(1-R)}{C_\mathrm{p}}\right) = (1-R)\times\left(\frac{V_\mathrm{c}}{V_\mathrm{f}}\right) \tag{2.13}$$

Both Equations 2.12 and 2.13 are applicable only when the rejection of a solute remains constant throughout a CD process, which is usually not the case. The ratio $V_\mathrm{c}/V_\mathrm{f}$ is termed the number of diavolumes removed (Lewis, 1996b).

2.2.5 *Parameters affecting flux and rejection*

Several parameters affect the separation between different solutes, their rejection characteristics, and the permeate flux during a membrane filtration process. First, the size and shape of the molecules to be separated are very important in determining their rejection characteristics, since they relate to the primary separation mechanism of the filtration processes, the sieve effect. Choosing a membrane based on the molecular-weight cutoff (MWCO) data supplied by manufacturers is often inadequate as far as the separation of two solutes is concerned. When two protein molecules, for example, have the same molecular weight, this does not necessarily mean that they also have the same volume. Hence, the MWCO data alone are insufficient to predict the separation characteristics of a particular membrane, and when molecules of a completely different nature are considered, the problem becomes even more complicated. As a general rule, in order to attain a satisfactory separation between two different components, at least a ten-fold difference in their molecular weights is required (Cheryan, 1998).

Another parameter affecting the separation characteristics of a membrane is the chemical nature of the material used for its manufacture, since it affects membrane/solute interactions. The membrane material also determines the porosity of membranes, hence giving rise to higher or lower permeate fluxes (Cheryan, 1998).

Liquid feeds that are fractionated by membrane separations usually contain mixtures of different components. This can give rise to unexpected rejections for some solutes, since interactions between different components may cause changes to their molecular structure. Aggregation of components may change the apparent size. Also, it is quite common for large, non-permeable solutes to form secondary dynamic membranes, which inhibit permeation of smaller molecules and also affect the total permeate fluxes. In addition, the chemical properties of the feed (especially factors such as pH and ionic strength) have an important role in separations. This is especially apparent with the rejections for protein solutions, where pH values in relation to the isoelectric point have a marked effect on separation characteristics (Cheryan, 1998). This effect is also observed for selective separations of amino acids with NF membranes, in which the surface charge of the membranes and the charged state of the amino acids, determined by the pH, can be used as a tool to modify the separation (Timmer *et al.*, 1998).

Operating variables, such as trans-membrane pressure, turbulence near the membrane surface, temperature and feed concentration, also affect the quality of the separations and the permeate fluxes. The dependence of permeate flux on the applied pressure is easy to understand, since pressure is the driving force of separation. Flux increases with increasing pressure, usually up to a limiting value above which the permeate rate becomes pressure-independent, due to the boundary layer or the gel-permeation layer on the membrane surface. For this reason, separation characteristics are distinguished as those observed in the pressure-dependent or pressure-independent regions. Increased solvent fluxes give rise to higher concentration polarization and consequently gel-boundary layer generation, which directly affect the permeability of solutes.

Pressure also causes membranes to undergo the phenomenon known as compaction (Lewis, 1996a). Compaction reduces the membrane thickness, which would normally be expected to lead to increased solvent flux and reduced solute rejection, but at the same time the pore diameter decreases, and this factor predominates, causing exactly the opposite effect on separation characteristics. Membrane compaction can be avoided by conditioning the membranes under the running conditions of the separation, in order to reach a steady state. However, and especially for flat-sheet membranes, compaction is not permanent and usually exhibits considerable reversibility, causing the separation characteristics of the membranes to change (Whu *et al.*, 2000).

Turbulence in the feed circulating above the membranes is another parameter that affects the formation of the boundary and the concentration polarization layers, and thus affects the permeate flux accordingly (increased feed flow rates result in increased turbulence which consequently leads to decreased concentration polarization and increased permeate fluxes).

Increasing temperature of processing generally causes the permeate fluxes to increase and the solute rejections to decrease (Tsuru *et al.*, 2000). The effect of temperature on permeate flux is related to the porosity of the membranes, the viscosity of the feed, and the diffusivity of the solvent and solute molecules. Increased temperatures generally reduce the viscosity of the feed causing the solvent adsorbed layer on the pore walls to become thinner, and thus increasing the effective pore diameter. The diffusivities of solvent and solute molecules also increase with temperature, because the available thermal energy increases, supplying them with the necessary energy to overcome the hydrodynamic drag (frictional) forces inside the pores (mostly for NF and RO).

The concentration of solutes is another key parameter that influences membrane separations. It is important, since it determines the extent of concentration polarization and consequently any fouling occurring. Generally, increased feed concentrations decrease the permeate flux, due to the higher levels of osmotic pressure exhibited, and increase the solute rejections due to lower convective flow and the secondary layers (of rejected solutes) forming on the membrane surface (Cheryan, 1998). However, there are occasions, as reported by Vellenga and Tragardh (1998), where the increased concentrations of solutes cause their rejections to drop. They reported that in combined salt and sugar solutions and using membranes too dense for sugars to permeate, rejections of the salts decreased as the total concentrations of the solutions increased. The explanation given was that as concentrations increased, concentration polarization on the membrane surface became more severe, hindering the back-diffusion of solutes that could permeate the membranes. It should be noted, however, that sugar and salt solutions do not cause severe fouling of the membranes, and similar phenomena may not be feasible with solutes that cause membrane fouling (i.e. attach irreversibly to the membrane).

2.3 Membrane classification, production methods and characterization

Membranes are classified as microporous or asymmetric according to their ultrastructure (Cheryan, 1998). Microporous membranes are further classified as homogeneous (isotropic) or heterogeneous (anisotropic). Homogeneous membranes possess a homogeneous pore structure parallel and perpendicular to the membrane's surface with the membrane being composed from the same material and having the same pore characteristics throughout its structure. Heterogeneous membranes have a heterogeneous structure usually composed of a thin 'skin' on the membrane's surface, responsible for the membrane's rejection characteristics, and a second much thicker layer with large voids, whose main function is to support and protect the rigidity of the surface 'skin'. Asymmetric membranes (considered to be heterogeneous) are also characterized by a thin dense skin on a porous membrane support. Rejection on these membranes occurs at the surface, and solutes larger than the pores of the membrane do not enter the membrane structure. These membranes are widely used for UF, NF and RO applications (Osada and Nakagawa, 1992; Cheryan, 1998).

A variety of materials are used in the manufacture of membranes, depending on the application and desired properties. Some of the most commonly used materials are either polymeric, e.g. poly-(vinylidene fluoride) (PVDF), poly-propylene (PP), poly-ethylene (PE), poly-carbonate (PC), cellulose acetate (CA), polysulfone (PSF) and poly-(ether sulfone) (PES), or inorganic, ceramic materials such as zirconium oxide or alumina. The first commercial membranes were manufactured from cellulose derivatives, which are still used for some applications. Polymeric membrane materials consisting of engineering plastics have, however, been developed and currently account for the vast majority of commercial membrane systems. Most polymeric asymmetric UF and RO membranes are prepared by the phase inversion method, which is a process where a polymer solution inverts to a swollen three-dimensional macromolecular network or gel (Osada and Nakagawa, 1992). This is accomplished by dissolving the polymer in a suitable volatile solvent, adding a swelling agent, casting the solution and subsequently regulating solvent evaporation in order to form a high polymer concentration surface skin layer. Subsequently, the remaining solvent evaporates slower due to reduced contact with air, and forms together with the polymer a concentrated phase that is interrupted by the swelling agent phase. As solvent evaporation proceeds, the polymer structure aggregates around the swelling agent, eventually forming polyhedra that give rise to an open-celled structure as both solvent and swelling agent completely evaporate.

Cellulose acetate is a classic membrane material, generally linear and rather inflexible. It is widely used for manufacture of 'skinned' membranes (a thin 0.1–1.0 μm 'skin' supported by a much thicker porous support). Cellulose acetate membranes are low-cost, relatively easy-to-manufacture, hydrophilic membranes with a wide variety of pore sizes. Their main disadvantages are their narrow pH (2–8) and temperature (<40°C) operating range, their poor resistance to chlorine, the loss of their properties due to compaction and their high biodegradability.

Polysulfone membranes, with their diphenylene sulfone structure, overcome some of the problems of cellulose acetate membranes, as they can operate at fairly high temperatures, over a wide pH range, and show a reasonable chemical resistance. Their main disadvantages are their low pressure tolerance and also their hydrophobicity that makes them susceptible to extensive fouling.

A new emerging generation of membranes are the composite or thin-film composite membranes that have much better characteristics than cellulose acetate and polysulfone membranes (Cheryan, 1998). Composite membranes consist of an ultrathin barrier layer on the surface and a porous support underneath. The two separate layers can be composed of the same or different materials, and the porous support may consist of more than one layer.

Inorganic membranes are generally manufactured from powdered inorganic materials (e.g. carbon, zirconia, alumina) with narrow particle-size distributions, following a thermal sintering process of an extruded paste of powder, and finally a coating step of the resulting membrane. Inorganic membranes overcome most of the disadvantages associated with polymeric membranes, since they are very resistant to extreme operating conditions. Their major limitations are the very high cost and the narrow range of pore sizes available (Cheryan, 1998).

Track-etched and dynamic membranes are fundamentally different in structure from the above. Track-etched membranes are essentially dense polymer films that are punctured by cylindrical holes. Microfiltration membranes are produced by this method, in which a polymer film is irradiated with alpha particles or neutrons. Dynamic membranes are produced by intentionally depositing material onto porous structures prior to filtration, thus creating a dynamic deposited layer over the porous structure (Gutman, 1987).

Due to the wide range of membrane pore-sizes available, there are different ways of characterizing them. Microfiltration membranes have measurable pore sizes and are usually characterized directly according to their pore diameter (maximum or mean) and pore size distribution. For UF membranes, it is difficult to measure the pore size directly, and they are characterized by their MWCO. The MWCO defines the size of the solute that would be almost completely (90%) retained by the particular membrane. The MWCO of ultrafiltration membranes is measured by testing the permeability of different-molecular-weight proteins (preferentially globular) or dextrans. This characterization is not perfect, since it is subject to the many other influencing factors (molecular shape and charge, interactions between feed components, membrane material, etc.). Moreover, there are a range of pore diameters within any membrane, which causes the separation to be diffuse. Generally, MWCO data are an indication for the preliminary selection of membranes. Reverse osmosis and NF membranes are usually characterized from their rejection values against mono- or divalent salt solutions. Sometimes, MWCO data are also provided for NF membranes.

2.4 Modules and modes of operation of pressure-driven membrane filtration processes

In practice, membranes are configured into modules, the design of which must incorporate the following:

(1) The membrane must be supported to withstand the required hydraulic pressures.
(2) The membrane should have a high surface area – preferably compact in volume.
(3) Established flow streams must be present in which the correct hydrodynamic conditions can prevail (e.g. flow rate, turbulence, pressure drop).
(4) There must be hygienic conditions, ease of cleaning and membrane replacement.

There are four different membrane configurations principally used today in the food industry: tubular, plate-and-frame, spiral-wound and hollow fibre membranes (Fig. 2.4). Tubular membrane modules (Fig. 2.4a) are prepared by direct casting of the membranes on to porous stainless steel, cardboard or plastic tubes. These modules

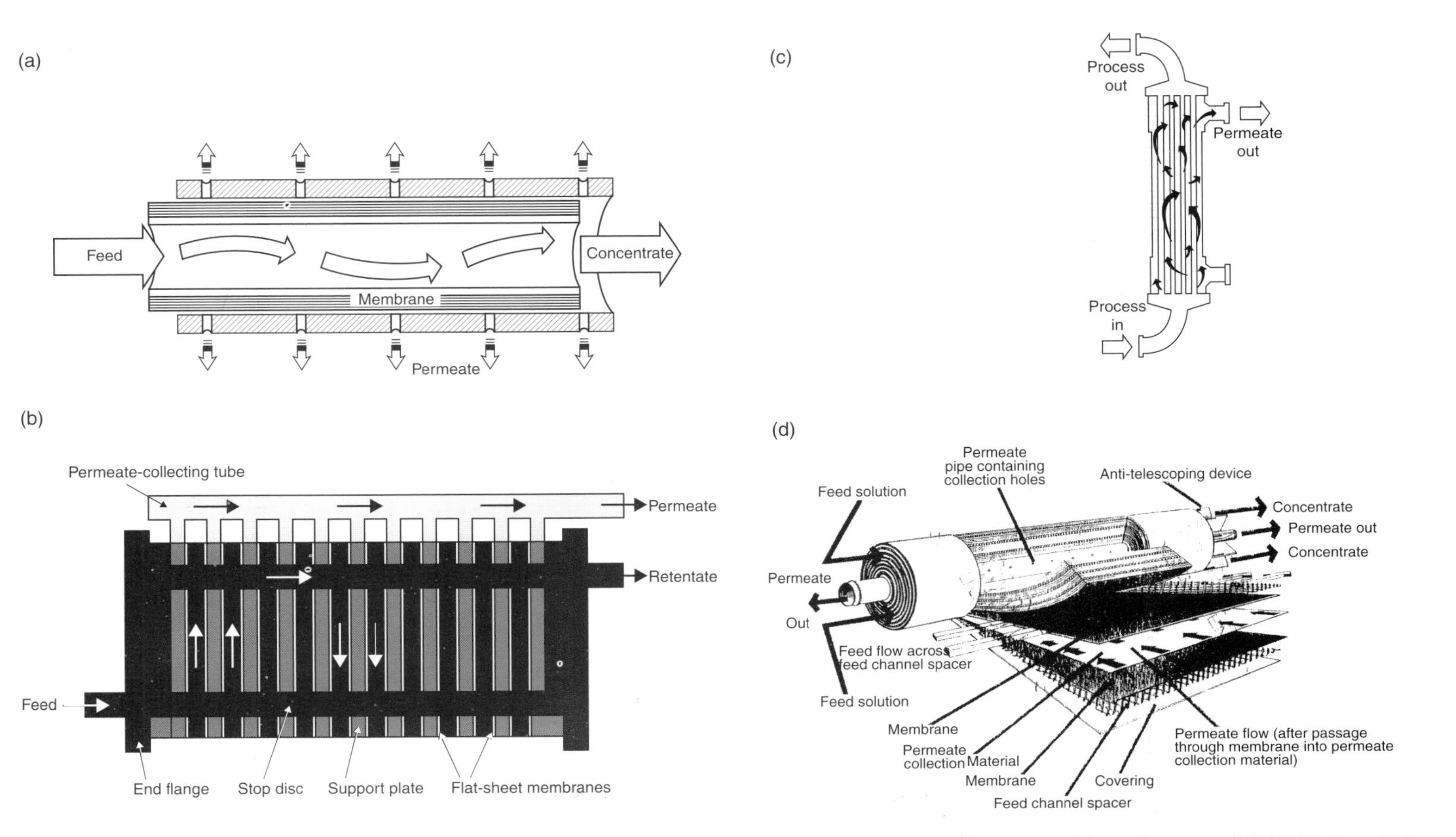

Fig. 2.4 Membrane module configurations and operating principles (a) Tubular system (courtesy of ITT Aquious). (b) Plate-and-frame (courtesy of DDS Silkeborg AS). (c) Hollow fibre (courtesy of Koch Membrane systems). (d) Spiral wound (courtesy of Koch Membrane Systems).

offer easy adjustment of the feed flow velocity and easy mechanical cleaning, but they occupy a relatively large space per unit surface area and have high hold-up volumes (Lewis, 1996a; Cheryan, 1998). Plate-and-frame membrane modules (Fig. 2.4b) consist of membranes based on porous membrane support materials and spacers providing larger membrane area per unit volume compared with tubular membranes. They are easy to handle, but care must be taken in order to avoid leakage and contamination of the different product and feed streams. Spiral-wound membrane modules (Fig. 2.4c) are in principle plate-and-frame systems rolled up. They are composed of two flat-sheet membranes separated by a spacer with their active sides facing away from each other. The membranes are glued together on three sides, and the fourth free side is fixed around a perforated centre tube. They are one of the most compact and inexpensive membrane designs available today, offering large membrane areas with low hold-up volumes, but they suffer from severe membrane fouling and dead spots in spaces behind the spacers (Osada and Nakagawa, 1992; Cheryan, 1998). Hollow-fibre membrane modules (Fig. 2.4d) consist essentially of a large number of tubular membranes that have an asymmetric structure in the radial direction and act as self-supporting membranes with no separate support or backing. They offer the highest membrane area per unit volume of all geometries, and they allow very efficient cleaning. Their main disadvantages are the high cost and the low pressures that they can sustain. All these modules have different limitations and advantages varying with the separation process (e.g. RO, UF) and the desired use (e.g. protein, carbohydrate solutions) (Cheryan, 1998).

The basic requirements for a membrane filtration process are a membrane module, feed tank, pump, flow control and pressure-retaining valves, along with appropriate flow rate, temperature and pressure monitoring devices. It is normal to include heat exchangers to control the feed temperature. The choice of pump depends upon the pressure and flow rate required. Generally, centrifugal pumps are sufficient for UF and MF, while the higher pressures used in RO require positive displacement pumps. Delicate particulate or viscous feeds may require more specialized pumps.

Membrane filtration processes operate as dead-end or cross-flow depending on the direction of the feed above the membrane (Fig. 2.5). In the dead-end mode, the feed is pumped directly towards the filter, and only the permeate stream leaves the membrane. The term cross-flow refers to the direction of the feed stream tangentially over the surface of the membrane, in order to sweep rejected solutes away from the membrane. In this mode of operation, the feed stream enters the module, and two streams are removed from the membrane, i.e. permeate and retentate. The cross-flow mode of operation prevents concentration buildup of rejected solutes on the membrane surface, thus reducing concentration polarization and membrane fouling. The flow of the feed stream above the membrane influences back-transfer of the accumulated solutes into the bulk of the feed, thus maintaining very thin boundary layers (Gutman, 1987; Osada and Nakagawa, 1992; Cheryan, 1998). Cross-flow filtration is utilized commercially, whereas dead-end systems are restricted to small laboratory operations.

Industrially, filtration processes are operated in batch or continuous modes of operation. In batch operations, the feed is circulated in the plant until the necessary degree of purity, concentration or separation is achieved. Following this, the process

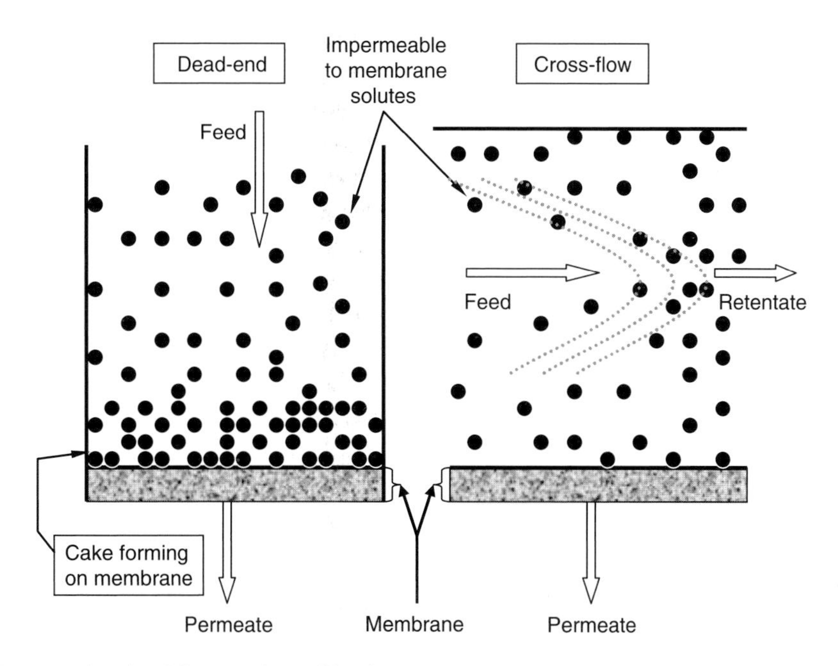

Fig. 2.5 Dead-end and flow membrane filtration processes.

is discontinued, and the desired product is collected. In practice, such operations are suitable for processing of small volumes that can be wholly contained within a plant and are more suited to small-scale or pilot-plant operations (basic requirements are shown in Fig. 2.6a).

On the other hand continuous operations can be carried out in single-stage internal recycle (feed and bleed) systems (Fig. 2.6b). In these systems, the feed is pumped around an inner loop consisting of the membrane module and circulation pump. The plant is first operated until the desired level of concentration is achieved. At this point, part of the retentate is bled off, the remainder returning to the circulation pump where fresh feed is added from the feed pump to maintain the volume of retentate. Fresh feed is continuously added at the same rate as concentrate is removed. This system may be operated for quite long periods in commercial plants and holds major advantages over batch systems in that the retention time in the plant is low (minutes rather than hours), and the internal loop is closed so the entire pressure is not lost on each recycle, which is energetically favourable. The major disadvantage is that the continuous operation is carried out at a maximum product concentration so that in most cases, the permeate flux is at the lower levels.

Alternative designs for large-scale processing are multi-stage plants, which may contain several hundred square metres of membrane. A single pump cannot provide the necessary power to drive these systems, so plants are made up of repeating units, each with its own pump, arranged in series. Within each unit, the modules may be arranged in parallel or in series. Figure 2.6c shows a system with three units, although more may be possible in practice. The size of successive units would decrease in practice as the volume of feed diminishes.

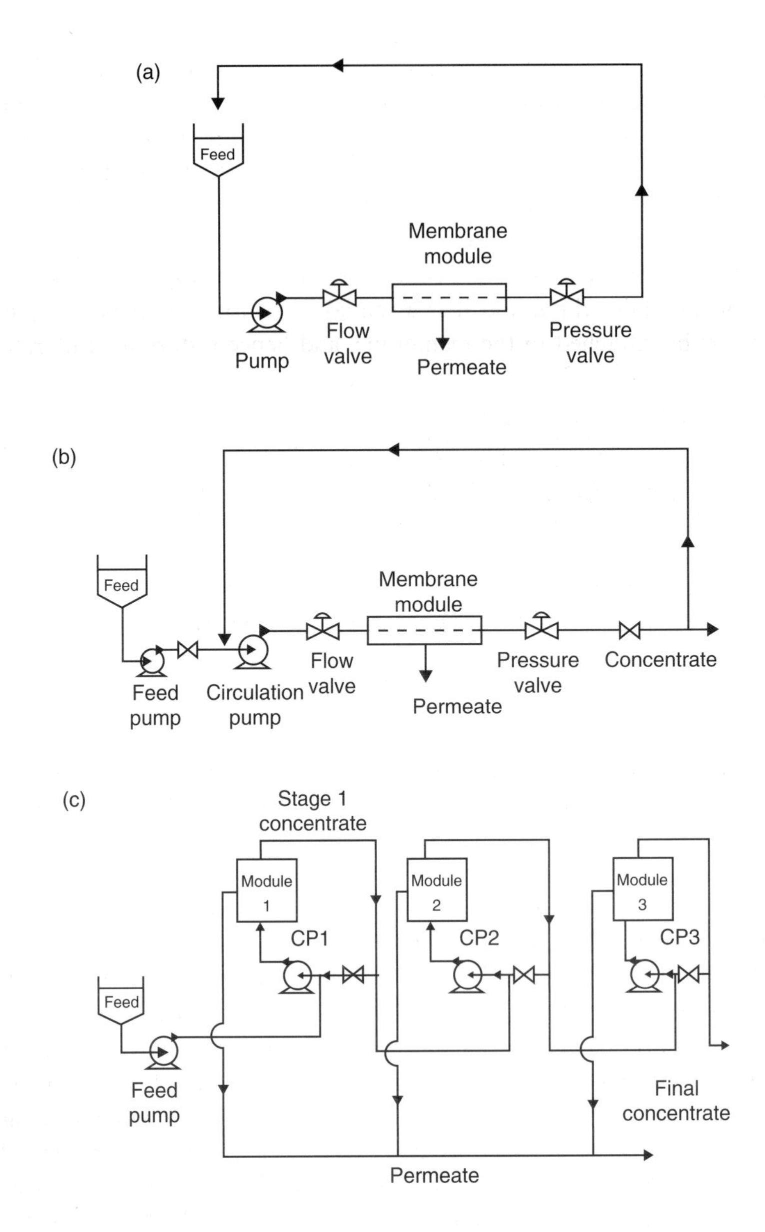

Fig. 2.6 Membrane operating systems: (a) batch operation; (b) continuous internal recycle (feed and bleed); (c) continuous operation (multi-stage plant). CP: circulation pump.

2.5 Hygiene and cleaning

In any membrane process, all microorganisms will be completely rejected by the membranes, and microbial numbers will therefore increase in the retentate in line with the concentration factor. It is also likely that some microbial growth will occur during processing, depending on the temperature and residence time within the plant. In food-processing operations with long residence times, it is advisable to operate at temperatures above 50°C or below 5°C to prevent growth of micro-organisms. Heat treatment before and/or following membrane processing may also be considered. The permeate stream, in contrast, should be sterile on production but may form a good medium for microbial growth. Hence, membrane systems should be designed for cleaning and sterilization on both retentate and permeate sides.

In addition to the requirement for good hygiene, regular cleaning is required to maintain the efficient operation of membrane systems. Fouling will develop during treatment of any dairy fluid, and it is essential to remove completely any material that is irreversibly attached to the membrane, and hence restore the flux rate to the original. This is commonly assessed by measuring the water flux before processing and after cleaning. The precise cleaning regime depends on the nature of the feed and the resilience of the membrane. Developments in membrane materials have generally led to membranes which can survive harsher cleaning regimes. In dairy applications, the main foulants are proteins and minerals (especially calcium phosphates), although fat may become associated with the deposited layer. It is normal to clean by circulation of caustic solutions to remove fat and protein, followed by acid washing to remove minerals. Proteolytic enzymes and ion-sequestering chemicals may also be used.

2.6 Composition and properties of dairy fluids for membrane processing

Milk is a complete food for the young mammal, containing nutrients in proportion to the dietary requirements. Different species of mammals produce milk with widely varying composition. Obviously cows' milk accounts for the vast bulk of UK dairy production, but goats, sheep and even buffalo are milked commercially. The latter two species provide milk with higher solids than cows' milk, while the gross composition of goats' milk is quite similar. Within a species, different breeds give a characteristically different milk composition; for example, Channel Island breeds produce milk with higher solids than the more numerous, high-yielding Friesian/Holstein. Cows' milk from the same breed will vary due to changes in diet, stage of lactation and a range of other environmental and genetic factors. Hence, milk composition is subject to seasonal and even day-to-day variation. Typical UK cows' milk contains about 87% water, 4% fat, 4.6% lactose, 3.5% protein and 0.9% other solids (including minerals, vitamins, non-protein nitrogen compounds, etc.). The fat may be separated by centrifugation to produce skim milk (<0.1% fat) which itself may be subject to some membrane-processing applications. Depending on hygienic quality, unprocessed milk contains numbers of microorganisms in vegetative or spore forms, as well as somatic cells (derived from the mother).

The fat fraction is in the form of globules with clearly defined membranes. Most of the fat is contained within globules 1–5 μm in diameter. The two protein fractions of milk are caseins (75–80%) and whey proteins (20–25%), each consisting of a number of different protein species. The caseins are associated with calcium phosphate in colloidal structures (micelles) in the size range 0.03–0.3 μm, while the whey proteins are dissolved in milk serum. Lactose is a disaccharide which is completely dissolved in the milk. The milk serum contains low concentrations of many other small chemical species in solution, e.g. urea, amino acids, water-soluble vitamins, lactic acid. Hence, milk forms the basis for a wide range of potential membrane separations, many of which are described in more detail in Section 2.7.

Whey is the liquid byproduct of cheese-making, and its composition varies depending on the type of cheese produced as well as the composition of the milk (and hence variation due to season, species, breed etc. as described for milk). As a general guide, bovine whey contains about 65 g kg^{-1} total solids, of which 50 g kg^{-1} is lactose, 6 g kg^{-1} protein, 6 g kg^{-1} ash, 2 g kg^{-1} non-protein N and 0.5 g kg^{-1} fat. A major distinction in whey composition depends on the method of manufacture, broadly in whether the whey results from enzymic or acid coagulation of the milk. Sweet whey is produced from chymosin-coagulated cheeses (or rennet casein) and has a relatively low acidity (pH 5.8–6.6), whereas acid whey, resulting from the manufacture of acid-curd cheeses (or acid casein), has a much higher acidity (pH 4.0–5.8) as a result of lactic acid fermentation or direct addition of food-grade acids. Furthermore, the acid development causes the release of minerals (especially calcium phosphates) from the micelles into the serum, and hence the mineral contents of acid and sweet whey differ markedly, the former having much higher levels of calcium and phosphorus (e.g. acid whey contains around 1.6 g kg^{-1} of Ca and 1.0 g kg^{-1} of P, whereas sweet whey contains about 0.6 and 0.7 g kg^{-1} of Ca and P, respectively). There are other, less obvious differences; for example, sweet whey will contain the soluble glycomacropeptide portion of the κ-casein, which is absent in acid whey.

Whey composition will also depend on milk treatment before cheese-making. Some cheeses are made from skim milk or milk with adjusted fat levels, which will affect the fat levels in whey. Whey from full-fat milk contains up to 0.5% fat and this is frequently removed by centrifugation prior to further processing. High heat treatments cause whey proteins to attach to the surface of the casein micelles. This interaction is minimal under standard pasteurization conditions, but whey produced from milk heated more severely will contain reduced protein levels.

In the manufacture of hard cheeses such as Cheddar, approximately 90 kg of whey is produced from 100 kg of milk, while for soft cheeses the figure is lower, e.g. approximately 80 kg from 100 kg of skim milk during the manufacture of cottage cheese. In the past, whey was considered to be a waste product, but more recently it has come to be viewed as a raw material for processing. Membrane processes have contributed greatly to this development.

The majority of membrane processes in the dairy industry involve the use of milk (including fermented milk) or whey as the feed. However, there are other possibilities, e.g.

- *Buttermilk* is the aqueous phase produced from churning butter. It is quite similar in composition to skim milk but contains more fat, including particularly high concentrations of phospholipids derived from the fat-globule membranes, which could act as valuable emulsifiers.
- *Brines* used in cheese-making may be recovered by UF.
- *Wastewaters* from cleaning dairy plant.

General guidelines for the behaviour of components of milk and other dairy fluids during the major membrane processes are as follows:

(1) *Reverse osmosis* would produce practically pure water as permeate, while all other components would be concentrated in proportion to their concentration in the feed. Only a small proportion of the smallest ions permeate the membrane, with rejection of most minerals being >0.99.
(2) *Ultrafiltration* membranes retain the fat completely, and the protein fractions almost completely, depending on the membrane characteristics. Lactose and other small dissolved species will permeate the membrane quite freely. Rejection values for lactose may be up to 0.1, while unbound ions have values approaching 0. However, the situation with respect to calcium, phosphorus and other materials associated with the protein fraction is more complex. Approximately two-thirds of the calcium and one-half of the phosphorus are incorporated in casein micelles in milk and hence will be retained during ultrafiltration, while the remainder is freely permeable. However, acidification of milk releases these minerals from the colloidal phase, so that as pH is reduced, the proportion passing through the membrane will increase. This is particularly important during the manufacture of fermented products, such as soft cheeses, with UF, where concentration is frequently carried out on fermented milks so that the concentrations of soluble minerals and citrate in the retentate are reduced, which is beneficial to the flavour of the products. In cheese whey, apart from a small amount of protein-bound mineral, the minerals are free to permeate through UF membranes. Water-soluble vitamins generally permeate through UF membranes, while fat-soluble vitamins are associated with the fat fraction and will remain in the retentate.
(3) *Microfiltration* provides a more variable picture depending on the pore size. Fat globules and any cells will be retained, while proteins may or may not permeate the membrane depending on their molecular weight, the characteristics of the membrane, and other factors such as whether they exist in true solution or as micelles, or other aggregates. Lactose and other small dissolved components will permeate the membranes freely.

2.7 Applications of membranes in the dairy industry

2.7.1 Reverse osmosis

Applications of RO in the dairy industry have been reviewed by El-Gazzar and Marth (1991). Full cream or skim-milk may be concentrated by RO to a factor of 2–3 times at pressures of 2–8 MPa, the permeate essentially being pure water. The concentration factor is much less than that attainable by evaporation as it is limited by the osmotic pressure of the feed and deposition of minerals (mainly calcium phosphate) in the pores of the membrane. There is also a danger of damaging the fat-globule membranes during RO, which could lead to release of free fat in the product. The latter problem would not arise with skim milk. Flux rates at the beginning of the process may be around 40 l m^{-2} h^{-1}, skim milk giving slightly higher flux rates than full-cream milk. Hence, RO is used as a preliminary step to increase the capacity of evaporation plants in the manufacture of milk powders. Another possibility is to concentrate milk on-farm to reduce transportation costs, although this is unlikely to be widely carried out in the UK where transport distances are small.

Skim milk concentrated by RO can be used as an alternative to fortification with milk powders in the manufacture of yoghurt and ice cream. A concentration factor of about 1.5 is used to give approximately 15% total solids in the retentate. Membrane concentration is less expensive than the use of powders.

Incorporation of RO technology into cheese-making is not as attractive as UF, because the substantial yield increases, resulting mainly from whey protein retention in the curds (see Section 2.7.3.1), are not available. However, RO milk concentrates can be used to increase the capacity of cheese vats and reduce the requirement for chymosin and starters.

Concentration of cheese whey by RO to a factor of up to 4 times, and hence about 28% total solids, is possible. The chief aims are to reduce transport costs and as a pre-concentration step before drying. Permeate flux values during RO follow the order: sweet whey > acid whey > milk. The difference between sweet and acid whey probably relates to the higher calcium levels in the latter, which act as a foulant.

Wastewaters from cleaning the dairy plant may contain considerable quantities of milk solids which could be concentrated by RO, dried and used in animal feed (Grandison and Glover, 1994). Rinse-water RO permeate could also be used as a source of water.

2.7.2 Nanofiltration

Nanofiltration can potentially be used for partially reducing the mineral content of milk or whey, while retaining other dissolved components. Typical rejection values are 0.95 for lactose and 0.5 for dissolved salts (Brennan *et al.*, 2006). Guu and Zall (1992) reported that nanofiltration of permeate gave improved lactose crystallization. Other possibilities could be to improve the heat stability of milk by reducing soluble calcium, or generally reducing salts for use in infant foods.

A promising application is the fractionation of whey-protein hydrolysates by nanofiltration (Groleau *et al.*, 2004) which could be particularly important in the development of large-scale manufacture of bioactive peptides.

Further, detailed information on the production of dairy derived carbohydrates by nanofiltration is provided in Section 2.7.7.

2.7.3 Ultrafiltration

Ultrafiltration has the potential to increase the concentration of protein and fat in milk, which can be useful in the manufacture of a number of dairy products, including cheeses, yogurt and other fermented products. Whole milk can be concentrated by a factor of up to five times, while skimmed milk can be concentrated up to seven times (Kosikowski, 1986). Diafiltration can be incorporated into UF processing of milk in the manufacture of milk protein concentrates, or to reduce the lactose content.

2.7.3.1 Manufacture of cheese and fermented products

The use of UF technology in cheese manufacture is reviewed in detail by Mistry and Maubois (1993). In cheese-making, there is a distinction between concentrating retentates by a factor up to two times, and concentrating to higher levels (Lawrence, 1989). Up to twofold concentration, there are benefits in terms of standardizing the protein (and fat) concentration of the retentate such that consistent yields of cheese are obtained. Seasonal and other factors lead to variation in milk composition which, in turn, leads to variation in the yield of cheese. Hence, standardization is essential in continuous cheese-making processes, for example in the commercial manufacture of Camembert using the Alpma coagulator or similar equipment, where fixed quantities of UF retentate are used to provide individual cheeses of predetermined weight. In addition, these levels of concentration will provide economic benefits in terms of reduced requirement for coagulating enzymes and starters, as well as increasing the yield of cheese per vat. However, no overall increase in cheese yield is obtained up to twofold concentration.

At higher concentration factors (VCF >2), there is a further major advantage in terms of cheese yield. In conventional cheese-making, 20% of the protein (i.e. the whey protein fraction) is lost in the whey. By concentrating the milk prior to cheese-making, and hence reducing or eliminating the curd syneresis (whey drainage), some or all of the whey protein is retained in the curd. These proteins have high water-binding capacities, and so more water and water-soluble components of the milk may also be retained in the curd. In some cases, UF is actually carried out on fermented milk to reduce the retention of minerals and citrate in the cheese, and consequently give a better flavour, as described in Section 2.6.

There is a distinction between processes where retentates are concentrated to an intermediate level in which some syneresis still takes place, but some whey proteins are retained (intermediate concentrated retentates), and concentration to the final composition of the cheese, where no syneresis takes place, and maximum yield benefits occur – the liquid pre-cheese concept (Mistry and Maubois, 1993). In either case, modifications to the manufacturing methods are required to achieve the correct cheese quality. Also, it has generally been found that cheese made from UF retentates ripens more slowly than traditional varieties. This is attributed to the presence of whey proteins which are more resistant to proteolysis and may even inhibit proteolysis during ripening, as well as modifying the buffering capacity. The proteolytic products from whey proteins may also alter the flavour of cheeses.

Intermediate concentrated retentates (milk concentrated in the range two- to five-fold) have been applied to a wide variety of cheeses. Reported yield increases of 6–8% with Cheddar and 14% with Feta cheese have been achieved while maintaining good-quality products (Mistry and Maubois, 1993).

The liquid pre-cheese concept has the advantage that there is no whey drainage, and processes can be designed without the need for cheese vats. Yield increases of up to 30% have been reported (Grandison and Glover, 1994). This approach has been generally successful in fresh, unripened cheeses such as quarg, ricotta and cream cheese, and in soft and semi-hard cheeses, including Camembert, Feta, Mozzarella and Saint Paulin.

However, the use of liquid pre-cheese to manufacture hard cheese varieties is much more challenging because it is difficult to reproduce the texture of traditional varieties. On one hand, the retained whey protein acts as a filler and binds more water than caseins, which results in softer cheeses. Also, the absence of whey drainage means that elastic texture, as is developed during the Cheddaring process, is not obtained.

Cheese powders can be used for exporting to countries where milk production is very low. The principle is to manufacture powder of the appropriate composition by UF and drying. The importing country need only add starters and rennet to reconstituted powders to produce cheese without whey production.

Incorporation of UF (VCF approximately 1.5) into manufacture of yoghurt, and related fermented products such as labneh and koumiss, has been described (Tamime and Robinson, 1999). The aim is to increase the protein levels without addition of expensive powders. As distinct from cheese-making, there is no yield improvement because the heat treatment (typically 80–85°C for 30 min) causes complete denaturation of whey proteins that become attached to the surface of the casein micelles. The resulting yoghurt gels retain water, and there is little or no syneresis. Hence, there is no scope for increasing whey protein retention. In practice, RO is often considered to be a better option for these products.

2.7.3.2 Whey processing

Whey contains only about 6 g kg^{-1} protein (see Section 2.6), but the whey proteins are highly nutritious and soluble, and possess excellent functionality. Use of UF in the production of whey protein concentrates, which can be dried to valuable high protein powders, is desirable. These products have wide applications in the manufacture of foods of high nutritive value as well as exploiting both their foaming properties and gelling properties in the baking, confectionery, meat processing and other industries. The initial flux during UF of whey is generally about twice that with milk under similar conditions, largely due to the lower protein concentration. Also, whey can be concentrated to much higher concentration factors than milk. Whey retentates start to become very viscous at concentration factors around 20, and the limit of concentration is at a factor of about 30 depending on the exact composition of the whey. The protein content of whey protein powders obviously depends on the degree of concentration during the UF process. Diafiltration can be incorporated to obtain very high protein concentrations. Typical values are as follows (Brennan *et al.*, 2006):

(1) VCF = 1; protein content (dry weight) 10–12%;
(2) VCF = 5; protein content approximately 35%;
(3) VCF = 20; protein content around 65%;
(4) VCF = 20 plus diafiltration; protein content up to about 90%.

2.7.3.3 Use of permeate

The permeates from UF of milk or whey are quite similar, containing about 5% total solids, which are predominantly lactose with some minerals and other minor soluble components of the milk. Large volumes of permeate with high BOD are produced during processing of milk or whey, which cannot be discarded to waste. Further processing and utilization of permeate is desirable and can contribute to the economic feasibility of the UF process. Three uses of permeate have been suggested following further processing: animal feed, human food or industrial fuel (Grandison and Glover 1994). Permeate can be concentrated by RO prior to further processing.

Liquid or dried permeate can be used directly, or in fermented forms as animal feeds. Lactose can be crystallized out from concentrated permeate, possibly following demineralization, but unfortunately lactose is of limited direct use. One solution is to convert lactose to glucose and galactose using β-galactosidase, for use as a sweetener in the confectionery and brewing industries. Fermentation of the lactose can be used in the manufacture of alcohol, lactic acid or antibiotics. Alternatively, fermentation of lactose to produce methane as an industrial fuel is possible.

2.7.4 *Microfiltration*

Microfiltration is the oldest membrane technology but has been slow to develop (Grandison and Finnigan, 1996). Recent development of cross-flow systems has led to novel applications of MF in the dairy industry involving removal of micro-organisms, clarification and protein separations.

2.7.4.1 Removal of microorganisms

Microfiltration is a well-established laboratory technique for the removal of micro-organisms (both vegetative and spore forms), and hence the production of sterile fluids, without the application of heat. The advent of cross-flow MF has enabled the same concept to be applied on a commercial scale.

Microfiltration of milk can be used with the aim of producing liquid milk with extended shelf-life either as a replacement for, or in addition to, heat treatment. Alternatively, MF milk with reduced bacterial numbers can be used as an alternative to pasteurized milk in the manufacture of cheese. The problem with milk processing is that the size ranges of fat globules and bacteria overlap, such that much of the fat will be removed along with bacterial cells in the retentate. However, effective removal of bacterial cells from skim milk is possible. Guerra *et al.* (1997) claim that by use of careful control with 0.87-μm-pore-size membranes, bacterial spores can be reduced by a factor of 10^4–10^5 with 100% transmission of casein at excellent permeate flux rates. Homogenization of milk leads to reduction of fat globule size, which may permit half-fat or full-fat milks to be effectively treated, if the globules are small enough to permeate the membrane. A product currently available in the UK is Cravendale 'Purfiltre' milk (marketed by Arla Foods, DK), which is a semi-skimmed product from which it is claimed 99.7% bacteria are removed by MF. The product is also pasteurized and has an extended shelf-life of 20 days. The combination of MF and pasteurization reduces bacterial numbers by 99.99% (4D) (Grandison and Finnigan, 1996). It is feasible that 'cold-pasteurized' liquid milks produced by MF alone will be available in future.

Removal of bacteria from skim milk for cheese-making by MF, using ceramic membranes on a commercial scale, has been described (Malmbert and Holm, 1988).

Permeate may be recombined with heat-treated cream in the manufacture of full-fat cheese. There are a number of potential advantages. In particular, the flavour characteristics of raw milk cheeses may be retained. The 'Bactocatch' system (marketed by Tetra Pak, Vernon Hills, IL) is claimed to hold further advantages in that bacterial contamination is reduced by 99.6% in either heat-treated or non-heated cheese milks; thus the quality is improved generally, and the incidence of blowing and the need to add nitrates to some cheeses are avoided. Reducing the bacterial load of cheese whey by MF prior to further processing is also a potential advantage.

2.7.4.2 Removal of fat globules

Hanemaaijer (1985) demonstrated the removal of residual fat from cheese whey by cross-flow MF producing lipid-rich or lipid-depleted (clarified) fractions. The presence of lipids greatly alters the functionality of powders manufactured from the clarified products, allowing 'tailoring' of functional whey protein powders for specific purposes. Recovery of fat from buttermilk has also been described (Rios *et al.*, 1989).

2.7.4.3 Fractionation of macromolecules

A promising application of MF is to the fractionation of macromolecules. Fractionation of milk proteins has usually involved separation of the proteins in the colloidal micellar phase from those in true solution. Hence, the casein micelles and whey proteins may be separated. For example, Brandsma and Rizvi (2001) demonstrated improved quality Mozzarella cheese manufacture from MF milk depleted in whey protein. This approach can be taken further if different proteins are induced to associate with or diffuse from the micelles. For example, Pouliot *et al.* (1993) described dissociation of β-casein from micelles of a phosphocaseinate suspension, which can be microfiltered to produce β-casein-enriched and β-casein-depleted fractions. Mehra and Kelly (2004) demonstrated fractionation of whey proteins into a retentate rich in immunoglobulins, bovine serum albumin and lactoferrin, and a permeate rich in α-lactalbumin and β-lactoglobulin. The same authors also reported separation of the latter permeate into α-lactalbumin- and β-lactoglobulin-rich streams using large-pore-size UF membranes (MWCO 30–100 kDa). It is feasible that, in this way, milk and whey could be tailored for specific products or dietary needs.

Fat may also be subfractionated by MF. Gouderanche *et al.* (2000) separated milk fat into globule size categories, which imparted different textural and organoleptic properties on the products.

2.7.5 Electrodialysis and electro-membrane filtration

Electrodialysis is used in Europe and Japan, mainly as a desalting process in the dairy industry. Monopolar membranes are employed for whey and skim milk demineralization, and amino acid and protein separations (Bazinet, 2005). The pH of these solutions affects the demineralization rate with the optimum obtained at the isoelectric point of the proteins, because the neutral global net charge of the proteins is not involved in the ion transport (although it should be noted that protein coagulation may occur). Purification of propionic and lactic acids from whey permeate fermentations has also been studied using ED, and the process was significantly improved when bipolar membranes were used. Electromembrane filtration has been used for the isolation of positively charged peptides with antimicrobial activity from α_{S2}-casein hydrolysates (Bargeman *et al.*, 2002).

2.7.6 Membrane bioreactors

An emerging application of membrane technology in the food and biotechnology industries, but not widely commercialized at present, is the use of membrane filtration systems as enzymatic or fermentation membrane bioreactors. Bio-catalytic conversions constitute processes where certain transformations (e.g. formation or modification of complex molecular structures etc) take place as part of the catalytic action of an enzyme or the metabolic pathway of a micro-organism. The advantages of bioprocesses over conventional and physical processes include: milder reaction conditions; use of renewable resources; reduced hazardous and environmental impact; greater specificity of catalysts and hence greater yield and process efficiency; less expensive process equipment and consequently lower capital investment; and use of recombinant technology to develop new processes. Disadvantages associated with bioconversions include the generation of complex mixtures requiring extensive downstream processing, usually at low product concentrations, low specific rate constants and inherent instability for biological processes and energy-intensive sterilization requirements (Belfort, 1989).

These processes can be carried out in batch or continuous modes of operation. Batch processes have several limitations compared with continuous processes: inherent lower efficiency due to their start-up and shutdown nature; high capital cost compared with their productivity; batch-to-batch product variations; need to inactivate and separate the biocatalyst from the product, which also leads to decreased productivity; and long operating periods for completing the reactions related to substrate depletion, product or substrate inhibition and low biocatalyst concentrations (Cheryan, 1998; Prazeres and Cabral, 1994).

These problems can be solved by operating the processes continuously, and for that purpose, immobilization or confinement of the biocatalyst is necessary. Immobilization refers to the chemical or physical attachment of the biocatalysts to solid surfaces or otherwise physically confining or localizing them in a defined region of space with retention of their catalytic properties (Cheryan, 1998; Cheryan and Mehaia, 1985). Although immobilized biocatalyst membrane reactors have several advantages, problems with losses of activity, steric hindrance, enzyme–substrate orientation, diffusional restrictions, high pressure drops, gas hold-up and the cost of immobilization mean that these processes have not been widely commercialized (Cheryan and Mehaia, 1985).

Membrane bioreactors are an alternative approach to conventional immobilization in the development of high-rate-conversion processes. Membrane bioreactors constitute an attempt to integrate bio-catalytic conversion, product separation and catalyst recovery into a single operation. Their basic concept lies in the separation provided by a synthetic semi-permeable membrane of the appropriate chemical nature and physical configuration, between the biocatalyst and the products, thus localizing the biocatalyst in a specific reaction area and allowing continuous removal of the reaction products (Prazeres and Cabral, 1994; Cheryan, 1998). Complete retention of the biocatalyst is the most important requirement for the successful operation of a membrane bioreactor, and there are two methods of organization: (1) the biocatalyst in its free form is confined to one side of the membrane by size exclusion, electrostatic repulsion or enlargement; (2) the biocatalyst is immobilized directly onto the membrane by chemical binding, physical adsorption, electrostatic attraction or entrapment within the membrane composite itself (Matson and Quinn, 1992; Prazeres and Cabral, 1994). Reaction products should pass through the membrane and be recovered in the permeate. The driving force for this separation is either diffusion (induced by a concentration gradient) or convection (usually induced by a pressure gradient). Pressure-driven permeation has the additional advantage of maximizing selectivity in sequential reaction by regulating the catalyst contact time with the substrate (Matson and Quinn, 1992).

Selection of membranes for the development of membrane bioreactors should take into account the size of the biocatalyst, substrate and product as well as the chemical nature of the membrane and the species in solution. Microfiltration membranes can be used for membrane fermenters where the biocatalysts are microbial cells, and UF membranes (1–100 kDa MWCO) can be used for enzyme membrane reactors. Particular attention should be given to the membrane material for enzyme membrane reactors, since it can significantly affect the enzyme stability (enzyme poisoning) (Prazeres and Cabral, 1994; Rios *et al.*, 2004). Membrane bioreactors can thus be classified into enzyme bioreactors or membrane fermenters depending on the nature of the biocatalyst (enzyme or microbial cell). More information on membrane fermenters can be found in Cheryan and Mehaia (1986), Chang (1987) and Belfort (1989).

2.7.6.1 Classification of membrane bioreactors and biocatalyst stability

For the classification of enzyme membrane reactors, criteria such as the type of configuration – hydrodynamics of the system or the mechanism of contact between biocatalyst and substrate – have been used. In both cases, two major categories exist: (1) the continuous stirred tank (or direct contact) membrane reactors and (2) the plug flow (hollow fibre, or diffusion) membrane reactors (Cheryan and Mehaia, 1985; Prazeres and Cabral, 1994; Cheryan, 1998).

Continuous stirred tank membrane reactors are operated in well-mixed conditions with pressure being the driving force for product permeation. Within this classification, two further subdivisions exist, i.e. dead-end and recycle membrane reactors (Figs. 2.7a and b). In dead-end membrane reactors, separation and reaction occur in the same compartment, and the reaction mixture is pressurized against the membrane, which is usually placed on the base of the system (Prazeres and Cabral, 1994) or submerged in the feed tank (Fane and Chang, 2002). These types of reactors are

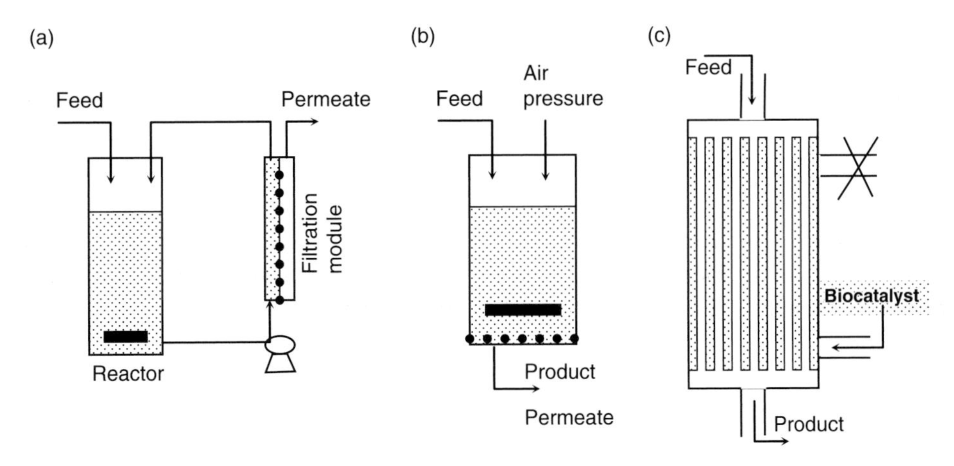

Fig. 2.7 General operation and configuration principles for the (a) recycle, (b) dead-end and (c) Plug flow membrane reactors.

associated with a major disadvantage, which is the necessity to compromise between the conditions necessary to obtain high conversions (reaction kinetics controlled) and maintain steady-state operation (governed by the performance characteristics of the membrane) (Cheryan, 1998). To that end, recycle membrane reactors, typically made up of a tank reactor coupled in a semi-closed-loop configuration via a suitable pump to a membrane module, offer the necessary conditions for adequate individual control of the reaction and separation (Cheryan and Mehaia, 1985; Cheryan, 1998).

In recycle membrane reactors, the membrane unit is physically separate from the reaction vessel. The reaction mixture is continuously pumped through the membrane module and recycled back, while the product permeate is removed and substituted with fresh substrate at the same rate. The completely mixed conditions prevailing in recycle membrane reactors (due to continuous feed circulation throughout the module) suggest that product and substrate concentrations (provided they are both permeable) are equal in the reaction vessel and permeate. Thus, this reactor type is more suitable for substrate-inhibited reactions than product-inhibited ones due to the equilibrium partitioning nature of membrane processes, but they suffer from low conversion rates due to low substrate concentrations (for the most common bioconversions) since they have a positive reaction order (reaction proceeds irreversibly) (Cheryan, 1998). In order to improve the conversion rates in these types of reactors, more than one can be connected in series (which can also facilitate multi-enzyme conversions), or a multistage increase of the reaction vessel volume can be used (in order to increase the residence time in the reactor). This bioreactor type is the most suitable for macromolecule hydrolysis, since the substrate availability does not depend on its molecular size (as discussed for plug flow membrane reactors) and also the membrane provides a means of controlling the size of the products in the permeate by its sieving characteristics (Cheryan and Mehaia, 1985; Cheryan, 1998).

Plug flow (or diffusion) membrane reactors use the membrane unit as bioreactor (Fig. 2.7c). Usually operated in hollow-fibre membrane cartridges, the biocatalyst is trapped on one side of the membrane, and the substrate is pumped across the other side (usually the shell and lumen sides, respectively). The substrate must diffuse across the membrane to undergo reaction, and the product and any unconverted substrate must then diffuse back into the flowing stream and be removed from the reactor. The concept of plug flow membrane reactor operation requires substrate and products to be of approximately the same molecular size, but much smaller (at least 10 times) than the pore size of the membrane, in order to facilitate diffusion-controlled mass transfer. The diffusion-controlled mass transfer of substrate in these types of reactors is their main disadvantage, since it limits the kinetic behaviour of the biocatalyst (Prazeres and Cabral, 1994). Several different modes of operation of plug flow reactors have been investigated, some of which are discussed by Kitano and Ise (1984) and Cheryan (1998).

Membrane bioreactors with catalytic active membranes are a new type of reactor where the biocatalysts are immobilized on the porous structure of the membranes (usually ceramic). The preparation of these catalytically active membrane types is a three-step process where an ultra-thin dynamic gel layer is generated on top of a membrane that is subsequently activated with cross-linking agents (e.g glutaraldehyde) and finally used as support for enzyme immobilization by covalent bonding. The expected advantages of this type of reactor include precise reaction control, minimized substrate and catalyst losses, faster reactions, higher yields, cleaner products and lower operating costs (Rios *et al.*, 2004). Membrane contactors are another recently developed concept, whereby enzymic reactions are performed in systems where membrane separation is coupled to a solvent extraction step. Further information on this subject can be found in Rios *et al.* (2004).

Inactivation of the biocatalyst and membrane fouling are important parameters determining the productivity and product consistency of membrane bioreactors, and consequently their application on a commercial scale. Several different parameters can lead to biocatalyst inactivation, including: dilution rate, thermal inactivation, loss of activators, membrane adsorption and poisoning, shear damage (associated also with interfacial inactivation, local heating, air entrainment) and inactivation due to reaction with other components present in the mixture (e.g. reducing sugars and browning reactions) (Deeslie and Cheryan, 1982; Prazeres and Cabral, 1994; Cheryan, 1998; Petzelbauer *et al.*, 2002). The reaction temperature is considered to be one of the major parameters causing enzyme inactivation. For that reason, it is frequently suggested that temperatures lower than the optimum should be used in membrane bioreactors in order to avoid loss of activity (Deeslie and Cheryan, 1982; Cheryan, 1998). Loss of activators is also a major reason for lower activities, because activators are frequently much smaller than the enzymes, and are thus lost in the permeate. Some membrane materials cause inactivation of enzymes on contact, but this problem can be solved only by trial and error during the development of any individual process (Cheryan and Mehaia, 1985).

Of all the parameters affecting enzyme activity, shear damage is the problem that most affects the operation of membrane bioreactors (especially continuous stirred tank recycle reactors). However, work carried out with a variety of enzymes (alcohol dehydrogenase, catalase, urease) has shown that shear was not the parameter directly causing inactivation, but other phenomena induced by shear, such as local heating, surface denaturation at cavities and most importantly the generation of gas liquid interphases, were actually the direct causative agents (Thomas *et al.*, 1979; Thomas and Dunnill, 1979; Virkar *et al.*, 1981; Narendranathan and Dunnill, 1982).

2.7.6.2 Basic theory characterizing membrane bioreactor operation

For continuous bioreactor operations, a series of different values characterizing the reaction conditions can be calculated using the data from sample analysis and permeate flux and volume measurements taken throughout the experiments.

The 'volume replacements' or number of diavolumes (dimensionless) is determined by the ratio V_c/V_f, where V_c is the total cumulative permeate removed during operation, and V_f the volume of the reaction mixture in the recycle membrane reactor.

The 'residence time' in a reactor is calculated as a function of the volume of the reaction mixture in the reactor divided by the permeate flow rate exerted during operation.

Bioreactor productivity, measured as weight of product per unit of biocatalyst used (e.g. mg U^{-1}), is calculated as 'instantaneous productivity' over a defined period of time and 'cumulative productivity' over the whole period of operation. The instantaneous productivity (P_i) is calculated using the following equation:

$$P_i = \frac{\bar{P} \times J_p \times \bar{t}}{E \times V} \tag{2.14}$$

where $\bar{P}$ is the average product output (e.g. mg ml^{-1}) in the time period $\bar{t}$, J_p is the permeate flow rate (e.g. ml min^{-1}), V is the reaction volume (e.g. ml) and E is the enzyme concentration (e.g. U ml^{-1}) (Sims and Cheryan, 1992). The cumulative productivity (P_c) can be calculated as the productivity over the full time period being considered, and is given by the equation

$$P_c = \sum P_i \tag{2.15}$$

The percentage 'substrate conversion' (dimensionless) is calculated using the following formula:

$$C = \frac{[S_0] - [S_t]}{[S_0]} \times 100 \tag{2.16}$$

where S_0 is the initial substrate concentration (e.g. mg ml^{-1}) in the feed-in stream and S_t the substrate concentration (e.g. mg ml^{-1}) in the product permeate stream leaving the reactor.

The 'rate of accumulation' (dimensionless) of any component (substrate, products, etc.) in a reactor is calculated from the rate of input into the reactor (substrate in feed, e.g. mg min^{-1} of carbohydrate introduced in the reactor) divided by the rate of output from the reactor (removed in the permeate, e.g. mg min^{-1} of carbohydrate removed from the reactor as product or unreacted substrate).

2.7.6.3 *Applications of membrane bioreactors and fermenters in the dairy industry*

Hydrolytic membrane reactors have been employed for the degradation of oligosaccharides including lactose (Cheryan and Mehaia, 1985; Cheryan, 1998). Galactosidases are versatile biocatalysts used for lactose hydrolysis (Prenosil *et al.*, 1987; Bakken *et al.*, 1990, 1992), to facilitate milk digestibility and improve the functional properties of dairy products, and galactooligosaccharide formation. Both plug-flow and recycle membrane reactors have been used for the enzymatic hydrolysis of lactose, since both substrate and product are of low molecular weight compared with the MWCO of the membranes used. A hollow-fibre enzyme reactor operated in plug flow, but with both substrate and enzyme phases recirculated with ultrafiltration swing, was developed by Park *et al.* (1985) for lactose hydrolysis. It was demonstrated that conditions (e.g. recirculation rate) favouring increased mass transfer also increased the conversion in the reactor.

Galactooligosaccharide manufacture using lactose as substrate is a fairly new and rapidly growing dairy-related process, due to the recent increased interest generated in functional foods. Galactooligosaccharides are prebiotic ingredients which may beneficially affect the colonic health of consumers. Applications of these products are found in commercial preparations for use by adults but also in infant formula preparations, since galactooligosaccharides are carbohydrates that are found in significant amounts in human milk. Continuous ultrafiltration stirred tank reactors for lactose hydrolysis and galactooligosaccharide synthesis have been developed by Petzelbauer *et al.* (2002) with hyper-thermophilic β-galactosidases, and Chockchaisawasdee *et al.* (2004) with a β-galactosidase from *Kluyveromyces lactis*. They identified membrane protein adsorption, recirculation rate and sugar protein interactions to be responsible for enzyme inactivation, and also showed variation in the degree of polymerization between the oligosaccharides produced in batch reactions and those produced with a membrane reactor.

Another dairy-based application of membrane bioreactors is in the purification of β-lactoglobulin from whey protein. The basis of the purification process is the enzymatic hydrolysis of contaminating proteins, α-lactalbumin and traces of serum albumin, by pepsin at conditions under which β-lactoglobulin is resistant to hydrolysis. The peptides generated would permeate the membrane leaving β-lactoglobulin in a purified form in the retentate stream, where it can be concentrated (Kinekawa and Kitabatake 1996; Sannier *et al.*, 2000). In the same context, casein-free whey and milk protein preparations, for use in infant formula preparations, can be prepared by enzymatic hydrolysis with the cell-envelope lactococcal proteinase in order to prevent immunological reactions to these products (Pouliot *et al.*, 1993; Alting *et al.*, 1998). These processes can be optimized to operate continuously. The same principle can be used for the production of bioactive peptides by fermentation in membrane fermenters with the microorganism *Lactobacillus helveticus* that possesses strong proteolytic activity. In this case, however, the desirable product would be in the permeate stream.

Applications of membrane fermenters in the diary industry have been aimed at the production of ethanol, lactic acid and the treatment of dairy effluents. In these cases, the lactose present in cheese whey UF permeate is used as substrate, which, after the necessary pre-treatment (concentration by RO, addition of nutrients and sterilization by MF), is fermented by different micro-organisms. Production of ethanol has been studied with *Kuyveromyces fragilis,* but other yeast strains can also be used if lactose is first hydrolysed to glucose and galactose. Lactic acid has been produced with *Lactobacillus* spp. (Börgardts *et al.*, 1998; Cheryan 1998).

2.7.7 Selective separations of dairy-derived carbohydrates by nanofiltration

Another novel dairy-related application of membrane technology in the functional food market is the use of nanofiltration to perform very selective purifications between carbohydrates with prebiotic properties (oligosaccharides), and other sugars with no such beneficial properties (monosaccharides). Such separations would have been considered unfeasible until recently due to the small differences in the size of the molecules to be separated; monosaccharides versus oligosaccharides (degree of polymerization varying between di- to hepta-saccharides). However, developments in the production methods of membranes have led to the manufacture of membranes with very well-defined permeability characteristics that made such separations attainable (Goulas *et al.*, 2002, 2003).

In low-molecular-weight sugar separations, the major parameters regulating solute permeability are the membrane characteristics, the feed concentration (and consequently the resulting concentration polarization), and the process operating parameters such as feed temperature and pressure. Most sugars are neutral molecules in aqueous solutions, whose mass transport through NF membranes is controlled by convection and diffusion (Tsuru *et al.*, 2000). Table 2.2 shows the rejection values of such sugars for two NF membranes. The differences between the rejection values of these sugars indicate the potential of some NF membranes for purifying oligosaccharides from monosaccharides. By regulating different filtration parameters, most importantly feed concentration and filtration temperature, optimization of the process

Table 2.2 Rejection values for raffinose, lactose and glucose during cross-flow NF of a model solution of the three sugars.

Membrane filtration temperature	Pressure (bar)	Rejection (–)		
		Raffinose	Sucrose	Fructose
NF-CA-50	6.9	0.80	0.56	0.10
25°C	13.8	0.90	0.73	0.26
DS-5-DL	6.9	1.00	0.99	0.54
25°C	13.8	1.00	0.99	0.72
DS-5-DL	6.9	0.99	0.94	0.28
60°C	13.8	1.00	0.97	0.48

NF-CA-50 cellulose acetate asymmetric membrane from Intersep Ltd (Wokingham, UK) and DS-5-DL thin-film composite membrane from Osmonics Desal (Le Mee sur Seine, France).

Table 2.3 Yield values from nanofiltration continuous diafiltration purification with a model solution and a commercial galactooligosaccharide mixture.

NF-CA-50, 13.8 bar at 25°C Yield (%)			**DS-5-DL**, 13.8 bar at 60°C Yield (%)		
Raffinose	Sucrose	Fructose	Raffinose	Sucrose	Fructose
81	59	15	98	89	14
Oligos[a]	Lactose	Glucose	Oligos	Lactose	Glucose
84	62	18	98	89	18

[a] The term 'Oligos' denotes all the galactooligosaccharide content of the mixture.

could be achieved. During continuous diafiltration purification of oligosaccharide mixtures, the developed method gave very good results, as shown in Table 2.3.

2.8 Future developments

Membrane technology has made significant advances in applications in the dairy industry in the past few decades. The relative cost of processing has fallen markedly over the past 15 years, making the technology much more attractive. However, the techniques have yet to reach their full potential due to several reasons such as

(1) the separations available with commercial membranes are often diffused so that the levels of purification are not good enough to compete with existing techniques;
(2) problems of low flux, or flux decline during processing, may render some operations non-economic;
(3) many sectors of the dairy industry are very traditional and slow to embrace the opportunities presented by novel membrane techniques.

Advances in membrane design and greater understanding of operating parameters are leading to sharper, more accurate separations. For example, it is now feasible to separate mono-, di- and oligosaccharides using NF (Goulas *et al.*, 2002), or to fractionate the proteins in milk using MF (Le Berre and Daufin 1996). These possibilities would have been considered unlikely some years ago.

Improving flux rates has been the goal of much research in recent years. Improvements in membrane design to increase porosity, or reduce tortuosity and thickness have gone some way to achieving this goal. There have been several approaches, which involve reducing the concentration polarization layer and rate of fouling, or removing deposits from the membrane. Optimizing pressures and feed flow rates too can be beneficial. The critical flux concept (Gesan-Guiziou *et al.* 2000), in which permeate flux is maintained at a level below that at which fouling occurs, is a promising idea, which has been demonstrated with milk (Youravong *et al.*, 2003). Backflushing of some permeate into the retentate stream at intervals is a well-established method of clearing blocked pores and hence recovering flux during MF. Turbulence promoters can be used to minimize concentration polarization. Other promising techniques include initiating toroidal vortices in the feed, the use of pulsatile flows or mechanically vibrating or electrically charged membranes. These techniques have met with success on a research level and should lead to more commercial applications.

Membrane processing, being a purely physical technique, fits well into the philosophy of producing minimally processed foods and will probably continue to replace more severe techniques of concentration or separation involving heat or chemical interactions in dairy processing. Membranes could be used to produce purified proteins with specific functional properties, purified oligosaccharides with prebiotic properties or even long-life milk without any heat treatment.

References

Alting, A.C., Meijer, R.J.G.M. and van Beresteijn, E.C.H. (1998). Selective hydrolysis of milk proteins to facilitate the elimination of the ABBOS epitope of bovine serum albumin and other immunoreactive epitopes. *Journal of Food Protection* **61,** 1007–1012.

Bakken, A.P., Hill, C.G. and Amundson, C.H. (1990). Use of novel immobilized β-galactosidase reactor to hydrolyse the lactose constituent of skim milk. *Biotechnology and Bioengineering* **36,** 293–309.

Bakken, A.P., Hill, C.G. and Amundson, C.H. (1992). Hydrolysis of lactose in skim milk by immobilized β-galactosidase (*Bacillus circulans*). *Biotechnology and Bioengineering* **39**, 408–417.

Bargeman, G., Houwing, J., Recio, I., Koops, G.H. and van der Horst, C. (2002). Electro-membrane filtration for the selective isolation of bioactive peptides from an s2-casein hydrolysate. *Biotechnology and Bioengineering*, **80,** 599–609.

Bazinet, L. (2005). Electrodialytic phenomena and their applications in the dairy industry: a review. *Critical Reviews in Food Science and Nutrition* **45**, 307–326.

Belfort, G. (1989). Membranes and bioreactors: a technical challenge in biotechnology. *Biotechnology and Bioengineering* **33**, 1047–1066.

Börgardts, P., Krinschke, W., Trosch, W. and Brunner, H. (1998). Integrated bioprocess for the simultaneous production of lactic acid and dairy sewage treatment. *Bioprocess Engineering* **19**, 321–329.

Brandsma, R.L. and Rizvi, S.S.H. (2001). Effect of manufacturing treatments on the rheological character of mozzarella cheese made from microfiltration retentate depleted of whey proteins. *International Journal of Food Science and Technology* **36**, 601–610.

Brennan, J.G., Grandison, A.S. and Lewis, M.J. (2006). Separations in food processing. In: Brennan, J.G. (ed.) *Food Processing Handbook*, pp. 429–511. Wiley-VCH, Weinheim.

Chang, H.N. (1987). Membrane bioreactors: engineering aspects. *Biotechnology Advances* **5**, 129–145.

Cheryan, M. (1998). *Ultrafiltration and Microfiltration Handbook*. Technomic, Lancaster, PA.

Cheryan, M. and Mehaia, M.A. (1985). Membrane bioreactors. *CHEMTECH* **16**, 676–681.

Cheryan, M. and Mehaia, M.A. (1986). Membrane bioreactors. In: McGregor, C.W. (ed.) *Membrane Separations in Biotechnology*, pp. 255–302. Marcel Dekker, New York.

Chockchaisawasdee, S., Athanasopoulos, V.I., Niranjan, K. and Rastall, R.A. (2004). Synthesis of galactooligosaccharide from lactose using β-galactosidase from *Kluyveromyces lactis*: studies on batch and continuous UF membrane-fitted bioreactors. *Biotechnology and Bioengineering* **89,** 434–443.

Deeslie, D.W. and Cheryan, M. (1982). A CSTR-hollow-fiber system for continuous hydrolysis of proteins. Factors affecting long-term stability of the reactor. *Biotechnology and Bioengineering* **24**, 69–82.

El-Gazzar, F.E. and Marth, E.H. (1991) Ultrafiltration and reverse osmosis in the dairy industry: a review. *Journal of Food Protection* **54**, 801–809.

Fane, A. and Chang, S. (2002). Membrane bioreactors: design and operational options. *Filtration and Separation* **39**, 26–29.

Gesan-Guiziou, G., Daufin, G. and Boyaval, E. (2000). Critical stability conditions in skimmed milk crossflow microfiltration: impact on operating modes. *Lait* **80**, 129–140.

Gouderanche, H., Fauquant, J. and Maubois, J.L. (2000). Fractionation of globular milk fat by microfiltration. *Lait* **80**, 93–98.

Goulas, A.K., Grandison, A.S. and Rastall, R.A. (2003). Fractionation of oligosaccharides by nanofiltration. *Journal of the Science of Food and Agriculture* **83**, 675–680.

Goulas, A.K., Kpasakalidis, P., Sinclair, H.R., Grandison, A.S. and Rastall, R.A. (2002). Purification of oligosaccharides by nanofiltration. *Journal of Membrane Science* **209**, 321–335.

Grandison, A.S. and Finnigan, T.J.A. (1996) Microfiltration. In: Grandison, A.S. and Lewis, M.J. (eds) *Separation Processes in the Food and Biotechnology Industries*, pp. 141–153. Woodhead Publishing, Cambridge.

Grandison, A.S. and Glover, F.A. (1994) Membrane processing of milk. In: Robinson, R.K. (ed.) *Modern Dairy Technology 1: Advances in Milk Processing*, 2nd edition, pp. 273–311. Chapman & Hall, London.

Groleau, P.E., Lapointe, J.F., Gauthier, S.F. and Pouliot, Y. (2004). Fractionation of whey protein hydrolysates using nanofiltration membranes. *International Dairy Federation Bulletin No. 389*, pp. 85–91.

Guerra, A., Jonsson, G., Rasmussen, A., Waagner Nielsen, E. and Edelsten, D. (1997). Low cross-flow velocity microfiltration of skim milk for removal of bacterial spores. *International Dairy Journal* **7**, 849–861.

Gutman, R.G. (1987). *Membrane Filtration. The Technology of Pressure-Driven Crossflow Processes*. Adam Hilger imprint by IOP Publishing, Bristol, UK.

Guu, Y.K. and Zall, R.R. (1992). Nanofiltration concentration on the efficiency of lactose crystallisation. *Journal of Food Science* **57**, 735–739.

Hanemaaijer, J.H. (1985). Microfiltration in whey processing. *Desalination* **53**, 143–155.

Kinekawa, Y.I. and Kitabatake, N. (1996). Purification of β-lactoglobulin from whey protein concentrate by pepsin treatment. *Journal of Dairy Science* **79**, 350–356.

Kitano, H. and Ise, N. (1984). Hollow fiber enzyme reactors. *Trends in Biotechnology* **2**, 5–7.

Kosikowski, F.V. (1986). Membrane separations in food processing. In: McGregor, W.C. (ed.) *Membrane Separations in Biotechnology*, Marcel Dekker, New York, 201–254.

Kulkarni, S.S. and Funk, E.W. (1992). Ultrafiltration. In: Winston Ho, W.S. and Sirkar, K.K. (eds) *Membrane Handbook*, pp. 393–453. Van Nostrand Reinhold, New York.

Lawrence, R.C. (1989). The use of ultrafiltration technology in cheese making. *International Dairy Federation Bulletin* **240**, 3–15.

Le Berre, O. and Daufin, G. (1996). Skim milk crossflow microfiltration performance versus permeation flux to wall shear stress ratio. *Journal of Membrane Science*, **117**, 261–270.

Lewis, M.J. (1996a). Pressure-activated membrane processes. In: Grandison, A.S. and Lewis, M.J. (eds) *Separation Processes in the Food and Biotechnology Industries*, pp. 65–97. Woodhead Publishing, Cambridge.

Lewis, M.J. (1996b). Ultrafiltration. In: Grandison, A.S. and Lewis, M.J. (eds) *Separation Processes in the Food and Biotechnology Industries*, pp. 97—140. Woodhead Publishing, Cambridge.

Malmbert, T.R. and Holm, S. (1988) Producing low-bacteria milk by microfiltration. *North European Food and Dairy Journal*, **1**, 1–4.

Matson, S.L. and Quinn, J.A. (1992). Membrane reactors. In: Winston Ho, W.S. and Sirkar, K.K. (eds) *Membrane Handbook*, pp. 809–832. Van Nostrand Reinhold, New York.

Mehra, R. and Kelly, P.M. (2004). Whey protein fractionation using cascade membrane filtration. *International Dairy Federation Bulletin, No. 389*, pp. 40–44.

Mistry, V.V. and Maubois, J.-L. (1993) Application of membrane separation technology to cheese production. In Fox P.F. (ed.), *Cheese: Chemistry, Physics and Microbiology*, 2nd edn, pp. 493–522. Chapman & Hall, London.

Narendranathan, T.J. and Dunnill, P. (1982). The effect of shear on globular proteins during ultrafiltration: studies of alcohol dehydrogenase. *Biotechnology and Bioengineering*, **24**, 2103–2107.

Osada, Y. and Nakagawa, T. (1992). *Membrane Science and Technology*. Marcel Dekker, New York.

Park, H.T., Kim, I.H. and Chang, H.N. (1985). Recycle hollow fiber enzyme reactor with flow swing. *Biotechnology and Bioengineering*, **27**, 1185–1191.

Pellegrino, J. (2000). Filtration and ultrafiltration equipment and techniques. *Separation and Purification Methods*, **29**, 91–118.

Petzelbauer, I., Splechtna, B. and Nidetzky, B. (2002). Development of an ultrahigh-temperature process for the enzymatic hydrolysis of lactose. III. Utilisation of two thermostable β-glycosidases in a continuous ultrafiltration membrane reactor and galacto-oligosaccharide formation under steady-state conditions. *Biotechnology and Bioengineering*, **77**, 394–404.

Pontalier, P.-Y., Ismail, A. and Ghoul, M. (1997). Mechanisms for the selective rejection of solutes in nanofiltration membranes. *Separation and Purification Technology*, **12**, 175–181.

Pouliot, Y., Gauthier, S.F., and Bard, C. (1993). Fractionation of casein hydrolysates using polysulfone ultrafiltration hollow fiber membranes. *Journal of Membrane Science*, **80**, 257–264.

Prazeres, D.M.F. and Cabral, J.M.S. (1994). Enzymatic membrane bioreactors and their applications. *Enzyme and Microbial Technology*, **16**, 738–750.

Prenosil, J.E., Stuker, E. and Bourne, J.R. (1987). Formation of oligosaccharides during enzymatic lactose hydrolysis and their importance in a whey hydrolysis process: Part II: experimental. *Biotechnology and Bioengineering*, **30**, 1026–1031.

Rios, G.M., Belleville, M.P., Paolucci, D. and Sanchez, J. (2004). Progress in enzymatic membrane reactors – a review. *Journal of Membrane Science*, **242**, 189–196.

Rios, G.M., Tarodo de la Fuente, B., Bennasar, M. and Guidard, C. (1989). Cross-flow filtration of biological fluids on inorganic membranes: A First State of the Art. In S. Thorne (ed.) *Developments in Food Preservation* – 5, pp. 131–175. Elsevier Applied Science, London.

Sannier, F., Bordenave, S. and Piot, J.M. (2000). Purification of goat β-lactoglobulin from whey by an ultrafiltration membrane enzymatic reactor. *Journal of Dairy Research*, **67**, 43–51.

Sims, K.A. and Cheryan, M. (1992). Continuous saccharification of corn starch in a membrane reactor. Part I: Conversion, capacity and productivity. *Starch/Starke*, **44**, 341–346.

Strathmann, H. (1992). Electrodialysis. In: Winston Ho, W.S. and Sirkar, K.K. (eds) *Membrane Handbook*, pp. 217–262. Van Nostrand Reinhold, New York.

Tamime, A.Y. and Robinson, R.K. (1999). *Yoghurt: Science and Technology*. Woodhead Publishing, Cambridge.

Thomas, C.R. and Dunnill, P. (1979). Action of shear on enzymes: studies with catalase and urease. *Biotechnology and Bioengineering*, **21**, 2279–2302.

Thomas, C.R., Nienow, A.W. and Dunnill, P. (1979). Action of shear on enzymes: studies with alcohol dehydrogenase. *Biotechnology and Bioengineering*, **21**, 2263–2278.

Timmer, J.M.K., Speelmans, M.P.J. and van der Horst, H.C. (1998). Separation of amino acids by nanofiltration and ultrafiltration membranes. *Separation and Purification Technology*, **14**, 133–144.

Tsuru, T., Izumi, S., Yoshioka, T. and Asaeda, M. (2000). Temperature effect on transport performance by inorganic nanofiltration membranes. *AIChE Journal*, **46**, 565–574.

Vellenga, E. and Tragardh, G. (1998). Nanofiltration of combined salt and sugar solutions: coupling between rejections. *Desalination*, **120**, 211–220.

Virkar, P.D., Narendranathan, T.J., Hoare, M. and Dunnill, P. (1981). The effect of shear on globular proteins: extension to high shear fields and to pumps. *Biotechnology and Bioengineering*, **23**, 425–429.

Whu, J.A., Baltzis, B.C. and Sirkar, K.K. (2000). Nanofiltration studies of larger organic microsolutes in methanol solutions. *Journal of Membrane Science*, **170**, 159–172.

Youravong, W., Lewis, M.J. and Grandison, A.S. (2003). Critical flux in ultrafiltration of skimmed milk. *Transactions of the Institution of Chemical Engineers* **81**, 303–308.

Chapter 3

Hygiene by Design

J. Ferdie Mostert and Elna M. Buys

3.1 Introduction

The quality of milk products, or any food product, can be defined against a wide range of criteria, including for example, the chemical, physical, microbiological and nutritional characteristics. Food and dairy product manufacturers aim to ensure that the safety and quality of their products will satisfy the highest expectations of the consumers. On the other hand, the consumers expect to a large extent that the manufacturer has ensured that the product:

- is safe for human consumption with respect to both chemical and microbial contamination;
- conforms to any regulations enshrined in law, or statutory requirements laid down by health or local authorities;
- is capable of achieving a specified shelf-life without spoilage; and
- has the highest possible organoleptic standard that can be achieved within the existing constraints of manufacture or marketing (Tamime and Robinson, 1999).

The most important raw material used in dairy product manufacture is milk. Milk, in addition to being a nutritious medium, presents a favourable physical environment for the multiplication of microorganisms and, being an animal product, is subjected to widely differing production, handling and processing methods, resulting in its contamination by a broad spectrum of microbial types, chemical residues and cellular material. The overarching principles applying to the production, processing, manufacture and handling of milk and milk products are:

- From raw material production to the point of consumption, all dairy products should be subject to a combination of control measures. Together, these measures (good agricultural practice – GAP and good manufacturing practice – GMP) should meet the appropriate level of public health protection.
- Good hygienic practices should be applied throughout the production and processing chain so that the milk product is safe and suitable for its intended use.
- Wherever appropriate, hygienic practices for milk and milk products should be implemented following the Annex to the *Codex Recommended International Code of Practice – General Principles of Food Hygiene* (IDF/FAO, 2004).

Food hygiene can be defined in many ways and from different perspectives, for example, as:

- all conditions and measures necessary to ensure the safety and suitability of food at all stages of the food chain (Anon., 1997);
- all measures necessary to ensure the safety and wholesomeness of food stuffs (Anon., 1993); and
- practices designed to maintain a clean and wholesome environment for food production, preparation and storage (Marriott, 1999).

The concept of hygiene is, thus, very broad and includes (a) all stages of the supply chain, from milk harvesting through to consumption, (b) any measure designed to prevent contamination of food at any stage of production, (c) maintaining a clean work environment during food processing and (d) those elements which make effective cleaning/sanitization possible, e.g. good plant, process and equipment design as well as issues such as correct working practices for personnel regarding handling of food, control of pests, etc. (Lelieveld, 2003a). These and other related aspects will be dealt with in more detail in this chapter.

3.2 Maintaining a clean working environment in dairy plant operations

3.2.1 Introduction

Factors that influence the quality, safety and shelf-life of dairy products during the manufacturing and distribution chain include the number and specific microorganisms of the raw milk supply, the design and effectiveness of the processing plant, cleaning, sanitation and maintenance programmes and refrigeration during transportation, retail distribution and consumer possession (Carey *et al.*, 2005). The sources of contamination that have to be controlled in and around the dairy processing plant, as well as the relevant regulations and guidelines that address these aspects, will be dealt with in the ensuing sections.

3.2.2 Regulations

Hygiene measures include measures intended to provide high-quality products and to control the hygiene and food safety. There is a steady increase in the involvement of regulatory and advisory bodies in the area of food process hygiene. The reason for this is the highly published incidents such as BSE, Foot and Mouth outbreaks in Europe, and numerous problems with individual products (Cocker, 2003). Many of these incidents are given a high profile because they involve international brands.

Most of the new legislation is based on the internationally developed United Nations Food and Agricultural Organization (FAO) *Codex Alimentarius*, contributing to a national and international trend towards harmonization. Trading blocs as well as individual nations, such as Europe and the USA, may exert an influence on hygiene legislations beyond their jurisdiction due to the fact that they have highly developed legislation in the area of food safety (Cocker, 2003).

The dairy industry was one of the earliest food sectors to develop legislation on hygienic practice and product safety, nearly 50 years ago. The International Dairy Federation (IDF) (www.fil-idf.org) and Joint FAO/WHO Committee of Government Experts compiled the Code of Principles concerning Milk and Milk Products in 1958. In 1993, the compositional standards of the IDF gained regulatory status when the IDF, FAO and WHO convened a group of experts, the Milk Committee, which was fully integrated into the Codex system as the Codex Committee on Milk and Milk Products (CCMMP) (www.codexalimentarius.net). Interestingly, if one regards the enhanced concern with food safety, most of the standards are concerned with the composition of the dairy products. The only standard concerned with the hygienic practice, specifically for milk and milk products, is the code of Hygienic Practice for Milk and Milk Products (CAC/RCP, 2004).

For compositional standards, revised or devised after 1997, a section on hygiene is included in the standard recommending that the product 'be prepared and handled in accordance with the appropriate sections of the Recommended International Code of Practice – General Principles of Food Hygiene (CAC/RCP 1-1969, Rev. 4-2003) (cited in CAC/RCP, 2004), and other relevant Codex texts such as Codes of Hygienic Practice and Codes of Practice'. It is also stated in selected compositional standards that the 'products should comply with any microbiological criteria established in accordance with the Principles for the Establishment and Application of Microbiological Criteria for Foods (CAC/GL 21-1997)' (cited in CAC/RCP, 2004).

A proposed draft guideline on the 'Application of General Principles of Food Hygiene to the Control of *Listeria monocytogenes* in Ready-to-eat Foods (ALINORM 05/28/12)' (CAC/RCP, 2004) will also be applicable to dairy processing.

In the EU, laws applied by the national authorities have been harmonized at EU level. In recent years, the food policy at international level has been moving in a new direction, towards industry taking responsibility for the control of the foodstuffs it produces, backed by official control systems. The European food industry has been at the forefront of the development of preventive food-safety systems, such as Hazard Analysis and Critical Control Point (HACCP), which requires the industry itself to identify and control potential safety hazards. The national authorities check that the controls are adequate. Although initially introduced by industry and employed in a non-mandatory manner, the success of this approach has led to it being included in several directives (Cocker, 2003). The EU directives that impact on food hygiene include:

- 93/94EEC Food hygiene;
- 89/392 EEC Machinery Directive and its amendments 91/368/EEC, 93/44, 93/68;
- EEC 92/59/EEC Council Directive Concerning General Product Safety;
- EEC 93/465/EEC Conformity Assessment and Rules for Affixing the CE mark;
- EEC 93/68/EEC Amending Directive on CE Marking: 87/404/EEC, 88/378/EEC, 89/106/EEC, 89/336EEC, 89/392/EEC, 89/686/EEC, 90/85/EEC, 90/385/EEC, 90/396/EEC, 91/263/EEC, 92/42/EEC, 73/23/EEC;
- EEC 94/62/EEC Packaging and Packaging Waste – Amended by 97/129/EEC and 97/138/EEC;
- Directive 98/83/EEC 'Potable Water';
- 90/679/EEC Worker Safety Pathogenic Organisms; and
- 90/220/EEC Deliberate Release of Genetically Modified Organisms (Cocker, 2003).

The European Hygienic Engineering and Design Group (EHEDG) develop criteria and guidelines for equipment, buildings and processing. The EHEDG (www.ehedg.org) is an independent group dealing specifically with issues related to design aspects of the hygienic manufacture of food products (Cocker, 2003). A series of guidelines have been published, and extended summaries are published in *Trends in Food Science and Technology*.

In the United States, the safety and quality of milk and dairy products are the responsibility of the FDA, the ultimate regulatory authority, and the USDA and

50 state regulatory agencies. In dealing with the safety and quality of milk and dairy product, two grades of milk are identified: Grade A and Grade B, which is used for the manufacturing of dairy products and has less stringent sanitation requirements. The '*Grade A Pasteurized Milk Ordinance* (PMO)' (www.cfsan.fda.gov) facilitates uniform sanitary standards, requirements and procedures. The FDA has the responsibility under the Food, Drug and Cosmetic Act, the Public Health Act, and the Milk Import Act to make sure that processors of both Grade A and Grade B milk take remedial action when conditions exist that could jeopardize the safety and quality of milk and dairy products (Cocker, 2003). In turn, the USDA will inspect dairy manufacturing plants to determine whether good sanitation practices are being followed, specified in the *General Specifications for Dairy Plants Approved for USDA Inspection and Grading Service of Manufactured or Processed Dairy Products* (7CFR 58.101 to 58.938) (Cocker, 2003). An arrangement exists between the FDA and USDA under which the USA will assist individual states in assisting safety and quality regulations for Grade B milk. These are set out in the *Recommended Requirements for Milk for Manufacturing Purposes and Its Production and Processing*. The 3-A Sanitary Standards, referenced in the PMO, are used throughout the USA as a source of hygienic criteria for food and dairy processing, packaging and packaging equipment (Cocker, 2003).

3.2.3 Sources of contamination

In any food-processing operation, particularly highly perishable dairy products, maintaining a clean working environment will contribute largely in assuring the safety and shelf-life of the products produced. There are, however, various sources of contamination, physical, chemical and microbiological, that have to be controlled (IDF, 1992, 1997a; Lelieveld, 2003a). Some pose a higher risk than others and the mechanism of contamination distribution differs from source to source, for example.

- Non-product contact surfaces, such as floors, walls and ceilings are important reservoirs of microbial contamination and can also be a source of physical and chemical contamination (Robinson and Tamime, 2002; Lelieveld, 2003a).
- Environmental factors. Distribution of condensation by contact, sprays and splashes are distributed as airborne droplets or aerosols and the risk to product safety is high. Powder, dust particles and fresh air pose a medium risk (IDF, 1997a).
- Equipment and processing. Contact of the food with surfaces leaves residual food debris; this contributes to the proliferation of microorganisms on the processing surfaces and impacts on product safety and quality (Lelieveld, 2003a).
- Personnel and their working practices distribute contaminants by contact and airflow and pose a high risk to product safety (IDF, 1997a; Robinson and Tamime, 2002).

3.2.3.1 Non-product contact surfaces

3.2.3.1.1 Location of the building. The location of the dairy plant is essential for ensuring a high hygienic level of dairy production (IDF, 1994, 1997a; Robinson and Tamime, 2002). For instance, the location of the building should ensure appropriate product flow through the plant. The separation of rooms with different pollution risks may, however, require extra space (IDF, 1992, 1997a). Poorly designed and maintained buildings can be a source of contamination (Lelieveld, 2003b). If a plant is well designed, also from a hygienic point of view, there will be no unnecessary critical control points (CCPs) when the plant is new.

Achieving the optimum degree of compartmentalization may not be easy, since old buildings are often not amenable to conversion (Robinson and Tamime, 2002). Indoor climate protection such as temperature regulation, dust protection and vermin protection will be influenced by the location's climatic conditions.

Since neighbouring industries may present a hygienic risk, this must not affect dairy production. Water quality and uninterrupted electricity supply are vital for a clean working environment. For instance, interruptions or fluctuations in power supply can cause failures in hygiene practices. There should be adequate facilities for the disposal of effluent and/or for the establishment of an effluent treatment plant in such a location in relation to the prevailing wind and at such a distance, so as to avoid pollution of processing and storage areas (IDF, 1994). Dairy wastewater may require purifying before disposal due to remains of products and cleaning agents. Sufficient ventilation is also essential to avoid mould growth (IDF, 1992, 1997a).

All traffic entering the factory site should be considered to be a potential source of contamination. This is particularly the case for raw-milk tankers, which will be significantly soiled by the farm environment. Driveways should be of smooth-finish concrete, asphalt or other similar sealing material to keep dust or mud to a minimum, with suitable slopes to prevent the accumulation of water. In areas where vehicles discharge milk to silos, ground surface areas should be regularly inspected and residues cleared up, since raw milk from tankers is a potential vehicle for salmonellae (IDF, 1994). A checklist for maintaining a clean working environment is given in Table 3.1.

3.2.3.1.2 Layout of the building. Since outdoor environments pose a high contamination risk, all areas where milk and milk products are received, treated, processed, handled and stored must be placed within one building. Post- and cross-contamination can be minimized by keeping products with different microbiological risk profiles separate. Milk must be received and processed in separate areas, and cleaning-in-place (CIP) systems for equipment used for storing, transporting or processing raw milk and CIP systems for pasteurized products should be separate. Staff driving pallet movers, for example, should not operate in areas with different hygiene requirements, i.e. raw milk area vs. processing of heat-treated milk. One of the most obvious means of avoiding aerial contamination involves the physical separation of specific processes, e.g. dry mixing from wet mixing operations (IDF, 1992; Robinson and Tamime, 2002). Adequate ventilation in dry areas, i.e. areas where mixing of dry material takes place, is essential, and water which will be in direct contact with milk for milk products must be of drinking-water quality. A system must be in place that will ensure the frequent removal and storage until the safe disposal of disposals such as packaging materials, chemicals, solid food, contaminated products, product spillages and CIP liquids (IDF, 1992, 1997a).

Table 3.1 Maintaining a clean working-environment checklist.[a]

Environmental and non-product contact surfaces
√ Approval by competent authorities
√ No existence of cross-paths: ingredients/processing/end product/storage
√ Water-supply treatment
√ Open windows: fine mesh to keep out insects
√ Sufficient ventilation to minimize odours/vapours/prevent water condensation
√ Air-quality/filter management
√ Lights in processing areas equipped with proper covers
√ Materials used in construction – easy cleaning
√ Selection and separation of solid waste/frequent removal
√ Waste treatment
√ Presence of garbage cans
√ Walls and floor material conservation
√ Ceiling, walls and floor hygiene level
√ Doors seal and hygiene level
√ Pest-control plan by specialized enterprise
Plant and manufacturing equipment
√ Cleaning and sanitizing regime
√ Maintenance
√ Design
Human activities
√ Establishment in compliance of a Code of Hygiene Practice
√ Health
√ No existence of cross-paths: raw material processing/heating/high care
√ Hand-washing facilities adequate
√ Effective hand-cleaning/sanitizing preparations
√ Presence of hand-drying devices
√ Existence of foot-washing device
√ Toilets and rest areas are kept clean
√ Protective clothing

[a] Source: Adapted from Fadda *et al.* (2005).

Any operation presenting a hygienic risk must be kept separate from stored non-milk products, open processing and products vulnerable to post-manufacturing contamination. High-contamination-risk areas where dust, vapour and aerosols are created require adequate ventilation and strict hygiene control to prevent recontamination of the final product (IDF 1997a). Laboratories and workshops may also pose hygiene risks related to the transfer of contaminants and should, therefore, be located as far away as possible from production areas (IDF, 1994, 1997a). All these aspects indicate that the flow of material through the factory should be thoroughly considered (CAC/RCP, 2004). Robinson and Tamime (2002) noted the general principles as follows:

- Ingredients should enter at the one end of the building and be held in designated stores or silos.
- The processing and cooling stages should follow.
- The flow of end products should then meet with a flow of packaging materials coming from a separate store.
- The packaged food should then enter a finished product store for bulk packaging and dispatch.

3.2.3.1.3 Roofs and ceilings. The roof cavity can be a collection area for contaminants and pests and provides one of the easiest means of entry for contaminants to the processing area. Entry of these areas to pests should be made impossible. Ceilings are often the least accessible area for cleaning, and every aspect of ceiling design from light fittings to ventilator grills must be given high priority with respect to hygiene. Ceilings should be sloped and form a barrier, and the ceiling materials should not support the growth of mould and the accumulation of dirt and condensation (IDF, 1997a; Mostert and Jooste, 2002).

3.2.3.1.4 Walls. Care should be taken that pests cannot gain access to the processing area through pipes, wire and ducts that pass through external walls. Cracks and crevices may form in walls due to thermal shock and fluctuation; this will cause ingress of dust and moisture. Mould growth or accumulation of dirt or condensation must not be possible. For similar reasons, there should be no ledges or protrusions; if a ledge is unavoidable, a downward sloping surface of at least 45° should avoid accumulation of dust and other debris (Mostert and Jooste, 2002; Robinson and Tamime, 2002). The wall material and surface finish should be resistant to deterioration by milk, milk products, decomposition products such as lactic acid and vapours from the CIP solutions condensing and drying on walls. All surfaces must be easily cleaned without damaging the walls and joint, and intersections must not allow the ingress of liquids from the cleaning process (IDF, 1997a).

3.2.3.1.5 Windows and doors. Windows must not be constructed from hollow sections, since this can pose a hygiene risk. External windows should be fixed, and ledges should be minimal or sloped to prevent the roosting of birds. Windows that open must be fitted with insect screens. To avoid the problem of droplets or condensation, carrying microorganisms into food, windows should be double-glazed, easily accessible for cleaning and, in processing and packaging areas, made of clear polyvinylchloride. Broken glass is difficult to detect in a retail food item, and it provides a common source of consumer complaints (Robinson and Tamime, 2002). Doors are the hygienic seal between the exterior and interior openings in walls, and care should be taken that doors do not become a source of contamination, themselves. Therefore, internal pressure should close doors automatically to ensure that barriers between areas are maintained. No door should open directly to the outside from a processing area; an exception would be emergency exit doors (IDF, 1997a).

3.2.3.1.6 Floors. The floor is also a consistent source of contamination and should be constructed of high-quality concrete with a topping that will ensure a hygienic flooring system and, since many food production areas are wet, a high degree of slip resistance to ensure operator safety. The floor should be designed to withstand the loads imposed on it by equipment and operations undertaken in the area, since contaminants can collect in cracks in floors, which will result in an unhygienic flooring system. Equipment should also be fixed to the floor in such a way that crevices and cracks are avoided. The floor should be easy to clean and sloped to prevent collection of liquids. For instance, the construction materials selected should prevent liquid ingress and erosion by the CIP liquids (IDF, 1994, 1997a). Resin flooring provides seamless covering that is produced by a chemical reaction between liquid components; they exhibit excellent durability and resistance to chemicals, but are intolerant of high temperatures, and the floor may retain a residue of volatile material long after application. Quarry tiles are impervious to water and chemicals, and if embedded in and grouted with a similarly resistant mixture, they can provide a long-lasting alternative to resin floors. Each type of flooring has advantages and disadvantages. For instance, the tiles are durable, but, if the grouting between the tiles becomes loose, the resulting cavities will provide a reservoir for potential microbial contamination. Synthetic finishes can crack or peel, particularly if laid by inexperienced contractors; and if carelessly drilled after laying, water can slowly migrate along the concrete/resin interface until the entire floor begins to lift (Robinson and Tamime, 2002). Drainage systems need to remove all liquids rapidly and prevent odours from entering the processing area. Unfortunately, these systems can be neglected and become areas of contamination, and care should be taken that they are maintained in a sanitary condition (IDF, 1997a).

3.2.3.2 Environmental factors

3.2.3.2.1 Air supply. Movement of people, products, packaging and machinery creates unavoidable airborne contamination within the factory and transfer contamination from outside the factory to inside (IDF, 1997a; Mostert and Jooste, 2002; Robinson and Tamime, 2002; Den Aantrekker *et al.*, 2003). Although microorganisms do not multiply in air, this is an effective method of distributing bacteria to surfaces within a dairy plant. In chilled rooms, evaporator fans draw large quantities of air over the evaporator cooling coils and distribute it around the room. Any contaminants in the air are likely to pass over the evaporator surfaces, some will be deposited, and, if conditions are suitable, attachment, growth and further distribution of airborne contaminants may occur. In addition to being a potential source of contamination, the development of a microbial biofilm on evaporator cooling coils may affect heat-transfer rates of the equipment (Evans *et al.*, 2004).

Air-control systems are effective in controlling airborne contamination, such as endospores of *Bacillus* spp., various species of non-spore-forming bacteria and yeasts, as well as a range of mould spores, transmitted by untreated air being drawn into the area, condensation, blow lines from equipment and compressed air lines from packaging equipment. These control systems are, however, not as effective in controlling contamination from other sources, for example, personnel, traffic, buildings and raw materials (IDF, 1997b; Robinson and Tamime, 2002; Lelieveld, 2003b). Chemical taints can, for instance, enter the production area through airborne transmission.

For products where air quality does not limit shelf-life and safety, the air quality should be controlled so that it does not become a limiting factor. Air control will be important in controlling the risk of contamination for products that are minimally processed, and where growth of microorganisms is controlled by preservation systems

(IDF, 1994, 1997b). In high-care areas, air supply is critical, if the area can be physically isolated, air systems should provide a positive air pressure and further reduce the risk of casual microbial contaminants. This will prevent microbiological contamination from potentially contaminated areas to non-contaminated areas (IDF, 1997a). According to Robinson and Tamime (2002), ducting of incoming air through a primary filter to remove gross contamination (5.0–10.0 μm diameter), followed by a filter capable of removing 90–99% of particles above 1.0 μm, is essential for areas containing open vats of food. Special consideration should also be given to airflow in cheese factories, especially where mould, smear-ripened or soft cheeses are produced. Routine monitoring to maintain air quality is essential. Temperature, humidity, airflow and pressure and microbiological monitoring are all aspects that should be included (IDF, 1997a; Mostert and Jooste, 2002). The location of air intakes and outlets, the method and frequency of cleaning filter-holders, and routine for replacing filters are aspects that merit inclusion within the HACCP plan (IDF, 1994; Jervis, 2002; Robinson and Tamime, 2002).

3.2.3.2.2 Water supplies. Water is used as an ingredient, as a production-process aid and for cleaning, and may be supplied from external sources or recovered water. Consequently, the overall water requirements of a dairy plant are large, and water conservation and water management have become major issues (Robinson and Tamime, 2002). While not all supplies need to be of potable quality or better, if used as an ingredient, there exists the potential for microbial and chemical contamination (IDF, 1994; Mostert and Jooste, 2002; Lelieveld, 2003b). If a water company supplies water, the water will mostly comply with quality standards, suitable for dairy processing (IDF, 1997a). Only occasionally, toxic algae or protozoa (e.g. *Cryptosporidium* spp.) in reservoirs cause problems. Furthermore, *Escherichia coli*, *Listeria monocytogenes* or *Salmonella* spp. cannot grow in water and are sensitive to the levels of chlorine found in drinking water. However, these organisms may survive in water, and so it is important that water used for rinsing equipment, or curd, or flavouring components of cottage cheese, should be of a higher quality than mains water. The level of chlorination in potable water is usually ineffective against spoilage groups like *Pseudomonas* spp. Since these bacteria cause taints under aerobic conditions, bacterial standards have been proposed for bacterial groups and are given in Table 3.2. Consequently, chlorinating plant water supplies to >20 mg l^{-1} and monitoring that the level is maintained throughout processing contributes significantly to assuring the safety and shelf-life of the products produced (Robinson and Tamime, 2002). If ground water is utilized, it must be pre-treated to comply with the mentioned safety and quality standards. According to IDF (1997a), water can be recovered from recycled, rinsing or cooling water. If the recovered water should be used for purposes where it comes into indirect or direct contact with food, it must not present a risk of microbiological contamination. Water from hand washing, unwanted water from steam and leaking pipes can all be vectors of contamination and stagnant water is especially hazardous (Lelieveld, 2003b). Water storage time should not exceed 24 h, and storage tanks should be constructed of stainless steel, easy to clean and closed.

Table 3.2 Microbiological specifications and significance of microorganisms in drinking and process water.[a]

Microorganism	European Union	WHO	Persistence in water supply	Resistance to chlorine	Animal reservoir
Total viable count					
22°C	<100 cfu ml^{-1}	–	–	–	–
37°C	<10 cfu ml^{-1}				
E. coli or Thermotolerant coliform bacteria	<1 cfu 100 ml^{-1}	ND[b] in 100 ml	Moderate	Low	Yes
Sulfite-reducing *Clostridium* spp.	<1 cfu 20 ml^{-1}	–	–	–	–
Pseudomonads	<100 cfu ml^{-1}	–	May multiply	Moderate	No
Salmonella typhi	–	–	Moderate	Low	No
Other *Salmonella* spp.	–	–	Long	Low	Yes
Cryptosporidium parvum	–	–	Long	High	Yes

[a]Sources: WHO (1994, 1996, 2002, 2005); Robinson and Tamime (2002).
[b]Not detectable.

3.2.3.3 Plant and manufacturing equipment

Factories that handle milk and milk products range from simple labour-intensive units through to highly automated plants, relying on sophisticated unit operations under computer control. Consequently, a wide range of processing equipment including heat exchangers, vats, mixers and containers may be needed; but whatever equipment used, it must be hygienically designed. Installation must be such that as far as possible, dead ends or dead spots in the equipment and milk pipelines, do not occur. Regular maintenance is essential, for instance, metal or plastic fragments, rust, or loose nuts or screws can pose physical hazards. Effective cleaning and disinfection of all equipment is essential, and equally important is the area behind or beneath an item of the equipment (Bénézech *et al.*, 2002; Robinson and Tamime, 2002; Lelieveld, 2003b; CAC/RCP, 2004).

Microorganisms can be established on surfaces in the processing area if the conditions for growth are favourable, particularly surfaces that provide stable environments for growth. Posing a high risk are surfaces exposed to air, unless frequently and effectively cleaned and sanitized by either physical (e.g. hand washing, high-pressure sprays) or chemical methods (e.g. hypochlorite, iodophores, quaternary ammonium compounds) (Hood and Zottola, 1995; Lelieveld, 2003b). Therefore, when choosing materials for processing line equipment, along with their mechanical and anticorrosive properties, is their hygienic status, i.e. low soiling level and/or high cleanability (Julien *et al.*, 2002). Stainless steel, which is widely used for construction of food-processing equipment, has been demonstrated to be highly hygienic. However, stainless steel can be produced in various grades and finishes, affecting bacterial adhesion. According to Julien *et al.* (2002), the stainless steel grade significantly affects the hygienic status, while the stainless steel finish is poorly linked to the hygienic status. However, the surface properties of material in contact with food have been taken into account by regulatory authorities when determining specifications for surface topography. For instance, Standard US 3-S Sanitary Standard 01–07 specifies that surfaces directly in contact with foods require a no. 4 grade finish and must be free of cracks and crevices. The EHEDG guidelines suggest an average roughness (R_A) of up to 3.2 μm that would be acceptable, provided flow is sufficient enough to remove soil from surfaces. Although it is still currently not clear what the relationship is between R_A values and bacterial adhesion, it does seem obvious that surface topography may play a major role in surface cleanability, even if not on bacterial adhesion then by protecting the cells from stress (Julien *et al.*, 2002).

In closed equipment, even if correctly designed, there are areas where product residues will attach to equipment surfaces, even at high liquid velocities. The microorganisms may multiply if present on the surface for an extended time period, and with the increase, the numbers of microorganisms washed away with the product will also increase, causing recontamination (Hood and Zottola, 1995; Den Aantrekker *et al.*, 2003). In dead spaces, the problem will of course be worse. Microorganisms also penetrate through very small leaks (Lelieveld, 2003b). Bacteria can also colonize a surface to form a biofilm (see Section 3.5). It has been suggested that attached microorganisms may be more resistant to sanitizing compounds than free-living cells, but there is no indication that proper sanitation will not be effective in reducing contamination. Controlling the adherence of microorganisms to food contact surfaces is an essential step in meeting the goals for a clean working environment (Hood and Zottola, 1995; Kumar and Anand, 1998; Den Aantrekker *et al.*, 2003). High contamination levels of bacterial cells may exceed the capacity of the subsequent process, e.g. heating or fermentation to reduce the initial levels of pathogens or spoilage bacteria to those resulting in a safe product with optimal shelf-life (Lelieveld, 2003b) (Table 3.1).

Irrespective of the capacity of the plant, cleaning to remove food residues on both the inside and outside of the processing equipment must have a high priority. Only successful cleaning and disinfection as a precondition of a successful CCP concept contribute to minimizing the risk factors by microbial recontamination of milk products (Orth, 1998). Overall, cleaning regimes should consider the following (Robinson and Tamime, 2002):

- identification of any points in the layout of the plant that may be especially prone to the build-up of food residues and subsequently biofilms; and
- whether the equipment needs to be dismantled for manual washing or whether it can be cleaned *in situ*; hand washing means using lower water temperatures, but the scrubbing and visual inspections involved may leave the surface sufficiently clean.

The nature of residues is important. Each major ingredient of a food reacts differently for cleaning agents. For example, sugars are easily soluble in warm water, but they caramelize in hot surfaces and become difficult to remove, whereas fats, although dispersible in warm water, are liberated more readily in the presence of surface-active agents.

3.2.3.4 Human activities

People are large reservoirs of microorganisms but can also be a source of physical hazards, such as hair, fingernails, etc. (Lelieveld, 2003b). If contamination from the environment is under control, and the food contact surfaces are clean, then plant operations remain the final avenue for the possible transfer of spoilage or pathogenic microorganisms (Robinson and Tamime, 2002). Pathogens on the hands or gastrointestinal infections can be transferred to the product or processing surfaces (Lelieveld, 2003b). The source of contamination can also be equipment; when a person then touches the equipment, the bacterial cells are transferred from the equipment to the hand and, by handling the product, to the product (Den Aantrekker *et al.*, 2003). Therefore, to control possible hazards, hands must be washed after touching the body, anywhere on the head or shaking hands with people; using a handkerchief; touching raw products; or touching bottoms of boxes that could be contaminated with foodstuff on the floor (CAC/RCP, 2004) (Table 3.1).

3.2.3.5 Animals and pests

One aspect in the dairy industry that often causes problems, regarding hygienic design, is pests. The occurrence of pests is a problem in the food industry, and their presence in dairy factories or in food products is unacceptable. Insects and mites occur at every point along the chain of food production, which starts with the production and storage of raw materials and continues through processing, packaging and distribution, and even in the household of the consumer. Their presence causes concern, not only in their appearance and the direct spoilage potential, but also because of the potential microbiological contaminants and pathogens they may carry, as well as the possible allergenic reactions that they may cause. There has also been increasing difficulty in achieving effective control of mites with conventional pesticides (Bell, 2003). A list of the most commonly encountered Arthropod pests associated with stored food is given by Bell (2003), while important critical biological data to understand invertebrate pest risk to products and packaging have been outlined by Kelly (2006). Particular care needs to be taken that birds and rodents do not enter food-production areas, as they are a source of microbial hazards (Lelieveld, 2003b; IDF, 2004). The data in Table 3.3 indicate which microbial hazards and problems may be associated with animals.

Table 3.3 Examples of problems associated with animal vectors[a].

Group	Result of attack
Rodents	Loss by consumption and partial eating of product
	Contamination with droppings, hair; health hazard from bacteria
	Indirect losses from damaging containers
	Possible blockage of pipework and considerable contamination if carcass accidentally macerated during product handling
Birds	Consumption and contamination by droppings and feathers
Insects	Limited losses but high risk of bacterial contamination

[a]Source: Robinson and Tamime (2002).

The rodents mostly responsible for contamination of foodstuffs are usually rats and mice. Bird droppings may contain *Salmonella* or other pathogens.

3.2.3.5.1 Biology of insects and mites. Most insect or mite species could be traced to natural habitats such as bird and animal nests, dried plant matter and dried animal carcasses. Increased storage of food for longer periods, and international trade, contributed to the establishment of many species of tropical origin in food manufacturing premises and buildings in temperate areas. Insect and mite species are generally well-adapted to survive periods at high and low temperatures and to establish themselves in various environments (Kelly, 2006).

Insect and mite storage pests have different abilities to invade and penetrate packaging, and there must be food debris, waste or food odour, and sufficient time to infest and contaminate raw material and stored food products. Some species, packed within the food at the factory, remain with it, being incapable of chewing out, while others enter and leave packages almost at will. Some prefer dampness, mouldy and poor hygienic ecosystems, and some are specifically primary food pests. Many beetles in their immature stages may be less capable of penetrating packaging materials, but as they grow they develop stronger mandibles. Common adult storage moths cannot penetrate packaging materials, but they lay eggs, which are very small and virtually invisible among pallets and cartons. They develop into larvae, many of which will become very effective packaging penetrators. Mites on the other hand are unable to chew through packaging material, but when fully grown they can often find and penetrate minute and tiny holes in folded packaging or faulty heat-sealed sachets. Some invader species can easily chew through paper, card board, waxed paper, cellophane, polyethylene, plastic films, metal and plasticized foils and foil-backed paper and card (Kelly, 2006).

The choice of packaging is important, but the infestation barrier properties should also be taken into consideration. Time is also an important factor because insects need time to chew through packaging material or to develop through to stages with chewing abilities. Conversely, time may also be used to our advantage to design and implement an effective pest-monitoring programme (Kelly, 2006). Different methods/measures to control insect and animal pests are discussed below.

3.2.3.5.2 Control of pests. Various physical, chemical and biological methods are available to control pest infestation in the food industry (Bell, 2003; Anon., 2005; Kelly, 2006). Pest control of insects and mites in the food industry has been thoroughly reviewed by Bell (2003).

It is advisable to follow a pragmatic approach in developing an effective hygiene and pest-monitoring/prevention-control strategy. Pest control is often, unintentionally, neglected. The aim should be to detect, at the earliest signs, any incipient infestation of raw materials, packaging materials, manufacturing equipment and storage facilities to avoid their establishment and proliferation. Since infestation can readily take place in conditions suited to their reproduction, general hygiene practices have to be implemented to avoid the establishment of a suitable environment for their development.

Buildings should be kept well maintained to prevent access of animals and eliminate possible locations for their reproduction. Inside the buildings, all potential refuge for pests, such as holes and crevices in walls and floors, obsolete material and equipment, etc., must be eliminated. The presence of food and water attracts pests and permits their reproduction.

Aspects which may also support a good preventative pest control programme/system include the following:

- insect traps (e.g. pheromone lures) – attracts and help detection of insects;
- infra-red early-warning systems – prevent infestations;
- electrical fly killers – attract and kill flying insects;
- regular inspections – ensure early warnings of insect presence; and
- regular disposal of waste and food debris, especially in sensitive areas.

An important factor in odour control is the odour barrier properties of the packaging material. It has been shown that preventing odours of food products from diffusing into an infested store keeps the products free from cross-infestation for relatively long periods. Thus, the bag in box is sometimes more effective than a plasticized card box, which is dependent on perfectly formed comer and flap folds. Time is also a critical factor. The longer a product remains in an infested warehouse, the more likely it is to become cross-infested. Insects and mites are more active at higher temperatures, which explain why sometimes product infestation and damage is greater in retail stores.

New developments in packaging which could help in this regard are insect-repellent treatment of base paper/card prior to manufacture and specially developed 'active packaging' materials which can affect the type and quantity of 'atmosphere' within the packaging. The active ingredients used in repellent packaging material are not widely approved for use on packaging in direct contact with food material. It is, however, interesting to see new developments in this regard especially in the field of physical control measures towards insect-unfriendly packaging design to prevent/minimize insect infestations.

Rodents are another pest problem that also poses a serious health risk and which may also cause damage to stock, machinery and structures. Preventative measures include

- physical exclusion of rats and mice from buildings;
- avoiding badly stored waste and food debris; and
- good management, housekeeping and an effective, preventative pest-control programme.

It is important to note that new technologies are now increasingly developed to reduce the widespread and long-term use of pesticides. The use of modern electronics to detect the level of pest activity and to monitor pest-control programmes is an innovative development with wide applications (Anon., 2005).

Birds and their nests also harbour insects and mites that can spread into the building and cause infestations. Birds could effectively be controlled by the use of bird wire, eco-spikes and netting to prevent them from using balconies, air conditioning vents and ledges.

3.2.4 Waste and effluent management

Wastage is defined as 'something expended uselessly, or that one fails to take advantage of, or that is used extravagantly' (Hale *et al.*, 2003). Water wastage occurs when water does not reach the point of use and is used excessively and is used for purposes where there are better alternatives and is not recycled. The general sources of milk, ingredient and product wastage are

- overfill (giving away products and components);
- spillage;
- inherent waste;
- retained waste;
- heat-deposited waste;
- product defects/returned product; and
- laboratory samples.

Overfill is of economic, but not environmental, significance. In modern well-designed dairy plants, spillage is minimal. Also, the use of welded pipelines has minimized the number of joints used. Leaks that do occur should be promptly rectified. Spillage, leakage or overflow that occurs routinely in dairy plants is an indication of poor equipment design, selection or maintenance. Collecting solid spillage and disposing of it as solid waste is more hygienically and environmentally preferable than disposing of it via the wastewater drainage system (Hale *et al.*, 2003). In some cases, as a result of unavoidable designs, some process equipment causes (inherent) waste. For example, in certain CIP cleaning cases at the end of production or at product changeover, the product is purged through the plant with water. Inevitably, the interface between the product and the water will not be perfect, and a quantity of waste will be produced. Retained product will make this problem worse. Retained waste is a liquid product or ingredient that does not freely drain to the next stage of the process. Possible causes may be

- dips in supposedly continuously falling pipelines such that trapped product will not drain either way; and
- where product rises in pipelines; in addition to purging, draining back product is an option but is likely to lead to further wastage.

With viscous products such as yoghurt, adhesion to the walls of pipelines and tanks is a significant source of retained waste. Unless such a product is mechanically removed, prolonged pre-rinsing is likely to be required to remove it before detergent circulation can commence. In addition to water wastage, the product will be very much diluted, making recovery or disposal more difficult. To minimize the amount of product adhesion, pipeline diameters should be as small as practically possible; pipelines should be more steeply sloped than for non-viscous products (Hale *et al.*, 2003).

When milk or milk products are heated, there is a likelihood of deposition of the product onto the heat exchange surfaces. These deposits on the plates or tubes in heat exchangers and on batch kettles will not rinse off and are lost in the waste stream when removed with detergent. Product is wasted, and a high strength and potentially environment polluting waste stream are generated. Products not meeting the required specification can be a major source of waste. Decisions on how to dispose of the product should be based on HACCP (Hale *et al.*, 2003).

3.3 Clean room design

3.3.1 Hygienic plant design

The primary aim of hygienic plant design should be to establish effective hygiene barriers between certain areas in the factory to limit the entrance of microbial and other contaminants. Within dairy plants, there will be areas with differing levels of hygiene design and operational requirements (IDF, 1997a; Wierenga and Holah, 2003):

- Non-production areas – These are areas where there is no risk of contaminating manufactured or pasteurized products, or areas where contamination is of minor importance. Such areas include raw-milk reception, cleaning facilities for pallets, return containers and toilets, storage facilities and service facilities, e.g. boiler rooms, CIP equipment, etc. Normal good hygienic practices should be maintained.
- 'Hygienic/low-risk' areas – In these areas, the risk of exposing the product to a contaminated environment is limited and where good hygienic manufacturing practices are required. Such areas are often situated adjacent to 'high risk/high care' areas, thus functioning as a hygienic 'barrier lock'. Examples of 'hygienic' areas are storage of packaging materials, laboratories, milk treatment area, etc. These areas should be separated, for example, wet mixing separated from dry mixing operations. In relation to the HACCP plan, these areas will often be regarded as areas where microbiological preventative measures are carried out.
- 'High risk' areas – A 'high risk' area is a well-defined, physically separated part of the factory which is designed and operated to prevent recontamination of the final product by the strictest hygiene requirements. Examples are processing, brining, ripening and packing rooms. Usually, there are specific hygiene requirements regarding building criteria, layout, standards of construction and equipment, flow of products, training and hygiene of the personnel and other operational procedures. Specific criteria for designing and operations in these areas are comprehensively reviewed in the *IDF Guidelines for Hygienic Design and Maintenance of Dairy Buildings and Services* (IDF, 1997a).

There are also other basic barriers to minimize/prevent contamination from

- non-product contact surfaces (location of building, floors, walls, etc.);
- processing environment (air and water supply);
- plant and manufacturing equipment;
- human activities; and
- animals and pests.

These potential sources of contamination and the control mechanisms to maintain a clean working environment have been outlined in Section 3.2.3.

3.3.2 Dealing with airborne contamination

Microbial air quality in processing and packaging areas is a critical control point in the processing of dairy products, since airborne contamination reduces shelf-life and may serve as a vehicle for transmitting spoilage organisms and, if pathogens are present, transmission of diseases (Anon., 1988; Kang and Frank, 1989a). Every precaution should therefore be taken to prevent airborne contamination of the product during and after processing (Kang and Frank, 1989b; Hickey *et al.*, 1993; Mostert and Jooste, 2002). Airborne microorganisms in dairy plants include bacteria, moulds, yeasts and viruses. Data on air counts and the generic composition of microorganisms in processing areas have been comprehensively reviewed by Kang and Frank (1989b). The composition and level of the microbial population can vary widely within and among plants, and on a day-to-day basis within the same plant. The variations can be attributed to differences in plant design, air flow, personnel activities and status of dairy plant hygiene. Installation of air filters, application of UV irradiation and regular chemical disinfection (bactericidal, fungicidal and viricidal agents) of air can be applied to critical areas to control airborne microorganisms (Homleid, 1997; Arnould and Guichard, 1999).

3.3.2.1 Sources and routes of airborne microorganisms

Airborne microorganisms can be attached to solid particles like dust, and they are present in aerosol droplets or occur as individual organisms due to the evaporation of water droplets or growth of certain mould species. The main sources of airborne organisms include the activity of factory personnel, dairy equipment, ventilation and air-conditioning systems, packaging materials and building materials (Heldman *et al.*, 1965; Hedrick and Heldman, 1969; Hedrick, 1975; Kang and Frank, 1989b; Lück and Gavron, 1990; Frontini, 2000; Brown, 2003). An increase in viable aerosols has been reported during flooding of floor drains (Heldman and Hedrick, 1971) and after rinsing floors with a pressure water hose (Kang and Frank, 1990; Mettler and Carpentier, 1998). Whenever possible, wet cleaning should not be used during the processing of milk products in areas in which the product is exposed and can be contaminated by aerosols. Once a high concentration of viable aerosol is generated, it can take more than 40 min to return to the normal background level. It has been shown that drains, floors, standing and condensed water may also be sources of pathogens in dairy plants (El-Shenawy, 1998). The potential contamination from bacterial biofilms is of major concern, since microbial cells may attach, grow and colonize on open exposed wet surfaces, e.g. floors, floor drains, walls and conveyor belts (Wong and Cerf, 1995; Carpentier *et al.*, 1998; Mettler and Carpentier, 1998).

Water used in open circulation systems is another significant source of airborne microbial populations, and the spray from cooling towers, if contaminated, may also be a possible source of certain pathogens (Hiddink, 1995) and, consequently, airborne contamination. Microorganisms and small particles are commonly found in the immediate vicinity of water surfaces (Al-Dagal and Fung, 1990).

3.3.2.2 Air-quality control

3.3.2.2.1 Dairy processing areas. The most practical approach to control microbial airborne contamination indoors is the removal of all potential contamination sources from sensitive areas where the product might be exposed to air. Good ventilation is necessary to remove moisture released during the processing of dairy products, and it will also prevent condensation and subsequent mould growth on surfaces. Attention is nowadays given to air cleaning in food plants, *inter alia*, by establishing air-flow barriers against cross-contamination from the environment (Jervis, 1992; Kosikowski and Mistry, 1997). The air entering processing rooms is normally chilled and filtered to remove practically all bacteria, yeasts and moulds. It is essential that filtered sterilized air be supplied to areas where sterile operations are to be carried out. Rigid-frame filters or closely packed glass fibres are available to achieve contamination-free air for culture transfer, and manufacturing and packaging of sterilized milk and milk products (Shah *et al.*, 1996, 1997). The use of high-efficiency particulate air (HEPA) filters will remove 99.99% of airborne particles 0.3 μm and larger, while ultra-low-penetration air (ULPA) filters remove 99.999% of particles as small as 0.12 μm (Shah *et al.*, 1997). Passage of air through a combined HEPA/ULPA filter is usually considered suitable for use where contamination-free work is to be carried out. Standard high-efficiency air filter systems allow more air into the room than normal, thereby establishing a positive air pressure. Upon opening a door, filtered air flows out, thus blocking the entry of untreated air and minimizing microbial contamination. For optimal ventilation, sufficient air changes have to be made to prevent the build-up of condensation on surfaces (Jervis, 1992). Specialist advice should be sought to choose the correct filter type/system on the basis of the air quality required for the specific operation in each controlled area. There is a difference between high- care areas, where the aim is to *minimize* air contamination, and high-risk areas that are designed to *prevent* recontamination (Brown, 2003).

Rooms in which direct exposure to outside air is inevitable can have air-flow barriers installed, mounted over open doorways to secure a significant downward velocity of air flow, preventing contamination from outside (Kosikowski and Mistry, 1997).

Compressed air can also contribute to contamination of products by dust and microorganisms and, in the case of lubricated compression systems, by oil fumes (Guyader, 1995; Wainess, 1995). Whenever air under pressure comes into direct contact with the product (pneumatic filling, agitation or emptying of tanks) or is directed at milk contact surfaces it should be of the highest quality. Sterile compressed air can be obtained by drying the air after compression in adsorption filters (e.g. chemically pure cotton, polyester or polypropylene) and to install a series of 0.2 μm-pore-size filters downstream, immediately preceding the equipment where the air is needed (Bylund, 1995; Guyader, 1995).

Air sanitation systems can also be applied to control airborne contamination and include (Brown, 2003)

- fogging to reduce the number of airborne microorganisms and also to disinfect surfaces that may be difficult to reach;

- UV light; and
- ozone treatment.

The use of clean-room clothing, head covering, masks and gloves largely eliminates the release of microorganisms and skin particles into the processing environment. A good hygiene-training programme for factory personnel will also contribute to reducing contamination by workers (Al-Dagal and Fung, 1990).

3.3.2.2.2 Outdoor environment. The control of airborne microorganisms in the immediate surroundings of dairy premises is more difficult than in closed, indoor environments where more controlled measures can be taken. One aspect that could be helpful in reducing the microbial load outdoors is the control of organic materials (Al-Dagal and Fung, 1990). UV light, humidity, temperature, wind direction and speed have a significant influence on the total number of airborne microorganisms in the outdoor atmosphere.

Air quality in processing areas, the factory environment (e.g. walls, floors, drains) and air used in the manufacturing of dairy products should be monitored regularly (see Section 3.6.1).

Future trends will be focussed on the local air control of production lines, instead of controlling production areas and also on the type of clothing that is used by personnel in food production areas (Brown *et al.*, 2002).

3.3.3 Hygienic equipment design

Milk and milk-product contact surfaces include all surfaces that are directly exposed to the product or all indirect surfaces from which splashed product, condensate, liquid, aerosols or dust may drain, drop or contaminate the product. This means that in the hygienic design of equipment, the immediate surrounding areas of the processing area must also be taken into consideration. The equipment/machinery used for the processing of dairy product must be designed and constructed so as to avoid microbiological, chemical and physical contamination and, ultimately, health risks. Lelieveld *et al.* (2003) gave a comprehensive review on the hygienic equipment design in the food industry. The key aspects in hygienic equipment design are, according to Shapton and Shapton (1991), the following:

- Maximum safety protection to the product. Good hygienic design prevents microbiological, chemical and physical contamination of the product that could adversely affect the health of the consumer (Holah, 2002) and also the image of the product and the manufacturer. Gaps, crevices, dead ends, etc., where microorganisms, especially pathogens, can harbour, grow, and subsequently contaminate the product, should be avoided.
- Contact surfaces that are readily cleanable and which will not contaminate the product.
- Plant hygiene/sanitation is dependent on the efficiency of cleaning (removal of residual soil from surfaces) as well as the effective destruction of most (sanitation) or all (sterilization) of the remaining microorganisms (Tamime and Robinson, 1999).

To be cleaned effectively, surfaces must be smooth without crevices, sharp corners and protrusions.
- Connections/junctures which minimize dead areas where chemical or microbiological contamination may occur. Hygienically designed equipment which is initially more expensive, compared with similarly performing poorly designed equipment, will be more cost-effective in the long term. Savings in cleaning time may lead to increased production. Complying with hygienic requirements may increase the life expectancy of equipment, reduce maintenance and, consequently, lower manufacturing costs. (Lelieveld *et al.*, 2003).
- Access for efficient cleaning, maintenance and inspection. Inspection, testing and validation of the design are important to check whether hygienic requirements have been met (Holah, 2002).

In many countries, there is legislation in place to ensure that manufacturing equipment is safe to operate. European legislation, for example, has for many years provided regulation to ensure that food-processing equipment is, additionally, also cleanable and hygienically safe. The basic hygienic design requirements include the following (Lelieveld *et al.*, 2003):

(1) Construction materials in contact with food must be durable, non-toxic, resistant to corrosion and abrasion, and easily cleanable.
(2) Product contact surfaces should be easily cleanable, smooth and free from cracks, crevices, scratches and pits which may harbour and retain soil and/or microorganisms after cleaning.
(3) All pipelines and equipment should be self-draining. Residual liquids can lead to microbial growth and microbial contamination or, in the case of cleaning agents, may result in chemical contamination of products.
(4) Permanent joints that are welded or bonded should be smooth and continuous and free from recesses, gaps or crevices. The product contact surface of welds must be smooth and the seams continuously welded. Improperly welded joints (e.g. misalignment, cracks, porosity, etc.) are environments which are difficult to clean and sanitize and which may harbour microorganisms. Dismountable joints, e.g. screwed pipe couplings, must be crevice-free with a smooth continuous surface on the product side. Flanged joints must be sealed with a gasket because metal-to-metal joints may still permit the ingress of microorganisms.
(5) Sharp corners, crevices and dead spaces should be avoided. Well-rounded (radiused) corners are important for easy cleaning. Crevices are mostly the result of poor equipment design and should be avoided because they cannot be cleaned. However, in some equipment where slide bearings are used in contact with the product (e.g. as bottom bearings or top-driven stirrers, etc.), crevices are unavoidable. Specific cleaning and sanitization procedures of these equipment parts should be specified in the cleaning procedures. Dead spaces (e.g. T-sections in pipes used for pressure gauges, etc.), dead ends and shadow zones that are outside the main product/cleaning agent flow provide areas where nutrients accumulate, and microorganisms grow. Such areas are hard to clean and sanitize, and should be avoided. If T-sections are used, they should be as short as possible to minimize decrease in flow velocity especially of liquid cleaning agents.

(6) Other equipment or parts of equipment that should also be hygienically designed include:
   - seals/gaskets (appropriate materials should be used and, wherever possible, mounted outside the product area);
   - conveyor belts (should be easily cleanable);
   - fasteners (nuts, bolts and screws must be avoided in product contact areas);
   - doors, covers and panels (should be designed to prevent entry/accumulation of dirt and soil); and
   - instrumentation and equipment controls (instruments should be constructed from appropriate materials, and controls should be designed to prevent the ingress of bacterial contamination and also easily cleanable).

3.4 Clean room operations

Cleaning procedures should effectively remove food residues and other soils that may contain microorganisms or promote microbial growth. Most cleaning regimes include removal of loose soil with cold or warm water followed by the application of chemical agents, rinsing and sanitation (Frank, 2000). Cleaning can be accomplished by using chemicals or combination of chemical and physical force (water turbulence or scrubbing). High temperatures can reduce the need for physical force. Chemical cleaners suspend and dissolve food residues by decreasing surface tension, emulsifying fats and peptizing proteins (Chmielewski and Frank, 2003).

3.4.1 Objectives of plant cleaning

The primary objective of maintaining effective dairy plant hygiene is to ensure the consistent production of excellent products. Dairy manufacturers aim to ensure that the safety and quality of their products will satisfy the highest expectations of the consumer. On the other hand, consumers expect to a large extent, unconditionally, that the manufacturer has ensured that the product

- is safe for human consumption with respect to both chemical and microbiological contamination;
- conforms to regulations enshrined in law, or statutory requirements laid down by health or local authorities;
- is capable of achieving a specified shelf-life without spoilage; and
- has the highest possible organoleptic standard that can be achieved within the existing constraints of manufacture or marketing (Tamime and Robinson, 1999).

The production of excellent products with good keeping qualities is governed by a multiplicity of various interrelated factors such as

- the hygienic quality of the product which, in turn, is dependent, for example, on the microbiological quality of the milk, raw materials, effective heat treatment and the care which is exercised during storage, handling and distribution;

- the cleanliness of milk and product contact surfaces, e.g. processing equipment, pipelines and packaging materials; and
- various other aspects, i.e. hygienic plant and equipment design, hygienic manufacturing practices and the level of training and attitude of dairy personnel (IDF, 1980, 1984, 1985, 1987a, b, 1991, 1992, 1994, 1996, 1997a, b, 2000; Wainess, 1982; Vinson, 1990).

The provision of hygienic processing equipment is an essential factor in this regard. The nature of contaminants from surfaces coming into contact with the product could be physical, chemical or biological. Contamination from these sources can be minimized by the following fundamental approach (Swartling, 1959; Dunsmore *et al.*, 1981a, b; Dunsmore, 1983):

- removal of milk and other residues which can provide nutrients for microorganisms remaining on equipment surfaces;
- cleaning and sanitization/sterilization of equipment by removal and destruction of microorganisms which survived the cleaning step;
- keeping equipment under conditions that limit the growth/survival of microorganisms when the equipment is not in use; and
- removal of residual cleaning agents which may contaminate the product.

It is important to note that processing equipment can be bacteriologically clean without necessarily being physically or chemically clean. However, it is for various reasons important to achieve physically cleanliness, i.e. removal of milk and other residues/dirt, before the cleaning and sanitization/sterilization steps are applied. It is evident that plant hygiene/sanitization is dependent on the efficiency of cleaning (removal of residual soil from surfaces) as well as on the effective destruction of most (sanitization/disinfection) or all (sterilization) of the remaining microorganisms (Tamime and Robinson, 1999). The principles of these steps will be outlined and discussed further.

3.4.2 Cleaning operations

3.4.2.1 Principles of the cleaning process

Residual dairy soil may be liquid milk and product residues, scale and unwanted miscellaneous foreign matter which have to be removed from plant surfaces during the cleaning process. The soil usually consists of organic compounds (e.g. lactose, protein, fat and non-dairy ingredients) and inorganic compounds, mainly from mineral origin. Most soils are, however, a combination of both organic and inorganic compounds. Milk stone, for example, is primarily a combination of calcium caseinate and calcium phosphate. Typical soil characteristics of soiling matter on dairy equipment surfaces are summarized in Table 3.4.

All processing equipment and pipelines should be emptied and properly rinsed to loosen and/or remove as much of the gross soil as possible before commencing the cleaning programme. Effective rinsing at this stage is a critical step in effective cleaning and may include scraping, manual brushing, automated scrubbing and pressure jet washing (Tamime and Robinson, 1999; Holah, 2003).

Table 3.4 Characteristics of different soil types and suggested cleaning procedures.[a]

Soil type on surface	Solubility characteristics	Ease of removal during cleaning: Without alteration by heat	Ease of removal during cleaning: Effect of alteration by heat	Suggested cleaning procedure
Lactose, starches and other sugars	Good in water	Good	Caramelization/ browning: More difficult to clean	Mildly alkaline detergent
Protein	Poor in water Good in alkaline solutions Slightly in acid solutions	Poor in water Better with alkaline solutions	Denaturation: difficult to clean	Mildly alkaline detergent
Fat	Poor in water, alkaline and acid solutions without surface-active agents	Good with surface-active agents	Polymerization: more difficult to clean	Use with surface-active solutions
Mineral salt, milk stone, heat-precipitated water hardness and protein scale	Poor in water and alkaline solutions Usually good in acid solution	Reasonable	Precipitation: difficult to clean	Acid cleaner–periodically

[a]Adapted from: IDF (1979); Romney (1990); Tamime and Robinson (1999).

3.4.2.2 Selection and functional properties of detergents

The cleaning process necessitates the use of certain chemical compounds, referred to as detergents, with specific functional properties. The basic properties/functions of detergents are (Tamime and Robinson, 1999)

- establishing intimate contact with the soiling matter through their wetting and/or penetrating properties;
- displacement of the soil, for example by melting/emulsifying the fat by wetting, soaking, penetrating and peptizing the proteins, and by dissolving the mineral salts;
- dispersion or displacement of the undissolved soil by deflocculation and/or emulsification;
- preventing redeposition of the soil by maintaining the properties of the above factors, and by ensuring good rinsing; and
- miscellaneous, i.e. to be non-corrosive, to have no odour nor taste, to be non-toxic and non-irritable to skin.

Various formulations are used to achieve the above-mentioned properties and functions. Some of the compounds and their properties that can be employed are shown in Table 3.5. There are many different types of detergents available on the market, and the choice of a cleaning agent will depend on the type of processing equipment, type of soiling matter to be removed (Table 3.4), water quality and hardness. Water is used during all the cleaning cycles in a processing plant, and it is essential that good quality potable water be used. It is also important that the degree of hardness be taken into account, since detergents are formulated in relation to the degree of water hardness,

Table 3.5 Functional properties and characteristics of detergent constituents.[a]

Type	Detergent components		I	II	III	IV	V	VI	VII	General comments
Inorganic alkalis	1. Sodium hydroxide		E	P	P	P			E	These compounds can effect degree of alkalinity, buffering action and rinsing power of a detergent.
	2. Sodium orthosilicate		G	F	F	P			G	
	3. Sodium metasilicate		G	G–P	VG	G			F	For high-alkalinity preparations use alkalis (1) and (2), which can cause skin irritation; therefore, handle them with care.
	4. Trisodium phosphate		F	F–P	VG	G			F	
	5. Sodium carbonate		F	P	P	P			P	For removing heavy soil, alkalis (2), (3) and (4) are very effective.
	6. Sodium bicarbonate		G	P	G	F			F	For low alkalinity (mild or hand detergents) use alkalis (5) and (6).
Acids	Inorganic	Nitric acid								Acids are normally used for the removal of tenacious soil e.g. in UHT plants.
		Phosphoric acid								
		Sulfuric acid								These materials are corrosive and can cause severe skin burns; therefore handle them with care, and if incorporated in a detergent formulation, they may have to be used with corrosion inhibitors.
	Organic	Hydroxy acetic acid	G			G				
		Gluconic acid								
		Citric acid								
Surface-active agents	Anionic	Sodium alkyl aryl sulfonate								Classification is dependent upon how these compounds dissociate in aqueous solution, e.g. surface-active anions, cations, etc.
		Sodium primary alkyl sulfate								
		Sodium alkyl ether sulfate								Some of these compounds are also used as emulsifying agents.
	Non-ionic	Polyethenory compounds		E	E	E				Non-ionic agents do not ionize in solution. Surfactants tend to reduce surface tension of the aqueous medium and promote good liquid/soil/surface interfaces.
	Cationic	Quaternary ammonium compounds (QAC)	(see sterilizing agents below)							
	Amphoteric	Alkyamino carboxylic								
Sequestering and chelating agents	Sodium polyphosphates		F	P	G	F	G			They prevent water-hardness precipitation, are heat-stable and are used for formulation of combined detergent/sterilizer compounds.
	Ethylenediamine tetra acetic acid (EDTA) and its salts				VG	E	E	E	G	
	Gluconic acid and its salts						E	E		Their inclusion in formulations is to 'hold' calcium ions in alkali solution and prevent reprecipitation. The bacteriostatic property of EDTA is achieved by withdrawing trace metals from bacterial cellular membranes. Gluconic acid is a stronger chelating agent than EDTA in alkali solutions (2–5% strength).

Contd.

Table 3.5 (Contd.)

Type	Detergent components		I	II	III	IV	V	VI	VII	General comments
Sterilizing agents	Chlorine	Chlorinated trisodium orthophosphate							E	Their inclusion provides a balanced product for cleaning and sterilization (e.g. hypochlorous acid, QAC, iodine or peroxide).
		Dichlorodimethyl hydration							E	Consult list of brands, approved by the authorities concerned, that can be used as detergent/sterilizers as an alternative to steam or boiling water for the sterilization of dairy equipment.
		Sodium dichloro-isocyanurate						E		
		Sodium hypochlorite							E	
	QAC	Cetyl trimethyl ammonium bromide		VG	VG	VG			E	
		Benzalkonium chloride		VG	VG	VG			E	
	Iodine	Iodophors	G	G		G			E	
	Peroxide	Peracetic acid and Hydrogen peroxide								
Miscellaneous inhibitors	Sodium sulfite									Inhibitors minimize corrosive attacks by acids and alkalis on metal.
	Sodium silicate									The sulfites protect tinned surfaces, and silicates protect aluminium and its alloys from attack by alkalis.
Antifoaming agents										Antifoaming agents are sometimes incorporated in a detergent formulation to prevent foam formation which could be generated by pumping/jetting action during detergent recirculation. Fats and alkalis may form soaps by saponification and these antifoaming agents prevent foam formation.
Suspending agents										Sodium carboxymethyl cellulose or starch assist in maintaining undissolved soiling matter in suspension (= suspending agents).
Phosphates	Orthophosphates				G		G	G		Some polyphosphate compounds hydrolyse to orthophosphates in aqueous solution at a high temperature, but the presence of alkalis reduces the rate of hydrolysis.
	Polyphosphates				G		G	G		
Water softening										Precipitation of calcium and magnesium ions from hard water in order to avoid water-scale deposition on surfaces of equipment especially for the last rinsing step after cleaning.

[a](E) Excellent, (VG) Very good, (G) Good, (F) Fair, (P) Poor. I, Organic dissolving; II, wetting; III, dispensing suspending; IV, rinsing; V, sequestering; VI, chelating; VII, bacteriocidal.
After: Tamime and Robinson (1999).

and the presence of excess inorganic salts, mainly calcium and magnesium can reduce their effectiveness. In addition, these salts can leave deposits on the surfaces of equipment which are difficult to remove (Tamime and Robinson, 1999).

3.4.2.3 Methods for cleaning of dairy equipment

Cleaning of a dairy processing plant may involve manual cleaning, CIP and other miscellaneous methods. Virtually all cleaning procedures follow the same basic steps, and these usually take place in the following order:

(1) In the *preliminary rinse*, the processing plant, equipment, filling machines, churns, etc. are rinsed with water to remove the bulk of the milk residues from the equipment. For conservation purposes, the final rinse (see below) is recovered, especially in large automated plants and used for this preliminary rinse.
(2) In the *detergent wash*, alkali compounds are usually used (see Table 3.5 for specific applications), and during this stage the aim is to remove any adhering soil.
(3) The *intermediate rinse* removes any detergent residues from the surfaces prior to the operations acid wash and/or sterilization/sanitation that follow.
(4) The *acid wash* cleaning operation is optional and may be performed only once a week to clean the heat-processing equipment. It is important to note that acids are harmful to the skin, and hence an acid wash is normally used in CIP. Inorganic (nitric and phosphoric) and/or organic (acetic, gluconic, oxyacetic) acids may be used, since they have the ability to dissolve milk stone and remove hard water scale.
(5) The *intermediate rinse* is to remove any acid residues from the equipment prior to the sterilization/sanitation treatment.
(6) *Sterilization/sanitation treatment* of the plant and processing equipment must be effective before commencing production, and this aim is achieved by using one of the following:
    - nitric acid;
    - chemical compounds (QAC, chlorine and chloramines to achieve sanitization);
    - heat can produce sterile plant surfaces (live steam is limited in its application, but hot water circulation at 85–90°C for 15–30 min is a practical procedure, the temperature must be maintained on the return side of the plant and at the product outlet points); and
    - miscellaneous (refer to section on sterilization).
(7) In the *final rinse*, good-quality potable water is used to remove the sterilant residues from the processing plant.

3.4.2.3.1 Manual cleaning. No matter how sophisticated dairy plants become, there will always be some parts/equipment that can only be cleaned by hand. Equipment such as homogenizers, separators and filling machines, if not designed to be cleaned in place, have to be dismantled and cleaned out of place (COP). The manual cleaning procedures are as follows: disconnect and dismantle the equipment; pre-rinse with potable slightly warm water (20–30°C); prepare the mild/hand detergent solution at the recommended concentration in water at 40–50°C; brush/wash the specific parts; carry out an intermediate rinse with potable water; sterilize with a suitable chemical agent; and finally rinse with water.

The human element is important in these operations, and the following factors may influence the effectiveness of hand cleaning:

- inefficient scrubbing;
- low detergent concentration; and
- insufficient contact time.

Good management, supervision and training of personnel can help to achieve the desired aims as well as following the detergent manufacturer's recommendations. The COP method may also be used for cleaning pipelines in a small dairy processing plant, or in those parts of a factory where it may be difficult to provide a proper CIP system (Tamime and Robinson, 1999).

3.4.2.3.2 Cleaning-in-place. The CIP system is designed to clean processing equipment without dismantling and reassembling the different parts and, in addition, to minimize manual operations. This system offer several advantages, which include (Romney, 1990; Majoor, 2003)

- improved plant hygiene;
- better plant utilization;
- increased cost savings;
- greater safety for personnel; and
- minimal manual effort.

CIP is regarded as the method of choice for cleaning storage tanks, pipelines, pumps and valves. It is also used for cleaning vats, heat exchangers, centrifugal machines and homogenizers. CIP systems should be carefully designed for effective cleaning and also to avoid product contamination with soils and/or chemicals. A comparison of the practical features of the different CIP systems is shown in Table 3.6.

The different types of the CIP system have been well documented (Romney, 1990; Bylund, 1995; Tamime and Robinson, 1999; Majoor, 2003), and the three basic types are

- *Single-use systems* – These systems are generally small units and normally situated as close as possible to the equipment to be cleaned and sanitized. In this system, the detergent is used only once and then discarded. This system is ideal for small dairy plants and also for heavily soiled equipment because the reuse of the cleaning solution is less feasible.
- *Reuse systems* – In these systems, the detergent and/or acid solutions are recovered and reused as many times as practically possible, especially in parts of the dairy plant where the equipment is not heavily soiled, for example in the milk-reception area, standardization tanks and/or yoghurt fermentation tanks. The preliminary rinse of such equipment removes a high percentage of the soil, and, since the detergent solution circulated during the wash cycle is not heavily polluted, it can be reused several times.

Table 3.6 Comparison of different CIP systems[a].

Feature	Single-use	Reuse	Multi-use
Design type	Simple, often modular	Complex, 'one-off'	Complex, but modular
Addition to new equipment	Easy	More difficult	More difficult
Flexibility of detergent types	Flexible	Inflexible without modification	Flexible
Changes in detergent strength	Easy	Difficult without modification	Easy
Detergent make-up	As used	Stored hot	As used
Peak thermal load	High	Low	Moderate to high
Soil-type system can deal with	Heavy	Light to moderate	Moderate to heavy
Reuse of water and detergent	Use once then throw away	Reuse where possible	Some single-use, some reuse
Costs – capital	Low	High	Lower than reuse
Costs – detergent	High	Low	Lower than single-use
Costs – heat energy	High	Low	Lower than single-use

[a]After Majoor (2003).

- *Multi-use systems* – These systems attempt to combine all the features of both single-use and reuse systems, and are designed for cleaning pipelines, tanks and storage equipment. These systems function through automatically controlled programmes that entail various combinations of cleaning sequences involving circulation of water, alkaline cleaners, acid cleaners and acidified rinse through the cleaning circuits for differing time periods at varying temperatures (Majoor, 2003; Romney, 1990).

3.4.2.3.3 Miscellaneous cleaning methods. Other alternative cleaning methods can also be applied in a dairy plant (Chamberlain, 1983; Potthoff *et al.*, 1997; Tamime and Robinson, 1999), for example,

- *Soaking* – Processing equipment and/or fittings are immersed in a cleaning solution for 15–20 min and thereafter cleaned manually or mechanically.
- *Ultrasonic treatment* – The equipment is immersed in a cleaning solution and any soil dislodged by high-frequency vibrations.
- *Spray method* – This method involves spraying hot water or steam onto equipment surfaces. The cleaning solution is sprayed from special units to remove heavily soiled matter from equipment surfaces, before they are cleaned using one of the conventional methods.
- *Enzymatic treatment* – Enzyme-based cleaning preparations have also been developed with specific characteristics to remove soil from equipment surfaces. This method of cleaning does not employ conventional strong alkaline and/or acid solutions, and the cleaning process takes place at relatively low temperatures (50–55°C), a high pH (8.5–9.5) and at a low concentration of reagents (0.09%) (Potthoff *et al.*, 1997).

3.4.2.3.4 Factors affecting the efficiency of cleaning. The efficiency of cleaning is dependent on various well-documented factors (Romney, 1990; Floh, 1993; Timperley *et al.*, 1994; Bylund, 1995; Graßhoff, 1997; Holah, 2003).

(1) *Soiling matter* – The efficiency of all cleaning agents is primarily reduced by the presence of organic material. It is of the utmost importance to remember that efficient cleaning (and sanitization) can only be accomplished if all/most of the milk residues/organic matter are removed prior to the cleaning process.
(2) *Cleaning method* – It is advisable to implement a CIP system because crucial factors, such as concentration and temperature of the detergent, can effectively be controlled.
(3) *Contact time* – Sufficient contact time between the cleaning agent and the soil matter is important, since the functional properties of a detergent, for example, wetting, penetration, dissolving and suspending of the soil have a longer time to act.
(4) *Temperature and detergent concentration* – In general, the higher the temperature of the detergent solution, the more effective its cleaning action. While manual cleaning has to be carried out at around 45–50°C, the major sections of a dairy plant will be cleaned at 85–90°C using CIP; higher temperatures (e.g. 100–105°C) are used during the alkaline wash of UHT plants. A caustic soda solution of about 1% is sufficient for cleaning storage tanks, pipelines and fermentation tanks, while 1–2% is recommended for cleaning multipurpose tanks and heat exchangers, and 2–3% for cleaning UHT plants. Acid solutions are normally used in the region of <1% at 60–70°C.
(5) *Velocity or flow rate* – The flow characteristics of a liquid in a pipe can be either laminar or turbulent, and these configurations are influenced by such factors as pipe diameter, fluid momentum and fluid viscosity. A numerical presentation of the degree of turbulence in the fluid is referred to as its Reynolds number (*Re*), and the higher the number, the more disturbed the flow. Thus, the physical scrubbing action in a CIP system is greatly influenced by the flow rate of the fluid, and the effectiveness of the cleaning operation is greatly improved by increasing the velocity of the solution. The design and operation of a CIP system needs to ensure that the required mean flow rate of 1.5 m s^{-1} or $Re > 10^4$ (Timperley and Lawson, 1980; Kessler, 1981; Floh, 1993) is maintained.
(6) *Chemical composition of detergents* – Effective cleaning is also dependent on the formulation and functional properties of the detergent (Wirtanen *et al.*, 1997) (Table 3.5).
(7) *Plant design* – A dairy processing plant is constructed from a variety of vessels, pipelines, elbows, pipe couplings, valves, pumps and other parts. The efficiency of plant cleaning is, thus, dependent on the plant design, and a multiplicity of factors (Romney, 1990; Tamime and Robinson, 1999) for example,

   - corrosiveness of the stainless steel;
   - surface finish and surface grain;
   - type of pipe couplings;
   - type of welding;

- dead ends/pockets;
- type of pumps;
- type of valves; and
- specific plant layout.

A comprehensive account of recommended methods for cleaning and sterilization of different dairy processing equipment is given by Graßhoff (1997) and Tamime and Robinson (1999).

3.4.3 Sanitization and sterilization

3.4.3.1 Principles of sanitization and sterilization

Failure by cleaning procedures to adequately remove residual soil from milk product contact surfaces may have serious quality implications. Microorganisms remaining on equipment surfaces may survive for prolonged periods, depending on the amount and nature of residual soil, and relative humidity. Milk is a highly nutritious medium; hence any residue not removed can promote bacterial growth, bacterial adhesion to the surface and consequently biofilm development (Wong and Cerf, 1995; Mostert and Jooste, 2002). Routine cleaning operations are not 100% efficient, and over a period of multiple soiling/cleaning cycles, soil deposits and microorganisms will be retained (Holah, 2003). Effective sanitization of processing plant equipment is, according to Tamime and Robinson (1999), governed by

- maintaining the correct prescribed cleaning cycle prior to the sanitization stage;
- following the recommended sanitization method adopted, e.g. strength of chemical cleaning solution, required contact time and temperature;
- properly draining cleaning agents to prevent moist conditions of product contact surfaces;
- frequent dismantling of components such as joints, dead ends, 'blind' areas, rubber gaskets, etc.;
- efficiency of cleaning and/or the sanitization procedures; and
- use of legally approved disinfection agent(s) and/or compounds.

3.4.3.2 Methods of sanitization and/or sterilization

The following methods can be employed to achieve either sanitization or sterilization:

3.4.3.2.1 Heat. Elevated temperature is the best method for sanitization and/or sterilization, as it penetrates into surfaces, is non-corrosive, is non-selective to microbes, is easily measured and leaves no residues (Jennings, 1965). The type of heating that is mostly used is moist heat, for example,

- *Hot and/or boiling water* – Hot water circulation (85°C/15–20 min) is widely used and recommended for sanitizing processing plants. It is an effective, non-selective method for surfaces; however, bacterial spores and bacteriophages may survive. Boiling water (100°C) has limited applications but could be used for disinfection purposes; spores survive, but bacteriophages are inactivated. Free-flowing steam (100°C) is not more effective than boiling water and has limited applications. Steam under pressure may, however, be used to sterilize UHT plants.

- *Steam (free flowing)* – The use of hot water or steam is uneconomic, hazardous, corrosive to certain materials, difficult to control and therefore ineffective. The efficiency of sterilization or sanitization, using hot water or steam, is primarily dependent on three factors: the time–temperature combination (i.e. the temperature reached and the time for which the temperature is maintained), humidity and pressure (Tamime and Robinson, 1999; Holah, 2003).

3.4.3.2.2 Chemical agents. The efficacy of chemical preparations used for sterilizing purposes is controlled by the following factors:

- amount of residual soiling matter on equipment surfaces;
- temperature and pH of the chemical disinfectant;
- concentration of the chemical compounds in the sterilizing solution;
- contact time between the chemical disinfectant and the equipment surface;
- type(s) of microorganisms being inactivated;
- hardness of the water; and
- inactivation of its antimicrobial properties by residual detergent.

The chemical disinfecting agents which are commonly used in the dairy industry are (Tamime and Robinson, 1999):

(1) *Chlorine-releasing components* – Chlorine is a general disinfectant and is available as hypochlorite (or as chlorine gas) or in slow-releasing forms (e.g. chloramines, etc.). These compounds may be obtained in liquid or powder form, and their bactericidal effect is due to the release of chlorine, which is normally in the range 50–250 μg ml^{-1}, depending on the application. The use of ozone as a sanitizer has been investigated to replace chlorine but, in general, has not yet reached the market (Reinemann *et al.*, 2003).

(2) *Quaternary ammonium compounds (QACs)* – QACs are amphipolar, cationic detergents that are surface-active bactericidal agents. QACs are sometimes used as detergents/sterilants, but it should be noted that certain alkaline compounds (anionic wetting agents) can reduce the bactericidal action of QACs. The recommended concentrations vary between 150 and 250 μg ml^{-1} of QAC at >40°C for no less than 2 min.

(3) *Iodophores* – Iodophores are also a commonly used chemical disinfectant. The bactericidal compound is iodine, which has been combined with a suitable non-ionic surfactant to provide a usable product. The iodine complex is acidified with, for example, phosphoric acid for better stability and improved bactericidal effect.

(4) *Amphoteric surface-active agents* – These agents are known to have good detergent/sterilizing properties and may be used for manual cleaning, since they are non-corrosive and non-irritant to skin. They are not recommended for CIP application due to their high foaming properties.

(5) *Acidic anionic compounds* – These are formulations that consist mainly of inorganic acids (e.g. phosphoric acid) and an anionic surfactant. They are used as combined detergents/sterilizers, or as sterilizing agents *per se*.
(6) *Peracetic acid* – Peracetic acid provides a rapid broad spectrum effect. It is particularly effective against bacterial spores but is hazardous to use. It is considered to be toxicologically safe and biologically active.
(7) *Miscellaneous sterilizing agents*:
- Sodium hydroxide – Caustic soda has a bactericidal effect due to its high alkalinity. Concentrations between 15 and 20 g l^{-1} at 45°C for 2 min are sufficient to inactivate vegetative organisms.
- Mixed halogen compounds – Compounds such as chlorine and bromine can be employed as sterilants at lower concentrations than the individual elements.
- Formaldehyde is used for sterilizing and/or storage of membrane plants.
- Hydrogen peroxide.

3.4.3.2.3 Irradiation. Ultraviolet (UV) radiation (*c.* 250–260 nm), in particular, has been used with success to sterilize air entering a processing area or sterilizing packaging materials before filling.

3.4.3.2.4 Spraying, fogging or fumigation. Solutions containing certain chemical agent(s) can also be applied to spray/fog the atmosphere of enclosed rooms to inactivate aerial contaminants and also to destroy organisms on surfaces. Various methods, e.g. mist spraying, foams and gel techniques, are also available (Holah, 2003) in this regard.

Many different methods and combination of techniques could be used for sterilization/sanitization of processing plants and have been well documented (Tamime and Robinson, 1999; Holah, 2003). It should be realized that the ultimate choice of any method is mainly governed by the recommendations of the equipment manufacturer and by the degree of the hygiene required.

There are several environmental factors that are beginning to have an impact on cleaning practices in the world. The discharge of cleaning and sanitizing chemicals into the environment can be an issue for elements such as phosphorus and chlorine. Much work has been done on the development of enzymes as cleaners or additives, to conventional cleaning solutions. The success of enzyme preparations seems to be matching the enzyme to the specific type of soil (Reinemann *et al.*, 2003). The volume of water used in cleaning processes and discharged into the environment, as well as the energy used for heating and pumping water, also has environmental implications.

3.5 Dealing with biofilms

Biofilm formation in dairy processing environments is of special importance because it may have a huge impact on the hygiene, food safety and quality of milk and dairy products. A biofilm may be defined as a community of microorganisms attached to a surface, producing extracellular polymeric substances (EPS) and interacting with each other (Hall-Stoodley *et al.*, 2004; Lindsay and Von Holy, 2006).

3.5.1 Biofilm formation

Biofilm development is a dynamic process and commences when planktonic (free-living) microbial cells attach to a surface. Irreversible attached cells produce extracellular polymeric substances (EPS) which allow for cell-to-cell bridges and anchoring of the cells to the surface (Lindsay and Von Holy, 2006). Micro-colony development results from simultaneous aggregation and growth of microorganisms, accompanied by EPS production. A mature biofilm consists of microorganisms in EPS-enclosed micro-colonies interspersed with less dense regions of the polymer matrix that include water channels transporting nutrients and metabolites (Stoodley *et al.*, 1994). Individual cells of the biofilm may also be actively released into the surrounding environment to attach and colonize other surfaces (Parsek and Greenberg, 2005). It is important to note that cells within biofilms are physiologically distinct from their planktonic counterparts (Oosthuizen *et al.*, 2001; Parsek and Fuqua, 2004). Modern dairy processing plants support and select for biofilm-forming bacteria on milk/product contact surfaces due to highly automated systems, lengthy production cycles and vast closed surface areas in processing lines (Lindsay and Von Holy, 2006).

Areas in which biofilms most often develop are those which are the most difficult to rinse, clean and sanitize. Dead ends, gaskets, joints, pumps, grooves, surface roughness due to surface defects, by-pass valves, abraded equipment parts, sampling cocks, overflow siphons in filters and corrosion patches, etc. are hard-to-reach areas (Wong and Cerf, 1995). The presence of nutrients, or even microscopic food residues, and frequent stress conditions from cleaning, sanitizing or processing treatments may individually or collectively influence biofilm development and biofilm structure (Chmielewski and Frank, 2003). Biofilms may develop in environments that have a high microbial diversity (e.g. floor drains) or in environments dominated by one or a few microbial species, such as on plate heat exchangers.

Biofilm accumulation in the dairy environment, and especially the development on milk/product contact surfaces, is important. Biofilms in dairy processing environments have, for example, the following potential implications:

- Microorganisms in established biofilms are highly resistant to treatment with antimicrobial agents (e.g. antibiotics, disinfectants, etc.) (Costerton *et al.*, 1995; Lindsay and Von Holy, 1999). Lewis (2001) suggested that adhered cells in a biofilm can tolerate antimicrobial compounds at concentrations of 10–1000 times that needed to kill genetically equivalent planktonic bacteria.
- Biofilm cells have the ability to survive harsh environmental conditions such as fluctuating pH, extreme heat or cold, low nutrient concentrations, and they are highly resistant to exposure to UV light, chemical shock, starvation and dehydration (Wong and Cerf, 1995; Hall-Stoodley *et al.*, 2004; Hall-Stoodley and Stoodley, 2005).
- Post-pasteurization contamination, decreased shelf-life, or potential spoilage of products (Koutzayiotis, 1992; Koutzayiotis *et al.*, 1992; Austin and Bergeron, 1995).
- Attached cells become irreversibly adsorbed to the surface, which enables the organisms to resist mechanical cleaning procedures (Lundén *et al.*, 2000).

- Food-borne pathogens and spoilage organisms can attach to and produce EPS on food contact surfaces and other dairy environments (Chmielewski and Frank, 2003; Hall-Stoodley and Stoodley, 2005; Lehner *et al.*, 2005). *Listeria monocytogenes* is a well-adapted pathogen with the ability to proliferate in cold wet conditions that are ideally suited for biofilm formation in various environments. *Listeria* spp. have been isolated from wooden shelves in cheese-ripening rooms (Noterman, 1994), processing and packaging equipment, and especially wet, difficult-to-clean environments such as conveyor belts, floor drains, condensate, storage tanks, etc. (Charlton *et al.*, 1990; Nelson, 1990). The growth of *L. monocytogenes* in food plant biofilms increases the general contamination level in the plant and may be an indication of unsatisfactory cleaning/sanitization procedures. Outbreaks of listeriosis and salmonellosis have been implicated to post-pasteurization/processing contamination of milk, cheese and ice-cream as a contributing factor (Brocklehurst *et al.*, 1987; Hedberg *et al.*, 1992). Pathogenic bacteria can also coexist within a biofilm with other organisms; for example, *Listeria*, *Salmonella* and other pathogens have been found in established *Pseudomonas* biofilms (Jeong and Frank, 1994; Fatemi and Frank, 1999).
- Heat-resistant spore-forming organisms are commonly found in dairy processing plants (Oosthuizen *et al.*, 2001) and even in extreme environments such as in hot (80°C) alkaline solutions in reuse CIP systems (Swart, 1995). *Bacillus* and other thermoduric bacteria may form a biofilm if hot fluid continuously flows over a surface for 16 h or longer (Frank, 2000).
- Although the presence of *Salmonella* spp. is not well documented, various studies suggested that *Salmonella* can establish themselves in biofilms on food surfaces (Joseph *et al.*, 2001). The significance of the growth and activity of bacteria at solid–liquid interfaces on milk product contact surfaces is furthermore emphasized by Koutzayiotis *et al.* (1992). These authors suggested that proteolytic enzymes may be produced and released from established *Flavobacterium* biofilms. It has also been found that the production of catalase by attached populations of *Pseudomonas aeruginosa* biofilms may be partly responsible for increased resistance to sanitizers containing hydrogen peroxide (Steward *et al.*, 2000).
- Reduction in the efficiency of heat transfer occurs if biofilm accumulation becomes sufficiently thick at locations such as plate heat exchangers (Mittelman, 1998). Biofilm microorganisms may also be responsible for the corrosion of metal milk pipelines and tanks due to chemical and biological reactions.

3.5.2 Detection of biofilms

The most common methods presently available to detect biofilms include the following (see also Section 3.6.2):

(1) *Swab/swab–rinsing plate methods*. This method may also be supplemented by the bioluminescence test for total ATP (see ATP-bioluminescence test).

(2) *Agar contact plate methods*:
   - RODAC plate count;
   - agar slice methods; and
   - dry rehydratable film method.

The agar contact plate methods are simpler than swabbing, but it is not possible to sample irregular or rough surfaces that are indeed niches that harbour biofilms. In addition, microorganisms do not quantitatively adhere to the agar surface upon application, again resulting in selection for a specific micro-population or underestimating microbial numbers on the sampled surface (Chmielewski and Frank, 2003).

(3) *ATP-bioluminescence test*. The most rapid biochemical method to detect biofilms, or the effective removal thereof, can be monitored by the ATP-bioluminescence test (Chmielewski and Frank, 2003). This test is a biochemical method for estimating total ATP collected by swabbing a surface. Total ATP is related to the amount of product residues left behind on surfaces and also to microbial contamination, collected by the swab. Results can be obtained within 5–10 min and is also a rapid method to determine cleaning effectiveness and the state of hygiene of plant surfaces (Reinemann *et al*., 2003).

3.5.3 Biofilm control/removal

The most important factors that contribute to biofilm formation are inadequate removal of residual soil from surfaces (cleaning) and ineffective sanitation and sterilization of milk/product contact surfaces. Microorganisms remaining on equipment surfaces may survive for prolonged periods depending on the amount and nature of the residual soil, temperature and relative humidity. Milk is a highly nutritious medium, so any residue not removed can promote bacterial growth, bacterial adhesion to the surface and, consequently, biofilm development (Wong and Cerf, 1995; Frank, 2000).

It is not practical to clean and sanitize frequently enough to prevent attachment of microbes to surfaces, since cell attachment may occur within a few minutes to hours. It has, however, been suggested that removal of biofilms during cleaning is significantly enhanced by applying mechanical force to a surface, such as high-pressure sprayers and scrubbers. Non-aerosol-generating detergents, such as foam, as well as the use of sanitizers, will result in a higher bacterial kill when used in conjunction with mechanical methods (Meyer, 2003).

The formation of aerosols or small droplets is often found during washing and spraying of surfaces, floor and drains. Care should be taken not to contaminate clean areas or sanitized processing equipment. High-pressure, low-volume water is normally used to rinse surfaces; however, Gibson *et al.* (1999) reportedly found that flow above a pressure of 17.2 bar does not enhance biofilm removal.

Ideally, plant layout and equipment should be designed to prevent the accumulation of soil and water, and to allow for easy cleaning and sanitation operations. Problems often occur at locations such as dead ends, pumps and joints where gaskets must be used, and areas where surfaces may not receive sufficient exposure to cleaning and sanitizing chemicals (Kumar and Anand, 1998). In addition, the modification of equipment surfaces by anti-microbial coatings and new ideas to improve surface hygiene may ultimately aid in inhibition of biofilm formation (Carpentier *et al*., 1998; Kumar and Anand, 1998; Meyer, 2003).

Cleaning procedures should effectively remove food residues and other soils that may contain microorganisms or promote microbial growth (see Section 3.4). Most cleaning regimes include removal of loose soil with cold or warm water followed by the application of chemical agents, rinsing and sanitation (Frank, 2000). Cleaning can be accomplished by using chemicals or combination of chemical and physical force (water turbulence or scrubbing). High temperatures can reduce the need for physical force. Chemical cleaners suspend and dissolve food residues by decreasing surface tension, emulsifying fats and peptizing proteins (Chmielewski and Frank, 2003).

Problems such as corrosion and biofouling in cooling systems by microbial biofilms are normally prevented/controlled by chemical treatment (Mattila-Sandholm and Wirtanen, 1992; Hiddink, 1995).

Research concerning the complex molecular mechanisms that regulate the synthesis of EPS, the attachment of microorganisms, as well as the development and detachment of biofilms will ultimately lead to improved strategies for the control of biofilms.

3.6 Monitoring dairy plant hygiene

3.6.1 Air quality

The purpose of monitoring air in the dairy plant is to evaluate its quality and to obtain information about the hygienic condition in certain critical areas where microorganisms may contaminate the product directly, or indirectly. Samples are usually taken

- at openings of processing equipment that may be subjected to potential contamination by air currents;
- at selected points in a room, e.g. where products are filled and packed; and
- in areas where high numbers of employees are working (IDF, 1987a).

Various air samplers have been designed for sampling airborne organisms (Kang and Frank, 1989b; Al-Dagal and Fung, 1990); however, none of these sampling devices recover viable particles without some inactivation or losses during or after sampling. The effectiveness of air-quality monitoring depends on the type of sampler used, as well as the nature of air in the specific environment to be monitored (Kang and Frank, 1989a). The two main principles by which airborne microbes can be sampled are

- collection onto solid and semi-liquid media or filters; and
- collection into a liquid solution or medium.

The objective in each case is to determine the number of organisms on the plates, in the filter or in the liquid media. The sampling time for all collection methods is usually standardized at 15, 30 and 60 min (IDF, 1987b). The basic methods include techniques such as sedimentation (gravitation settling), impaction on solid surfaces, impingement in liquids as well as centrifugation and filtration (Kang and Frank, 1989b; Al-Dagal and Fung, 1990; Hickey *et al.*, 1993; Neve *et al.*, 1995). Comparative studies of air

Table 3.7 Suggested standards for air counts in dairy processing areas.[a]

	Plate count (cfu m^{-3})		Yeasts and moulds (cfu m^{-3})	
Processing area	Satisfactory	Unsatisfactory	Satisfactory	Unsatisfactory
Cultured milk, cream and cottage cheese	<150	>1500	<50	>1000
Milk and cream	<150	>1500	<50	>1000
Butter	<100	>1000	<50	>1000
Powdered milk	<200	>2000	<100	>1000
Ripened cheese	<200	>2000	<100	>1000

[a]Source: Adapted from Lück and Gavron (1990).

sampling devices have indicated that there is often no obvious choice of the correct sampler to use.

Microbiological standards for airborne counts in various dairy processing areas have been proposed by various authors (Kang and Frank, 1989b). The standards proposed by Lück and Gavron (1990) are relatively strict, but experience has shown that they are achievable in practice (Table 3.7).

3.6.2 Cleanliness of sanitized surfaces

Various methods and/or techniques have been devised to monitor the hygiene of dairy equipment surfaces (IDF, 1987b; BSI, 1991; Hickey *et al.*, 1993; Wong and Cerf, 1995; Tamime and Robinson, 1999), thus helping to maintain production of high-quality products and at the same time ensuring compliance with legal requirements. Whatever tests are employed, it is essential that they are applied routinely, for individual readings are in themselves meaningless; only when values for a typical high standard of hygiene have been established for a given plant, along with acceptable tolerances, do the results of any microbiological/hygiene test become valuable.

Enumeration of total bacterial counts, coliforms, yeasts and moulds are the most common microbiological examinations carried out to assess the hygiene of dairy equipment surfaces. The types of microorganisms present reflect to some extent the standard of plant hygiene (Tamime and Robinson, 1999). Selective and differential culture media may also be used to test specifically for given groups of organisms. Although a given method may not remove all the organisms, its consistent use in specific areas can still provide valuable information as long as it is realized that not all organisms are being removed (Jay, 1992). The most commonly used methods to monitor surface hygiene have been described by Mostert and Jooste (2002) and Chmielewski and Frank (2003). The conventional methods include

(1) Swab/swab–rinse plating methods. These methods are applicable to any surface, especially hard-to-reach areas such as surfaces with cracks, corners or crevices (Hickey *et al.*, 1993) that can be reached by hand. A moistened swab or sponge is rubbed over a designated area to remove the microorganisms from the surface.

The sample liquid, or decimal dilutions, if necessary, is then examined by the plate-count method. The reproducibility of the swab techniques is variable due to the unreliable efficiency of swabbing and the proportion of bacteria removed from the surface is unknown. Furthermore, it is time-consuming (results available within days) and highly operator dependent (Wong and Cerf, 1995). Despite their limitations, the swab methods are very useful and almost universally applied in the dairy industry (Tamime and Robinson, 1999). The swab–rinse method may also be supplemented by a bioluminescence test for total adenosine-5-triphosphate (ATP) (see ATP bioluminescence test).

(2) Agar contact plate methods. Flat or slightly bent surfaces which are smooth and non-porous can be sampled by pressing a solidified piece of appropriate nutritive agar against a surface. A number of commercial products are available in this regard:
   - RODAC plate count. The replicate organism direct agar contact (RODAC) method employs special commercially available plastic plates in which the agar medium protrudes slightly above the rim. The agar surface is pressed onto the test area, removed and incubated (Lück and Gavron, 1990; Jay, 1992; Hickey *et al.*, 1993).
   - Agar slice methods. Modified large syringes or plastic sausage castings can be filled with agar medium and a portion pushed out and pressed onto the test surface, cut off and incubated (Jay, 1992). Unless caution is taken to apply agar to the sample surface with constant pressure and time, reproducibility of sampling can be questionable (Wong and Cerf, 1995).
   - Dry rehydratable film method. This (Petrifilm aerobic count) method also provides a simple direct-count technique on both flat and curved surfaces. This procedure is less applicable for surfaces with cracks or crevices (Hickey *et al.*, 1993).

(3) ATP-bioluminescence test. Recent developments of ATP detection methods using bioluminescence have been proposed as a rapid method for assessing the effectiveness of sanitation in the dairy industry (Reinemann *et al.*, 2003). This test is the most rapid biochemical method to determine the state of plant surface hygiene and takes less than 5 min to perform. The method must be used carefully and with a sufficient number of tests to obtain meaningful results. The readings are not intended to correlate with the microbial count, but there is an excellent correlation between clean surfaces and low levels of ATP (Anon., 1995; Tamime and Robinson, 1999).

Table 3.8 Suggested standards for dairy equipment surfaces prior to pasteurization/heat treatment.[a]

Total colony (or coliform) count 100 cm^{-2}	Conclusion
500 (coliforms <10)	Satisfactory
500–2500	Dubious
>2500 (coliforms >100)	Unsatisfactory

[a]Adapted from Harrigan and McCance (1976); Tamime and Robinson (1999).

(4) Visual inspection. Inefficient cleaning usually results in a visual build-up of a residual film(s) on surfaces. Some of these films have a characteristic appearance which can help to determine the cause of the cleaning failure. Films containing fat are soft when wet and dry, while protein films are hard when wet or dry and have a light brown colour. Inorganic/mineral films are hard when wet or dry, usually have a rough porous texture and are invisible when wet and white when dry (Reinemann *et al.*, 2003).
(5) Other methods. Other methods are also described in the literature, for example the adhesive (sticky) tape method (Tamminga and Kampelmacher, 1977) and rapid methods for monitoring the hygiene of dairy equipment surfaces (Russell, 1997). Suggested standards for dairy equipment surfaces prior to pasteurization/heat treatment are shown in Table 3.8. Nowadays, with improved cleaning and sanitation programmes, a total colony count of 200 cfu 100 cm^{-2} would be expected, and below 50 cfu 100 cm^{-2} for equipment containing pasteurized products (Lück and Gavron, 1990).

3.6.3 Water quality

Water has many important applications in the dairy industry, but it may present a risk if the microbiological quality is not regularly monitored and appropriate water treatment applied (Hiddink, 1995). Spoilage of milk and milk products by water-borne organisms, e.g. psychrotrophs (Witter, 1961), can occur directly, through product contact with the water itself, or indirectly, by organisms metabolizing nutrient residues on improperly cleaned product contact surfaces (Hickey *et al.*, 1993).

The type of sample taken depends on the purpose of sampling. General bacteriological sampling (for *E. coli* and coliforms) involves the collection of relatively small volumes of water (0.5–3 l), whereas samples for detection of specific pathogens involve larger volumes (10–1000 l) (IDF, 1987b; Clesceri *et al.*, 1989; Fricker, 1993). Water samples should be examined as soon as possible after collection, preferably within 6 h. The methods for the microbiological examination of water are intended to give an indication of the degree of contamination and to ensure the safety of supply. In general, the tests are based on indicator organisms, the presence or absence of which provides an indication of the microbiological quality of the water. Detailed procedures for the sampling and testing of the microbiological quality of water are outlined in the 17th Edition of *Standard Methods of the Examination of Water and Waste Water* (Clesceri *et al.*, 1989). The microbiological specifications for drinking and process water are shown in Table 3.2.

References

Al-Dagal, M. and Fung, D.Y.C. (1990) Aeromicrobiology – a review. *CRC Critical Reviews of Food Science and Nutrition* **29**, 333–340.

Anon. (1988) Recommended guidelines for controlling environmental contamination in dairy plants. *Dairy and Food Sanitation* **8**, 52–56.

Anon. (1993) Council directive 93/43/EEC of 14 June 1993 on the hygiene of foodstuffs. *Official Journal of European Communities* **L175**, 1–11.

Anon. (1995) *HACCP and the Lumac Solution*. Lumac, Landgraaf, Netherlands.

Anon. (1997) Recommended international code of practice: General principles of food hygiene. In: *Codex Alimentarius Commission Food Hygiene Basic Texts: CAC/RCP 1-1969 Rev 3*. Food and Agriculture Organization of the United Nations, World Health Organization, Rome.

Anon. (2005) The never ending war on pests in the food sector. *International Food Hygiene* **16**, 5, 7.

Arnould, P. and Guichard, L. (1999) Disinfection of dairy factories using fumigation. *Latte* **24**, 44–46.

Austin, J.W. and Bergeron, G. (1995) Development of bacterial biofilms in dairy processing lines. *Journal of Dairy Research* **62**, 509–519.

Bell, C. H. (2003) Pest control: insects and mites. In: Lelieveld, H.L.M., Mostert, M.A., Holah, J. and White, B. (eds.) *Hygiene in Food Processing*, Woodhead, Cambridge, pp. 335–379.

Bénézech, T., Lelievre, C., Membre, J.M., Viet, A.-F. and Faille, C. (2002) A new test method for in-place cleanability of food processing equipment. *Journal of Food Engineering*, **54**, 7–15.

Brocklehurst, T.F., Zaman-Wong, C.M. and Lund, B.M. (1987) A note on the microbiology of retail packs of prepared salad vegetables. *Journal of Applied Bacteriology* **63**, 409–415.

Brown, K.L. (2003) Control of airborne contamination. In: Lelieveld, H.L.M., Mostert, M.A., Holah, J. and White, B. (eds.) *Hygiene in Food Processing*, Woodhead, Cambridge, pp. 106–121.

Brown, K.L., Wileman, R. and Smith, J. (2002) *Airborne Contamination Risk from Food Production Personnel. R&R Report 163*. Campden and Chorleywood Food Research Association, Chipping Campden, UK.

BSI (1991) *Methods of Microbiological Examination for Dairy Purpose, BS 4285: Part 4*. British Standards Institution, London.

Bylund, G. (1995) In: *Dairy Processing Handbook*. Tetra Pak Processing Systems A/B, Lund, Sweden.

CAC/RCP (2004) *Code of Hygiene Practice for Milk and Milk Products. CAC/RCP 57*, pp. 1–40.

Carey, N.R., Murphy, S.C., Zadoks, R.N. and Boor, K.J. (2005) Shelf lives of pasteurized fluid milk products in New York State: a ten year study. *Food Protection Trends* **25**, 102–113.

Carpentier, B., Wong, A.C.L. and Cerf, O. (1998) In: *Biofilms on Dairy Plant Surfaces: What's New?* Bulletin 329, pp. 32–35. International Dairy Federation, Brussels, Belgium.

Chamberlain, C.J. (1983) Opportunities for ultrasonics. *Food Processing* **52**, 35–37.

Charlton, B.R., Kinde, H. and Jensen, L.H. (1990) Environmental survey for *Listeria* species in California milk processing plants. *Journal of Food Protection* **53**, 198–201.

Chmielewski, R.A.N. and Frank, J.F. (2003) Biofilm formation and control in food processing facilities. *Comprehensive Reviews in Food Science and Food Safety* **2**, 22–32.

Clesceri, L.S., Greenberg, A.E. and Trussell, R.R. (1989) *Standard Methods for the Examination of Water and Waste Water*, 17th edn, pp. 9.1–9.227. APHA, Washington, DC.

Cocker, R. (2003) The regulation of hygiene in food processing: an introduction. In: Lelieveld, H.L.M., Mostert, M.A., Holah, J. and White, B. (eds.) *Hygiene in Food Processing*, pp. 3–18. Woodhead, Cambridge.

Costerton, J.W., Lewandowski, Z., Caldwell, D.E., Korber, D.R. and Lappin-Scott, H.M. (1995) Microbial biofilms. *Annual Review of Microbiology* **49**, 711–745.

Den Aantrekker, E.D., Boom, R.M., Zwietering, M.H. and Von Schothorst, M. (2003) Quantifying recontamination through factory environments – a review. *International Journal of Food Microbiology* **80**, 117–130.

Dunsmore, D.G. (1983) The incidence and implications of residues of detergents and sanitizers in dairy products. *Residue Reviews* **86**, 1–63.

Dunsmore, D.G., Thomson, M.A. and Murray, G. (1981a) Bacteriological control of food equipment surfaces by cleaning systems. III. Complementary cleaning. *Journal of Food Protection* **44**, 100–108.

Dunsmore, D.G., Twomey, A., Whittlestone, W.G. and Morgan, H.W. (1981b) Design and performance of systems for cleaning product-contact surfaces of food equipment: a review. *Journal of Food Protection* **44**, 220–240.

El-Shenawy, M.A. (1998) Sources of *Listeria* spp. in domestic food processing environment. *International Journal of Environmental Health Research* **8**, 241–251.

Evans, J.A., Russell, S.L., James, C. and Corry, J.E.L. (2004) Microbial contamination of food refrigeration equipment. *Journal of Food Engineering* **62**, 225–232.

Fadda, S., Aymerich, T., Hugas, M. and Garriga, M. (2005) Use of a GMP/GHP HACCP checklist to evaluate the hygienic status of traditional dry sausage workshops. *Food Protection Trends* **25**, 522–530.

Fatemi, P. and Frank, J.F. (1999) Inactivation of *Listeria monocytogenes/Pseudomonas* biofilms by peracid sanitizers. *Journal of Food Protection* **62**, 761–765.

Floh, R. (1993) Automatic recirculation cleaning in the 90's. *Dairy Food and Environmental Sanitation* **13**, 216–219.

Frank, J.F. (2000) Microbial attachment to food and food contact surfaces. *Advances in Food and Nutrition Research* **43**, 319–370.

Fricker, C. (1993) Water supplies: Microbiological analysis. In: Macrae, R., Robinson, R.K. and Sadler, M.J. (eds.) *Encyclopaedia of Food Science, Food Technology and Nutrition*, pp. 4859–4861. Academic Press, London.

Frontini, S. (2000) Clean air during production. *Latte* **25**, 28–34.

Gibson, H., Taylor, J.H., Hall, K.E. and Holah, J.T. (1999) Effectiveness of cleaning techniques used in the food industry in terms of the removal of bacterial biofilms. *Journal of Applied Microbiology* **87**, 41–48.

Graßhoff, A. (1997) Cleaning of heat treatment equipment. In: *Fouling and Cleaning of Heat Treatment Equipment – a Monograph*. Bulletin 328, pp. 32–44. International Dairy Federation, Brussels.

Guyader, P. (1995) Contamination of products: be careful with compressed air. *Dairy Science Abstracts* **58**, 208.

Hale, N., Bertsch, R., Barnett, J. and Duddelston, W.L. (2003) Sources of wastage in the dairy industry. In: *Guide for Dairy Managers on Wastage Prevention in Dairy Plants*, Bulletin 328, pp. 7–30. International Dairy Federation, Brussels.

Hall-Stoodley, L., Costerton, J.W. and Stoodley, P. (2004) Bacterial biofilms: from the natural environment to infectious diseases. *Nature Reviews Microbiology* **2**, 95–108.

Hall-Stoodley, L. and Stoodley, P. (2005) Biofilm formation and dispersal and the transmission of human pathogens. *Trends in Microbiology* **13**, 7–10.

Harrigan, W.F. and McCance, M.E. (1976) *Laboratory Methods in Food and Dairy Microbiology*, 2nd edn. Academic Press, London.

Hedberg, C.W., Korlath, J.A., D'Aoust, J.-Y., White, K.E., Schell, W.L., Miller, M.R., Cameron, D.N., MacDonald, K.I. and Osterholm, M.T. (1992) A multistate outbreak of *Salmonella javiana* and *Salmonella oranienburg* infections due to consumption of contaminated cheese. *JAMA* **268**, 3203–3207.

Hedrick, T.I. (1975) Engineering and science of aeromicrobiological contamination control in dairy plants. *Chemistry and Industry* **20**, 868–872.

Hedrick, T.I. and Heldman, D.R. (1969) Air quality in fluid and manufactured milk products plants. *Journal of Milk and Food Technology* **32**, 265–269.

Heldman, D.R. and Hedrick, T.I. (1971) Air-borne contamination control in food processing plants. *Research Bulletin* **33**, pp. 76–77. Michigan State University, Agricultural Experiment Station, East Lansing, MI.

Heldman, D.R., Hedrick, T.I. and Hall, C.W. (1965) Sources of airborne microorganisms in food processing areas – drains. *Journal of Milk and Food Technology* **28**, 41–45.

Hickey, P.J., Beckelheimer, C.E. and Parrow, T. (1993) Microbiological tests for equipment, containers, water and air. In: Marshall, R.T. (ed.) *Standard Methods for the Examination of Dairy Products*, pp. 397–412. American Public Health Association, Washington, DC.

Hiddink, J. (1995) In: *Water Supply, Sources, Quality and Water Treatment in the Dairy Industry*, Bulletin 308, pp. 16–32. International Dairy Federation, Brussels.

Holah, J.T. (2002) Foodborne pathogens: hazards, risk analysis and control. In: Blackburn, C. de W. and McClure, P.J. (eds.) *Hygienic Plant Design and Sanitation*, pp. 145–148. Woodhead, Cambridge.

Holah, J.T. (2003) Cleaning and disinfection. In: Lelieveld, H.L.M., Mostert, M.A., Holah, J. and White, B. (eds.) *Hygiene in Food Processing*, pp. 235–278. Woodhead, Cambridge.

Homleid, J.P. (1997) Spray-disinfection in the dairy industry. *Meieriposten* **86**, 111–112.

Hood, S.K. and Zottola, E.A. (1995) Biofilms in food processing. *Food Control* **6**, 9–18.

IDF (1979) *Design and Use of CIP Systems in the Dairy Industry,* Document No. 117. International Dairy Federation, Brussels.

IDF (1980) *General Code of Hygiene Practice for the Dairy Industry*, Document No. 123. International Dairy Federation, Brussels.

IDF (1984) *General Code of Hygienic Practice for the Dairy Industry and Advisory Microbiological Criteria for Dried Milk, Edible Rennet Casein and Food Grade Whey Powders.* Document No. 178, International Dairy Federation, Brussels.

IDF (1985) *Surface Finishes of Stainless Steels/New Stainless Steels.* Document No. 189, International Dairy Federation, Brussels.

IDF (1987a) *Hygiene Design of Dairy Processing Equipment.* Document No. 218, International Dairy Federation, Brussels.

IDF (1987b) *Hygiene Conditions – General Guide on Sampling and Inspection Procedures*, Standard 121A: 1987. International Dairy Federation, Brussels, Belgium.

IDF (1991) *IDF Recommendations for the Hygienic Manufacture of Spray-Dried Milk Powders,* Document No. 267. International Dairy Federation, Brussels.

IDF (1992) *Hygienic Management in Dairy Plants,* Document No. 276. International Dairy Federation, Brussels.

IDF (1994) *Recommendations for the Hygienic Manufacture of Milk and Milk Based Products,* Document No. 292. International Dairy Federation, Brussels.

IDF (1996) *Codex Standards in the Context of World Trade Agreements/IDF General Recommendations for the Hygienic Design of Dairy Equipment,* Document No. 310, pp. 26–31. International Dairy Federation, Brussels.

IDF (1997a) *IDF Guidelines for Hygienic Design and Maintenance of Dairy Buildings and Services,* Bulletin 324. International Dairy Federation, Brussels, Belgium.

IDF (1997b) *Implications of Microfiltration on Hygiene and Identity of Dairy Products/ Genetic Manipulations of Dairy Cultures,* Document No. 320, pp. 9–40. International Dairy Federation, Brussels.

IDF (2000) *Hygienic Design of Storage Tanks,* Bulletin 358, pp. 49–53. International Dairy Federation, Brussels.

IDF (2004) *Dairy Plant – Hygienic Conditions – General Guidance on Inspection and Sampling Procedures,* Standard 121:2004/ISO 8086. International Dairy Federation, Brussels.

IDF/FAO (2004) *Guide to Good Dairy Farming Practice*, pp. 1–2. Joint Publication of the International Dairy Federation and the Food and Agriculture Organization of the United Nations, Rome.

Jay, J.M. (1992) M*odern Food Microbiology*, 4th edn., Von Nostrand Reinhold, New York.

Jennings, W.G. (1965) Theory and practice of hard-surface cleaning. *Advances in Food Research* **14**, 325–459.

Jeong, D.K. and Frank, J.F. (1994) Growth of *Listeria monocytogenes* at 10°C in biofilms with microorganisms isolated from meat and dairy processing environments. *Journal of Food Protection* **57**, 576–586.

Jervis, D.I. (1992) Hygiene in milk product manufacture. In: Early, R. (ed.) *The Technology of Dairy Products*, pp. 272–299. Blackie, Glasgow.

Jervis, D. (2002) Application of process control. In: Robinson, R.K. (ed.) *Dairy Microbiology Handbook*, 3rd edn., pp. 593–654. John Wiley, New York.

Joseph, B., Otta, S.K., Karunasagar, I. and Karunasagar, I. (2001) Biofilm formation by *Salmonella* spp. on food contact surfaces and their sensitivity to sanitizers. *International Journal of Food Microbiology* **64**, 367–372.

Julien, C., Bénézech, T., Carpentier, B., Lebret, V. and Faille, C. (2002) Identification of surface characteristics relevant to the hygienic status of stainless steel for the food industry. *Journal of Food Engineering* **56**, 77–87.

Kang, Y-J. and Frank, J.F. (1989a) Evaluation of air samplers for recovery of biological aerosols in dairy processing plants. *Journal of Food Protection* **52**, 665–659.

Kang, Y-J. and Frank, J.F. (1989b) Biological aerosols: a review of airborne contamination and its measurement in dairy processing plants. *Journal of Food Protection* **52**, 512–524.

Kang, Y-J. and Frank, J.F. (1990) Characteristics of biological aerosols in dairy processing plants. *Journal of Dairy Science* **73**, 621–626.

Kelly, M. (2006) Insect unfriendly packaging design. *International Food Hygiene* **17**, 5–6.

Kessler, H.G. (1981) *Food Engineering and Dairy Technology*. Verlag A. Kessler, Freising, Germany.

Kosikowski, F.V. and Mistry, V.V. (1997) *Cheese and Fermented Milk Foods*, Vol. 2, 3rd edn, pp. 46, 308. F.V. Kosikowski, L.L.C., Westport, CT.

Koutzayiotis, C. (1992) Bacterial biofilms in milk pipelines. *South African Journal of Dairy Science* **24**, 19–22.

Koutzayiotis, C., Mostert, J.F., Jooste, P.J. and McDonald, J.J. (1992) Growth and activity of *Flavobacterium* at solid–liquid interfaces. In: Jooste, P.J. (ed.) *Advances in the Taxonomy and Significance of* Flavobacterium, Cytophaga *and Related Bacteria*, pp. 103–109. University Press, University of the Orange Free State, Bloemfontein, South Africa.

Kumar, C.G. and Anand, S.K. (1998) Significance of microbial biofilms in the food industry: a review. *International Journal of Food Microbiology*, **42**, 9–27.

Lehner, A., Riedel, K., Eberl, L., Breeuwer, P., Diep, B. and Stephan, R. (2005) Biofilm formation, extracellular polysaccharide production, and cell-to-cell signalling in various *Enterobacter sakazakii* strains: aspects promoting environmental persistence. *Journal of Food Protection* **68**, 2287–2294.

Lelieveld, H.L.M. (2003a) Introduction. In: Lelieveld, H.L.M., Mostert, M.A., Holah, J. and White, B. (eds) *Hygiene in Food Processing*, pp. 1–2. Woodhead, Cambridge.

Lelieveld, H.L.M. (2003b) Sources of contamination. In: Lelieveld, H.L.M., Mostert, M.A., Holah, J. and White, B. (eds) *Hygiene in Food Processing*, pp. 61–75. Woodhead, Cambridge.

Lelieveld, H.L.M., Mostert, M.A. and Curiel, G.J. (2003) Hygiene equipment design. In: Lelieveld, H.L.M., Mostert, M.A., Holah, J. and White, B. (eds.) *Hygiene in Food Processing*, pp. 122–166. Woodhead, Cambridge.

Lewis, K. (2001) Riddle of biofilm resistance. *Antimicrobial Agents and Chemotherapy* **45**, 999–1007.

Lindsay, D. and Von Holy, A. (1999) Different responses of planktonic and attached *Bacillus subtilis* and *Pseudomonas fluorescens* to sanitizer treatment. *Journal of Food Protection*, **62**, 368–379.

Lindsay, D. and Von Holy, A. (2006) What food safety professionals should know about bacterial biofilms. *British Food Journal* **108**, 27–37.

Lück, H. and Gavron, H. (1990) Quality control in the dairy industry. In: Robinson, R.K. (ed.) *Dairy Microbiology – The Microbiology of Milk Products*, Vol. 2, 2nd edn, pp. 345–392. Elsevier Applied Science, London.

Lundén, J.M., Miettinen, M.K., Autio, T.J. and Korkeala, H.J. (2000) Persistent *Listeria monocytogenes* strains show enhancement adherence to food contact surface after short time contact time. *Journal of Food Protection*, **63**, 1204–1207.

Majoor, F.A. (2003) Cleaning in place. In: Lelieveld, H.L.M., Mostert, M.A., Holah, J. and White, B. (eds.) *Hygiene in Food Processing*, pp. 197–219. Woodhead, Cambridge.

Marriott, N. (1999) *Principles of Food Sanitation,* 4th edn. Aspen, Gaithersburg, MD.

Mattila-Sandholm, T. and Wirtanen, G. (1992) Biofilm formation in the industry. A review. *Food Reviews International* **8**, 573–603.

Mettler, E. and Carpentier, B. (1998) Variations over time of microbial load and physicochemical properties of floor materials after cleaning in food industry premises. *Journal of Food Protection* **61**, 57–65.

Meyer, B. (2003) Approaches to prevention, removal and killing of biofilms. *International Biodeterioration and Biodegradation* **51**, 249–253.

Mittelman, M.W. (1998) Structure and functional characteristics of bacterial biofilms in fluid processing operations. *Journal of Dairy Science* **81**, 2760–2764.

Mostert, J.F. and Jooste, P.J. (2002) Quality control in the dairy industry. In: Robinson, R.K. (ed.) *Dairy Microbiology Handbook*, 3rd edn., John Wiley, New York, pp. 655–736.

Nelson, J. (1990) Where are *Listeria* likely to be found in dairy plants? *Dairy Food and Environmental Sanitation* **10**, 344–345.

Neve, H., Berger, A. and Heller, K.J. (1995) A method for detecting and enumerating airborne virulent bacteriophage of dairy starter cultures. *Kieler Milchwirtschaftliche Forschungsberichte* **47**, 193–207.

Noterman, S. (1994) The significance of biofouling to the food industry. *Journal of Food Technology* **48**, 13–14.

Oosthuizen, M.C., Steyn, B., Lindsay, D., Brözel, V.S. and Von Holy, A. (2001) Novel method for the proteomic investigation of a dairy-associated *Bacillus cereus* biofilm. *FEMS Microbiology Letters* **194**, 47–51.

Orth, R. (1998) The importance of disinfection for the hygiene in the dairy and beverage production. *International Biodeterioration and Biodegradation* **41**, 201–208.

Parsek, M.R. and Fuqua, C. (2004) Biofilms 2003: emerging themes and challenges in studies of surface-associated microbial life. *Journal of Bacteriology* **186**, 4427–4440.

Parsek, M.R. and Greenberg, E.P. (2005) Sociomicrobiology: the connections between quorum sensing and biofilms. *Trends in Microbiology* **13**, 27–33.

Potthoff, A., Serve, W. and Macharis, P. (1997) The cleaning revolution. *Dairy Industries International* **62**, 25, 27, 29.

Reinemann, D.J., Wolters, G., Billon, P., Lind, O. and Rasmussen, M.D. (2003) Review of practices for cleaning and sanitation of milking machines. In: *IDF Bulletin,* 381, pp. 3–11. International Dairy Federation, Brussels.

Robinson, R.K. and Tamime A.Y. (2002) Maintaining a clean working environment. In: Robinson, R.K. (ed.) *Dairy Microbiology Handbook,* 3rd edn, pp. 561–591. John Wiley, New York.

Romney, A.J.D. (Ed.), (1990) In: *CIP: Cleaning in Place*, 2nd edn. Society of Dairy Technology, Huntingdon, UK.

Russell, P. (1997) Monitoring hygiene. *Milk Industries International* **99**, 25–29.

Shah, B.P., Shah, U.S. and Siripurapu, S.C.B. (1996) Air-borne contaminants in dairy plants – a review. *Indian Dairyman* **48**, 19–21.

Shah, B.P., Shah, U.S. and Siripurapu, S.C.B. (1997) ULPA filter for contamination free air in dairy plant. *Indian Dairyman* **49**, 23–27.

Shapton, D. and Shapton, N. (1991) *Principles and Practices for the Safe Processing of Foods.* Woodhead, Cambridge.

Steward, P.S., Roe, F., Rayner, J., Elkins, J.G., Lewandowski, Z., Ochsner, U.A. and Hassett, D.J. (2000) Effect of catalase on hydrogen peroxide penetration into *Pseudomonas aeruginosa* biofilms. *Applied and Environmental Microbiology* **66**, 863–868.

Stoodley, P., De Beer, D. and Lewandowski, Z. (1994) Liquid flow in biofilm systems. *Applied and Environmental Microbiology* **60**, 2711–2716.

Swart, M. (1995) The presence and importance of microorganisms in alkaline detergents used in the dairy industry. MSc thesis, Potchefstroom University for CHE, Potchefstroom, South Africa.

Swartling, P. (1959) The influence of the use of detergents and sanitizers on the farm with regard to the quality of milk and milk products. *Dairy Science Abstracts* **21**, 1–10.

Tamime, A. Y. and Robinson, R.K. (1999) *Yoghurt Science and Technology*, 2nd edn. Woodhead, Cambridge.

Tamminga, S.K. and Kampelmacher, E.H. (1977) Bacteriology of dairy products. *Zentralblatt fur Bakteriologie Mikrobiologie und Hygiene I Abteilung Originale* **B165**, 423–426.

Timperley, D. and Lawson, G. (1980) Test rigs for evaluation of hygiene in plant design. In: Jowitt, R. (ed.) *Hygienic Design of Operation of Food Plant*, pp. 171–179, Ellis Horwood, Chichester, UK.

Timperley, D.A., Hasting, A.M.P. and de Goederen, G. (1994) Developments in the cleaning of dairy sterilization plant. *Journal of the Society of Dairy Technology* **47**, 44–50.

Vinson, H.G. (1990) Hygienic design of a dairy. *Journal of the Society of Dairy Technology* **43**, 39–41.

Wainess, H. (1982) Hygienic design and construction of equipment used in dairy plants. In: *IDF Bulletin,* Doc. No. 153, pp. 3–10. International Dairy Federation, Brussels, Belgium.

Wainess, H. (1995) Design, construction and operation of packaging equipment. In: *Technical Guide for the Packaging of Milk and Milk Products*, 3rd edn., Bulletin 300, pp. 40–46, 47–51. International Dairy Federation, Brussels.

WHO (1994) *Environmental Health Criteria 160*. World Health Organization, Vammala, Switzerland.

WHO (1996) *Guidelines for Drinking-Water Quality*. World Health Organization, Geneva.

WHO (2005) *Water Drinking Guidelines*. World Health Organization, Geneva. http://w3.whosea.org/techinfo/water.htm (accessed 2 December 2005).

Wierenga, G. and Holah, J.T. (2003) Hygienic plant design. In: Lelieveld, H.L.M., Mostert, M.A., Holah, J. and White, B. (eds.) *Hygiene in Food Processing*, pp. 76–105. Woodhead, Cambridge.

Wirtanen, G., Salo, S., Maukonen, J., Bredholt, S. and Mattila-Sandholm, T. (1997) *Sanitation in Dairies*, VTT Publication No. 309. Technical Research Center, Espoo, Finland.

Witter, L.D. (1961) Psychrophilic bacteria – a review. *Journal of Dairy Science* **44**, 983–1015.

Wong, A.C.L. and Cerf, O. (1995) In: *Biofilms: Implications for Hygiene Monitoring of Dairy Plant Surfaces*. Bulletin 302, pp. 40–44. International Dairy Federation, Belgium.

Chapter 4

Automation in the Dairy Industry

Evaggelos Doxanakis and Asterios Kefalas

4.1 Introduction

As our society transformed from a purely agricultural economy to an industrial economy, at the turn of the 20th century, food production, processing and distribution became gradually mechanized. The dairy sector followed the universal pattern of the evolutionary process of mechanization. In this process, each successive new step involves a conscious process of substitution of mechanical activities for human participation. Gradually, the small farm dairy, in which humans operated the entire chain of activities, from farm to market, became obsolete. From the first step in this process, milking the animal, to the last step, the delivery of the product to the customer, each stage became part of the agribusiness.

At the beginning, economic reasons, such as labour costs, were the primary drivers behind this mechanization. Gradually, safety considerations, mandated by legislatures, entered the scene. Food safety became the dominant factor in considering the participation of humans in the process of food production. The process of milking animals by humans, storing, processing and delivering the milk to the customer provided plenty of opportunity for contamination. It eventually became evident, via scientific investigations motivated by health incidents that at times reached pandemic proportions, that machines could perform some of these activities in a more sanitary manner than human beings. In addition, as the cost of the machinery used in the automation of dairy declined and at the same time labour costs increased, automation flourished.

4.2 A brief history of automation in the dairy

In everyday parlance, the process of performing an activity by a mechanical device is called automation. However, today it would be more appropriate to think of automation as an ongoing evolutionary process of non-human devices taking over human activities. In general, automation could be thought of as a process that started with machines doing the 'dirty and heavy' work, while humans maintained the function of telling the machines what to do. This is the *first wave: mechanization*. In mechanization, the machine is performing the physical execution of work while the human being maintains control. The next stage is the *second wave: automation*. Here, the machine is taking over a substantial portion of the routine control functions. The human being still maintains some, albeit insignificant, control functions such as turning the machine on and off. Finally, the process of machine takeover reaches its maximum in the stage of *the third wave: cybernation*. In cybernation, the 'system' is the ultimate controller. In this system, the human being may or may not have a decisive role to play.

The dairy industry followed the same path of automation of the Industrial Revolution era that began in Great Britain at the turn of the twentieth century. This started with the well-known automatic looms and knitting machines that led to the notorious revolt of the Luddites in nineteenth-century England, which caused the destruction of entire factories, the Industrial Revolution spread virtually undeterred, and even welcomed, all over the world. In the US, Henry Ford, in his famous assembly line, elevated the role of the machine to its maximum. Ford, in a very clever way, enticed workers to accept the machine as a partner. The assembly-line worker in the Ford factory saw the machine as a friendly device that took the physical-exertion part of the work off his shoulders. In addition, Ford shared with the worker the benefits of machine use, i.e. a higher productivity, by raising wages and lowering working hours. Finally, Ford's automation of the assembly line benefited both customer and society at large via lower prices for a product that became the most desirable acquisition.

During the first quarter of the twentieth century, Ford's concept of the assembly line, that later became known as Fordism, became identified with 'progress.' It seemed that everybody won. The world was, virtually overnight, divided into two parts: the industrialized or developed and the non-industrialized or underdeveloped. This process of substituting mechanical energy for animal and human muscle energy continued in the developed world until the middle of the twentieth century and is still going on in the so-called developing world. Although negative human reactions have not yet reached the level of the Luddite movement of the previous century, worker enthusiasm for the benevolent 'partner' has diminished considerably.

The 1950s ushered in the new industrial revolution. This new revolution is known by many names. Initially, it was named 'The Second Industrial Revolution or The Post Industrial Revolution', to differentiate it from the previous revolution which was known as The First Industrial Revolution. Later on, as it became apparent that this revolution was very different from the pervious one, it was renamed 'The Information Revolution', to emphasize the fact that this revolution had at its centre not energy but information. In short, it suddenly dawned on the experts that these new machines were not aimed at lifting the burden off the shoulders of the human being. Rather these new devices were aimed at lifting the burden off the 'minds' of the workers who were charged with keeping an eye on the energy-driven machines.

These new workers were buried in huge piles of paper containing voluminous reports on the time, costs and benefits of the various uses of these machines. After World War II, as industrialization was galloping, it became apparent to the captains of industry, governments and academics that these workers, and their bosses, were drowning in the vast rivers and lakes of data but had little useful information and knowledge. Gradually, but steadily, a new machine was developed that was exclusively designed to 'crunch' huge amounts of facts and tabulate the results in forms easily consumable by the human mind. These new facts came to be known as 'data', and the new machines were christened 'Electronic Data Processing' (EDP) equipment.

As knowledge of the miraculous capabilities of these machines increased, humans added an increasing number of 'mind like' functions to these machines. In a clever journalistic facts-based fiction, Kurt Vonnegard described his observations of a large American company's introduction of the new 'mind automation' portraying it as the new Luddite era. In his fictitious city, there were two kinds of people. On the one side of the river, were the managers, and on the other were the people. The people, fed up with the managerial delirious excitement with the new machines, formed The Ghost Shirt Society and, by following the footsteps of their distant predecessors of a century before, went around the city destroying the new machines. Some half a century later, this excitement with the new machine reached the level of its forefather, Fordism, and came to be known as Wienerism in honour of Norbert Wiener, the father of Cybernetics (Wiener, 1948).

4.3 Factors contributing to automation

As with all human endeavours, automation in its contemporary form is the result of the culmination of many related developments. In the early days, the cost of labour was the dominant factor. As scientific discoveries became marketable innovations, the cost of the machine diminished precipitously. In addition, the reliability, accuracy and ease of use of the mechanical devices skyrocketed. These cost and process considerations are primarily the so-called internal factors or drivers of automation. The entire dairy industry, from the isolated farmer in the rural areas to the organization executive in the big city suburbs, immediately saw the benefits of the machine. Automation meant that time was saved, and since time is money, that also meant that money was saved, and consequently profits increased.

These technological and economic developments were the facilitators of automation and were more or less the results of other more powerful forces. These are the external forces and are related to the shifts in the demographic and the accompanying changes in people's lifestyles. The rapid increase in population at the turn of the twentieth century combined with an unprecedented industrialization led to a considerable uncoupling of the place of production of dairy products and the place of consumption. In order to accommodate this huge movement of people from the places near the source of production, i.e. the animal, the dairy industry had to build factories for processing either closer to the animal or closer to the market. In either case, dairy products, raw, semi-finished or final products, must be transported over large distances. Given the fragile nature of these products, ways had to be invented of protecting and preserving their quality. One cause of contamination of dairy products is contact with humans. Thus, for safety reasons, the need to minimize human contact led to automation. Legislation was quick to recognize this and acted swiftly.

A host of other factors dictate that the dairy industry must continually strive to automate. For example, globalization and increasing competition, the global factory, the increase number and variety of products that each dairy unit is called to offer in the market and the need to customize products are some of the external drivers of automation. Siemens Logistics and Assembly Systems recently published a report entitled 'Improving Food Safety Through Automation', which is a free download from the company's website (www.usa.siemens.com/logisticsassemby). The report identified the following six emerging trends and also indicates how automation helps.

4.3.1 Six factors driving automation

In addition to food safety and security concerns, the food and beverage industry faces at least six additional emerging drivers that further accentuate the importance of implementing automation as soon as possible. Integrated automation can help food and beverage manufacturers address these six emerging trends while providing important safety benefits.

- Trend 1: consumer demands;
- Trend 2: customer demands;

- Trend 3: labour availability and reliability issues;
- Trend 4: labour laws;
- Trend 5: regulatory requirements;
- Trend 6: supply-chain management.

4.4 Benefits of automation

Some of the benefits of automation are rather obvious, while others need to be explained and even elaborated on. In general, the benefits of automation derive from the elimination of the possibility of contamination and the creation of health hazards for humans and animals. In addition, cost savings are passed on to the consumer in the form of lower prices and therefore higher availability to those less fortunate. At the micro- or factory level, automation serves several more specific purposes such as

- the production of consistent quality goods;
- the reduction in production costs;
- the flexibility to meet market demands;
- the adoption to constantly changing legal demands;
- it ensures high and consistent product quality;
- it increases production;
- it reduces losses;
- it guarantees a safe operation;
- it ensures manpower savings; and finally
- it allows for efficient production planning, execution and reporting.

Overall, the mission of automation in a dairy is to meet the company need to produce safe goods, in a cost-effective manner, with high and consistent quality and in the appropriate quantities to meet market demand. Successful execution of this mission calls for a holistic approach or a systems approach to automating the production of these highly sensitive and essential dairy products. In the rest of this chapter, we will present the conceptualization, design and implementation of an integrated control system for a dairy factory. Although this system incorporates all the traditional mechanical energy-driven devices, its real strength lies in the latest accomplishments of the science of information and telecommunication industries.

4.5 Conceptual framework of an automated system

We said in the previous paragraph that successful execution of the mission of the dairy industry calls for a holistic approach or a systems approach to automating the production of its highly sensitive and essential dairy products.

4.5.1 What is a system?

The concept of a 'system' (Fig. 4.1) has originally been borrowed from the management literature in the field of engineering. A generic definition is that 'a system is a set of

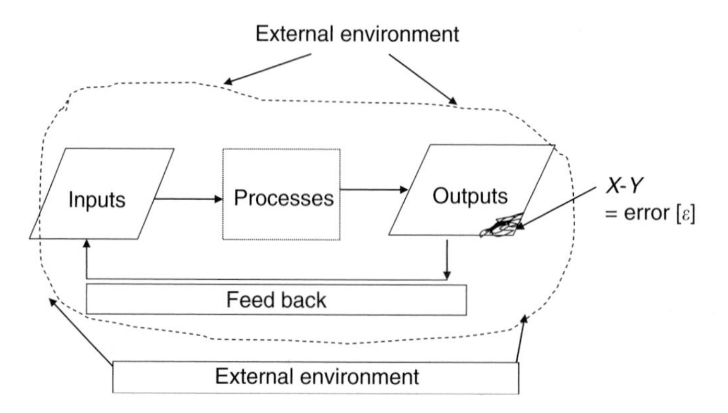

Fig. 4.1 System and components. Adapted from: Management Systems: Conceptual Considerations, Schoderbek, Kefalas and Schoderbek, R.D. Irwin, 1975, 1980, 1985, 1990, 1995. Permission of the author.

objects together with relationships between the objects and their attributes'. This definition was used as the basis on which a new, more pragmatic definition was developed. This definition adds the concepts of 'whole' and 'environment' to satisfy the condition sine qua non of the survival of biological systems, which is openness to the system's environment. A system is now defined as 'a set of objects together with relationships between the objects and their attributes connected or related to each other and to their environment in such as way as to form an entity or a whole' (Schoderbek, P.P, Kefalas, A.G., and Schoderbek, C.G., Management Systems Conceptual Considerations, Business Publications, Inc., Plano Texas, 1975, P. 30).

We shall briefly examine one by one the main components of the definition of a system. A system is defined as 'a set of *objects* together with *relationships between* the objects and their *attributes* connected or related to each other and to their *environment* in such as way as to form an entity or a whole.'

4.5.2 Objects

Objects are the components of the system. From the static viewpoint, the objects of a system would be the parts that constitute the system. From the functional viewpoint, however, a system's objects are the basic functions performed by the system's parts. Thus, the objects of a system are the inputs, processes, outputs and feedback control.

4.5.2.1 Inputs

Inputs to a system may be matter, energy, humans or simply information. Inputs are the start-up forces that provide the system with its operating necessities. Inputs are the energizing forces of a system. Inputs can be of three main kinds. Serial inputs are the outputs of another system with which the focal system is serially or directly related. These serial or online inputs are usually referred to as 'direct-coupled' or 'hooked-in' inputs. Random inputs represent sets of potential inputs from which the system can choose. The choice of a random input is determined by the probability that it satisfies the particular need of the focal system. In other words, the choice will be determined by the degree of correspondence between the input needs of the focal system and the attributes of the available inputs. The actual selection by the focal system is then based upon the probability distribution and the decision criterion of the focal system. Finally, feedback inputs are reintroductions of a portion of the system's outputs. These portions represent deviations, called errors (ε), between the system's goal (X) and its actual outputs or performances (Y).

4.5.2.2 Processes

Processes are the activities that transform inputs into outputs. As such, they may be machines, human muscular, mental or chemical activities, a computer or a combination of all of the above. White Box processes are activities that can be described in a detailed manner and when the potential outcome is predictable. Black Box processes are situations where these activities are indescribable in detail. In these processes, combinations of inputs may result in different output states. Processes may also be assemblies, where a number of different inputs are converted into one or a few outputs (e.g. a car assembly plant), or they may be disassemblies, where one input is converted into a number of outputs (e.g. a meat-packing plant).

4.5.2.3 Outputs

Outputs are the results of the processes. They can fall into one of the following categories: intended or planned, unintended but avoidable, unavoidable and feedback. Good systems designers pursue maximization of the intended or planned outputs and minimization of all the others.

4.5.2.4 Relationships

Relationships are the bonds that link objects and/or sub-systems together. Although each relationship is unique, and should therefore be considered in the context of a given set of objects, still the relationships most likely to be found in the empirical world belong to one of the three following categories: symbiotic, synergistic and redundant.

A symbiotic relationship is one in which the connected systems cannot continue to function alone. These types of relationships can be parasitic or unipolar (i.e. the relationship is running in one direction) or bipolar in which both systems derive a mutual benefit (i.e. the relationship is running in both directions). A synergistic relationship, though not functionally necessary, is nevertheless useful because its presence adds substantially to the system's performance. Synergy, in its narrow meaning, means 'combining energies' or 'combined actions'. In systems terminology, the term synergy means more than just combined action. Synergistic relationships are those in which the cooperative effort of semi-independent sub-systems taken together produces a total output greater than the sum of their outputs taken independently. A colloquial and convenient expression of synergy is to say that $2 + 2 = 5$ or $1 + 1 > 2$. Finally, redundant relationships are those that duplicate other relationships. The reason for this duplication is reliability. Redundant relationships guarantee that the system will operate all of the time and not just some of the time.

4.5.2.5 Attributes

Attributes are properties of the objects and their relationships. They manifest the way that something is known, observed or introduced in a process. Attributes are of two main categories: defining or accompanying. Defining attributes or characteristics are those without which an entity cannot be identified. Accompanying attributes are those whose presence or absence would not make any difference with respect to the use of the term describing it at a given moment in time and for a given purpose. However, these attributes may become very relevant when the circumstances have changed substantially.

4.5.2.6 Feedbacks

A feedback was defined above as an input that represents a deviation, called error (ε) between the system's goal (X) and its actual output or performance (Y). [Error (ε) = $X - Y$]. Feedbacks are of two kinds: negative and positive. It must be said at the outset that the terms feedback and negative and positive are used in a 'special' way and do not correspond to the colloquial meaning. Thus, giving a system a negative feedback does not mean giving 'bad News'. Negative feedback is the reintroduction of an error that carries a message to the system that it must reverse the actions that lead to the error. That is why it is called negative because it asks the system to do the opposite of what it has been doing thus far. The aim is to *minimize* the error. A positive feedback on the other hand tells the system to *maximize* the error. Negative or error minimization feedback is the backbone of control system. Positive or error maximization feedback is also called an amplifier and is the backbone of growth processes.

4.5.2.7 Boundary

Thus far, the main objects of the systems and their attributes have been explained. There remains only to explain the component of the definition called the environment. Before doing so, however, we must deal with 'that' which 'separates' the system from its environment within which it operates. A boundary is that which demarcates the system from its environment. A functional definition of a boundary is given as 'the line forming a closed circle around selected variable, where there is less interchange of energy (or information) across the line of the circle than within the delimiting circle' .Two things must be noticed when looking at Fig. 4.1. First, the line that demarcates the system from its environment is arbitrary (i.e. is determined by the observer or the designer of the system), and second the line is not solid (i.e. it is porous so that 'stuff' can go through it).

4.5.2.8 Environment

A system operates within an environment. As with the concept 'system', the concept environment is also misunderstood and misused. The vernacular use of the term evokes in one's mind images of green trees, flying birds, grazing cows, running rivers, and blue skies. In the systems language, this is clearly a case of mistaking the part for the whole.

In other words, the term environment is a much wider concept, which encompasses the popular term ecology. Environment is defined as all factors which satisfy the following two conditions: (1) they affect the functioning of the system; this is the relevancy condition; and (2) they are beyond the immediate control of the management of the system; this is the controllability condition. Thus things, events, happenings, and so on, which are 'out there' but do not relate to the system's functioning are not part of the environment. By the same token, things, events, happenings, and so on, which do relate to what the system does, but which the system does not control, are part of its environment. If the system does control them, then these become part of its resources.

4.6 Stages in automation in the dairy

Automation in the dairy industry followed the same process described at the opening section of this paper.

4.6.1 First wave: mechanization

The first point of automation was the feeding and milking places. Farmers adopted automatic feeders and milking machines easily, recognizing early on the time- and effort-saving features of these devices (Fig. 4.2).

Fig. 4.2 Progress in mechanization and transportation in the dairy: (a) mechanized feeding; (b) mechanized milking; (c) transportation by horse-driven cart; (d) modern transportation by truck.

The next stage in the production process is the transportation to the processing places. Because of the large distances products had to travel, farmers upgraded their horse-driven carts for fully automated refrigerator trucks to ensure safe transportation to the milk factory.

4.6.1.1 Summary

It is obvious from the very brief description of mechanization that the role of the human being is central and that the human is the decision maker. The machine is the 'slave', while the human is the 'master'. Despite the machine's important contribution to the process of milk creation and transportation, it still depends on the human being to tell it what to do.

4.6.2 Second wave: automation

The process of mechanization reaches its climax at the factory. It is here that one experiences the wonders of the human desire to eliminate his own presence from the process of creating the millions of dairy products, from the simple milk carton to the delicious ice-cream product. A sketch of a simplified automated dairy factory is given in Fig. 4.3. Inputs to the system come from the environment via refrigerated trucks. The content is deposited into a series of tanks where various processes are performed. The outputs of these processes are checked by a feedback system, and the results are fed back to the input for possible corrective actions.

4.6.2.1 A typical example

Figure 4.4 presents a simplified sketch of a typical automation system of a dairy plant. A typical system monitors and guides all the activities via computer programs. It is

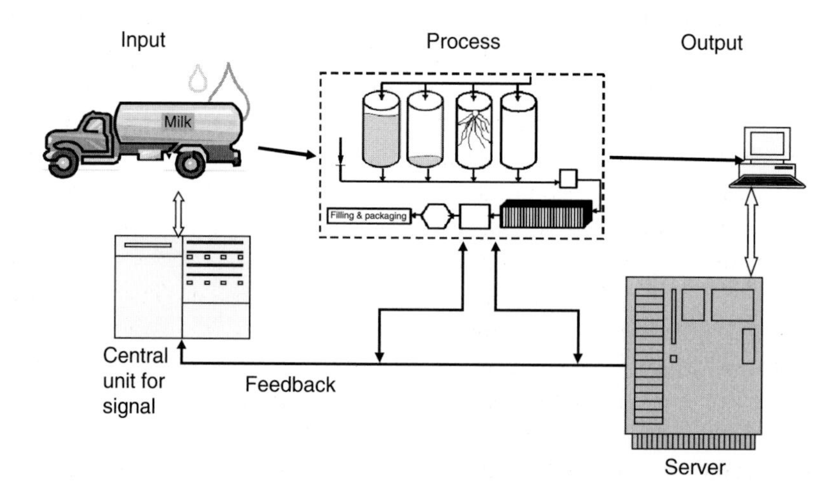

Fig. 4.3 Simplified automated milk processing system. Source: Adapted from *Dairy Handbook*, ALFA-LAVAL, Food Engineering AB, 5-22100, Lund, Sweden, Chapter 23, pp. 319–333.

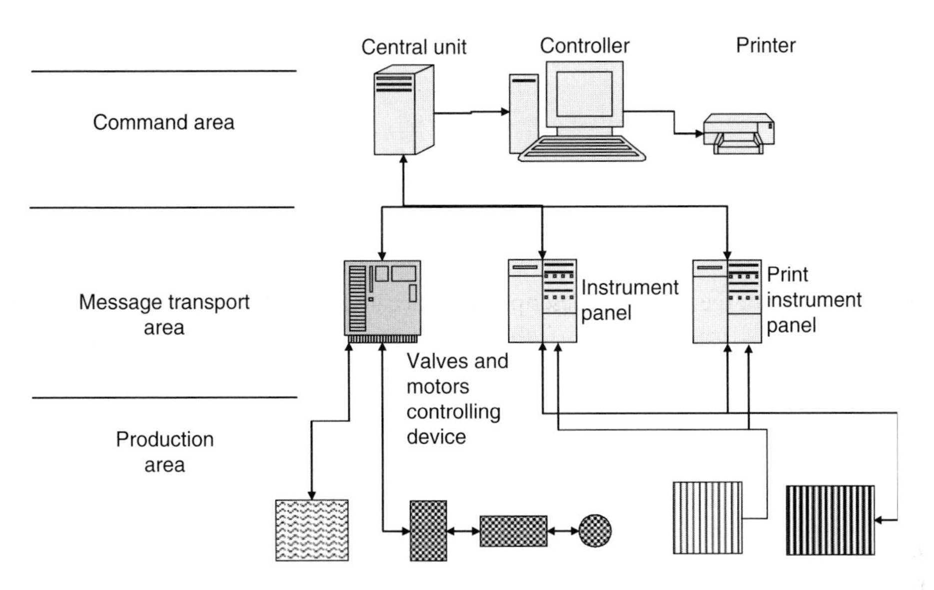

Fig. 4.4 Typical automated system in the dairy.

simple in its operation and user-friendly, and automatically detects any malfunction. It is composed of three areas

(1) The command area.
(2) The message transport area.
(3) The production area.

The first two areas are under the direct supervision of the operator who sits in the control room. The first area includes the central computer and constitutes the heart of the system. The computer recognizes the state of the conversion process and, with the help of input signals from the production area, processes the information, modifies it according to the program and finally sends out commands as output signals.

The first area includes a series of hardware units such as console, colour screen and printer. The operator receives information and issues commands regarding possible modifications in the process. The second area – area of transport and transformation of messages – is composed of instruments that connect the central control with the area of production. These instruments accept the signals from the area of production and change them to signals recognizable by the computer. Subsequently, the computer reverses these commands into analogue signals that can be recognized by the instruments in the area of production. Finally, these signals are changed into forms comprehensible to the operator. The instruments of the second area consist of the valve and motor-controlling devices, the IP (Instrument Panel) that contains various displays of information for the operator such as registration instruments and others, and the process interface that checks the binary and analogue signals.

Finally, the third area is the production area which has various mechanical equipments such as motors, the MCC (Motor Control Cabinet) where the reels and motors are assembled, valves and instruments that detect the state of treatment such as thermometers, manometers, conductometers, switches and meters of level.

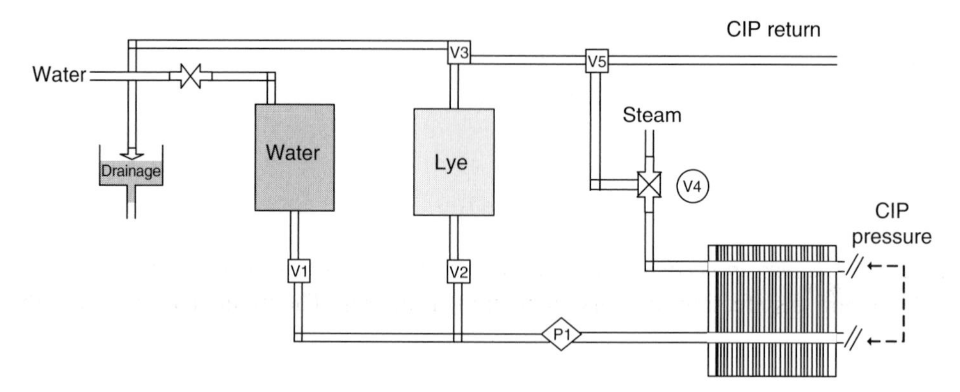

Fig. 4.5 Example of a typical operation: cleaning.

4.6.2.2 Two examples

The following two examples illustrate the different levels of the system. The first is the 'level of control of processes' and second is 'level of control of analogue signals'.

Figure 4.5 depicts a situation with two tanks. The first tank contains water and the second a solution of NaOH that is used in cleaning the product line. There is also a plate exchanger for heating the cleaning solutions.

In order to clean this line, a five-step process is followed:

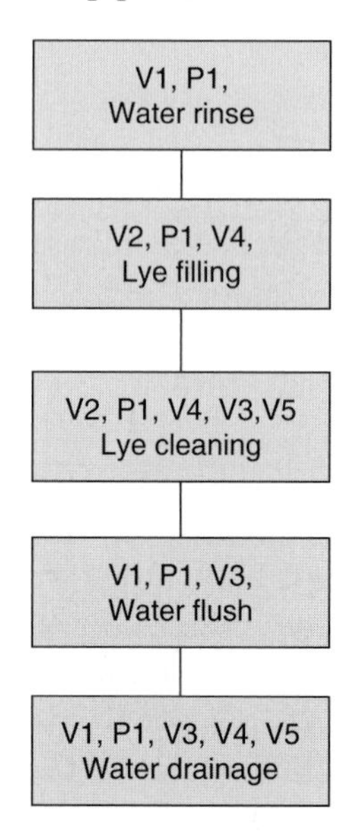

The second case concerns the control levels via analogue serial signals. Here, the case is a simple heat adjuster where, on one side of its plates, there is the product flow and, on the other side, the thermal source, which in this case is hot water. The product temperature is determined manually via the continuous monitoring of the thermometer (TT) that shows the product temperature and manual intervention of the valve that determines the hot-water flow. In the automated case, the indication of the TT is transported to the Controller (TIC). Here, this indication is immediately changed into an analogue signal and is compared with the set point. The message is subsequently transmitted to the valve which controls the flow of hot water to determine the appropriate flow and hence the temperature of product at the target level (Fig. 4.6).

Figure 4.7 provides an example of a more sophisticated system that shows in detail the various processes sketched in the previous two examples.

4.6.2.2.1 Cleaning-procedure automation: a detailed description. Cleaning procedures and disinfection are very important in dairies. The maintenance of a high

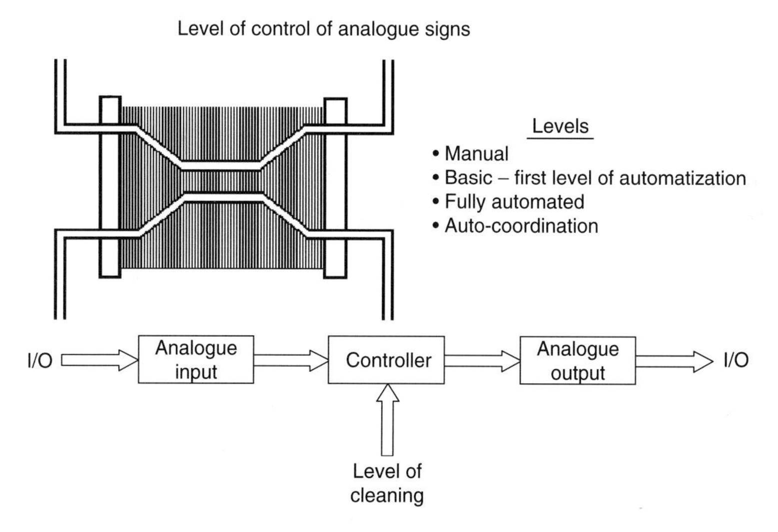

Fig. 4.6 Level of control with analogue signals.

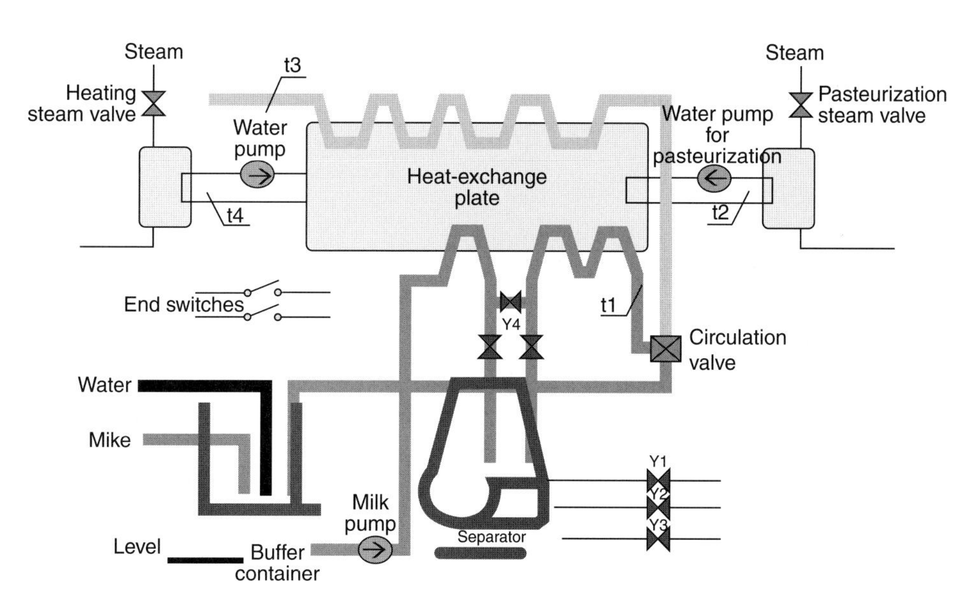

Fig. 4.7 A more sophisticated system. www.btc-automation.lv/.../big/food_10_1_en.jpg.

product quality depends, to a large extent, on appropriate cleaning and disinfection of the equipment and all factory areas. Milk, as the principal raw material for a dairy, is the best substrate in which bacteria can multiply rapidly. Milk, either as a raw material or as a produced good, flows through the whole mechanical machinery of the factory and comes into contact with milk receptors, pipes, tanks, pasteurizers, separators and filling machines. As a result, effective cleaning is required on a daily basis so that a high product quality can be guaranteed.

Even the smallest error can have catastrophic results for the company. Faulty cleaning can result in very large volumes of finished products being wasted. Also, faulty cleaning may have disastrous consequences in the area of public health, with equally catastrophic consequences for the dairy industry itself. For these reasons, the dairy industry has developed a number of sophisticated systems for cleaning, some of which are briefly addressed below.

These systems can be grouped into the following categories:

(1) The first category can be divided into two sub-categories, depending on whether any disassembly takes place during cleaning. Cleaning-In-Place Systems (CIP): the whole cleaning procedure is carried out in a circuit without equipment disassembly. CIP is an integrated system consisting of tanks, valves, pumps, detectors and automated equipment operated under precise procedures which control
   - temperature;
   - flow;
   - pressure;
   - density and contact time of the cleaning solutions;
   - Cleaning Out of Place (COP): This cleaning procedure works by disassembling the parts to be cleaned. These parts are then placed in specific containers that are cleaned in an open circuit.

(2) The second category is related to the automation level:
   - manual systems;
   - semi-automatic systems;
   - fully computerized systems.

The above-mentioned systems can be

- centralized;
- decentralized.

4.6.2.2.2 Example of a fully computerized system.

(1) Application: Automated cleaning in place of pipes, valves, tanks, heat exchangers and filling machines.
(2) Capabilities: Different cleaning programmes for different machinery and equipment, i.e. a different programme for cleaning the storage tanks and the heat exchangers.
(3) Equipment:
   - Storage tank for water equipped with:
      (i) level controllers (high-level–low-level electrode);
      (ii) automated feeding valve (input); and
      (iii) automated emptying valve (output).

- Storage tank for detergent solution equipped with
  - (i) level controllers (high-level–low-level electrode);
  - (ii) air vent;
  - (iii) sampling valve;
  - (iv) agitator;
  - (v) input and output valves;
  - (vi) level transmitter.
- CIP lines consisting of
  - (i) heat exchanger;
  - (ii) dosing pumps, the concentrated agent;
  - (iii) forward and returning pumps, the cleaning solution;
  - (iv) valves;
  - (v) electrical and electronic equipment.

Automation's contribution to CIP systems is the accuracy in the regulation of all those parameters that are essential for an effective cleaning. Those parameters are related to

- predetermined temperature of the cleaning solutions or detergents;
- processing time of each cleaning solutions or detergents;
- sequence in which the cleaning circuits will be carried out;
- security in opening and closing the cleaning valves;
- easy adjustment in the new demands (change of cleaning agents, time, sequence, etc.).

4.6.3 Third wave: cybernation

In the two previous waves of substitution of mechanical participation for human activities in the process of milk producing, transporting, processing and transforming into consumer products, machines were given the role of 'slave', while the human was granted the role of 'master'. Despite the fact that, as we saw in the Second Wave: Automation, machines were performing many decision-making functions, most of them were coupled or forced decisions. In other words, the system's designer, a human, had built in the decision. For example, the machine has no choice but to open a valve when the weight or the pressure has reached a certain level.

In The Third Wave: Cybernation, the system itself has a choice of selecting among many alternatives. The human is part of the system; not apart of it, as in the case of mechanization and conventional automation. An automated dairy factory is essentially an error minimization, closed loop, negative feedback system. These types of systems are called cybernetic systems. In cybernetic systems, the system maintains control in the act of and by the act of going out of control. It is a self-regulating system that depends on the existence of a difference between the goal and the actual performance, i.e. the error.

Figure 4.8 provides a sketch of a negative-feedback control system. The control object, that is, that which is to be controlled, must be maintained between an upper and lower limit. Performances over the upper limit (overshooting) or below the lower limit (undershooting) are brought back within the threshold by a process of detecting

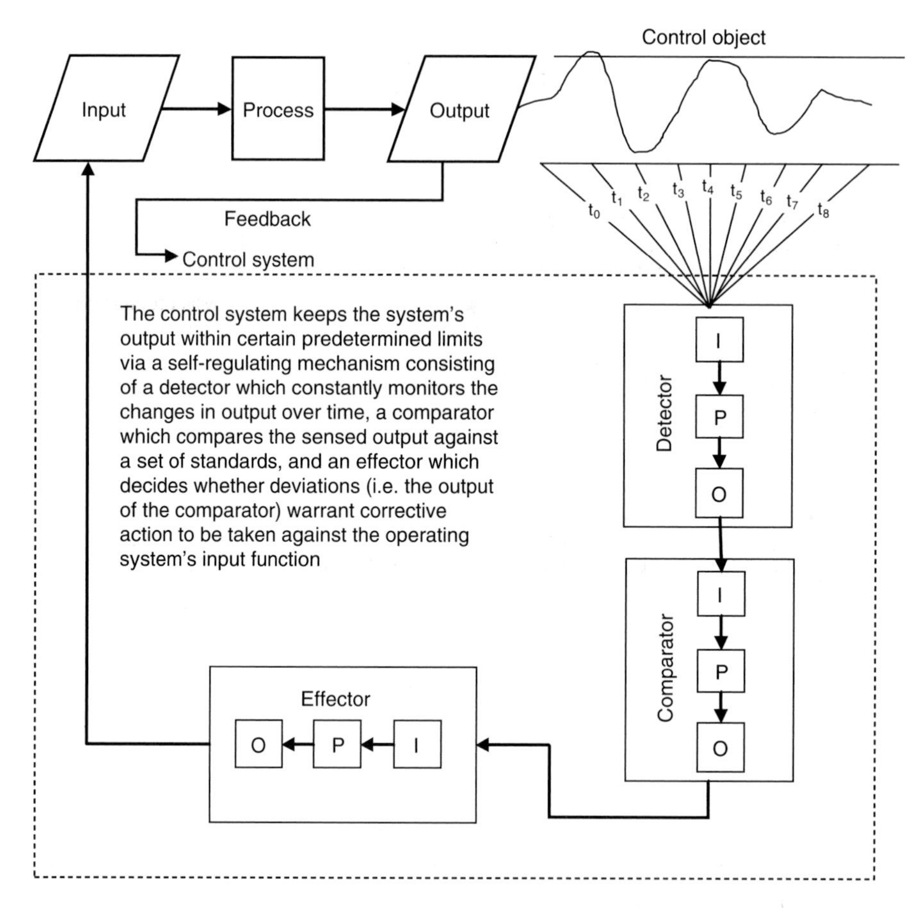

Fig. 4.8 Negative feedback control system. Adapted from: Management Systems: Conceptual Considerations, Schoderbek, Kefalas and Schoderbek, R.D. Irwin, 1975, 1980, 1985, 1990. Permission of the author.

the error, comparing it and then issuing a corrective action that changes the behaviour of the system.

4.6.3.1 Some new applications

In this final section, we offer two proprietary systems that deal with the latest regulatory requirements that impact the entire food and beverages industry. These systems are currently at the Alpha Test stage. The first system deals with the provision of information regarding the entire company's compliance with all safety requirements as well as with Knowledge Management and the Management of Crisis. The second system deals with the track-and-tracing requirements.

4.6.3.2 Lotus integrated food safety system

The first system is an integrated food safety system developed by Lotus Consulting, Athens, Greece (www.lotusconsulting.gr). Figure 4.9 provides a diagrammatical

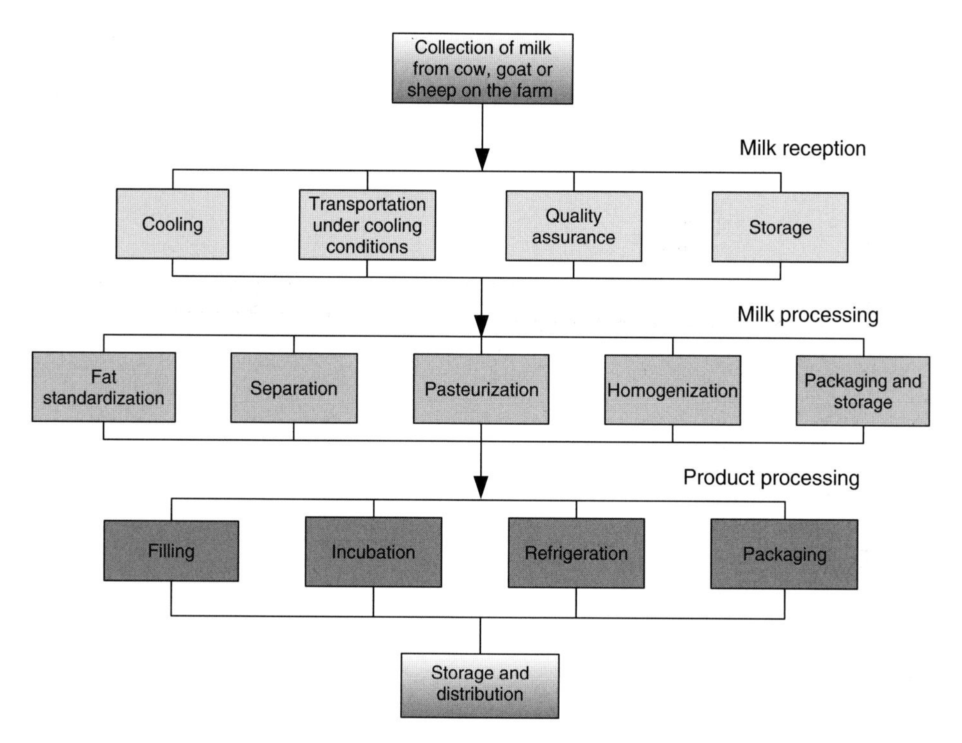

Fig. 4.9 Milk production from animal to market.

presentation of the various stages involved in the process of converting milk into consumable products. In every step in this process from the milk reception to storage, there are opportunities for automation. In some of the sub-processes, second-wave automation is used. In others, more sophisticated and more reliable systems are needed. Finally, the need for fully automated systems reaches its climax when the desired control function is elevated to the administrative whole enterprise level. The Lotus Integrated Food Safety System (LIFSS) is offered here as an example of a higher-level integration of the administration of the whole company. The system aims at satisfying all legally mandated safety requirements by maximizing knowledge management and crisis management.

There is little doubt that the issue of producing safe food safely is among the primary concerns of modern society. When humans lived in their rural homes, food gathering and preparation were the exclusive concern and obligation of the heads of the household. Conventionally, men gathered food, and women prepared it and served it to the family. Thus, food safety issues were confined mostly to a single family. The simplicity inherent in the traditional foods made the safety issue rather simple. Foods were consumed almost immediately or within a short period of time.

Today, the job is not easy. Opportunities for errors and mishaps exist in everything. The complexity inherent in food production, storage, processing and delivery is enormous. Thus far, dairies have employed a piecemeal approach. They did a little bit of this and a little bit of that, hoping to comply with laws and regulations. Lotus Consulting has learned from experience that firms need an integrated approach.

A systems approach, so to speak, where the various aspects involved in food production and safety are seen as a series of interrelated activities that form a whole, or a system, that is in a continuous interaction with its component sub-systems and with its external environment.

LIFSS takes a *holistic approach* to the process of producing and delivering safe and nutritious food. This approach sees the series of processes that are involved in food producing, processing and delivering as feeding on each other. The accomplishment of one step energizes the next. In other words, the output of one sub-system becomes the input to the other sub-system, which energizes it to create its own output that in turn becomes the input to the other and so on.

At the centre of the system is not the mechanical hardware that was the heart of the Second Wave Automation; rather, the basic ingredient of LIFSS is information. It is the creation and dissemination of accurate and timely information among the various levels and departments of a company that must be managed effectively and efficiently.

Before going into a detailed description of LIFSS, let us state some basic assumptions:

(1) We assume that the law must be strictly obeyed.
(2) We assume that things can go wrong and crises will arise.
(3) We assume that we, as an industry and in cooperation with the state, have made progress and created a number of programmes and systems such as ISO14001 or Hazard analysis and Critical Control Point (HACCP), finally.
(4) We acknowledge that a company's safety depends on the existence of people who have the necessary knowledge and skills and the will to secure the safety of the entire system.

Food enterprises do not usually develop an integrated system for safety. They usually implement a number of safety-management systems. While it is commonplace for businesses to be insured against known common hazards, such as fire or earthquakes, hazards derived from their own operations are neglected or ignored. These latter hazards could lead to disastrous results for the business itself. LIFSS aims to minimize the probability of occurrence of such hazards and their consequences by taking a systems approach to the issue of safety in the food industry.

4.6.3.3 Goal of LIFSS

The goal of LIFSS is to ensure that certain key control objects are kept within certain predetermined limits. These control objects belong to one of two categories.

(1) Category A: Mandatory requirements, the various concepts or 'things' that are prescribed by the national, regional and international legislation. These are known in our professional language Critical Control Points (CCP).
(2) Category B: Voluntary requirements, firm-specific things that have been identified by management as been necessary for the effective food production and distribution. These are called Critical Decision Points (CDP) at Lotus.

LIFSS attempts to integrate both of these two categories into a meaningful, fast, effective and user-friendly, computer-aided system.

4.6.3.4 Sub-systems

The main components of LIFSS are shown in Fig. 4.10.

4.6.3.4.1 Sub-system 1: Standardized Safety Management Systems (SSMS). This sub-system includes the 'standardized' management systems, based on worldwide approved standards, principles or guidelines. Such systems are ISO14001, HACCP, OHSAS 18001, GHP and others. Most of us are familiar with these systems, and most firms in our industry have in place most if not all of them. These system describe in great detail the various Hazard Critical Control Points and the specific activities that must be performed to ensure that specific variables are within certain prescribed limits, the frequency of verifications and the results with their likely consequences. The logic of the HACCP is applicable to the other systems. All these systems have all the necessary elements for easy and definite assurance of compliance. They have:

(1) concrete definitions;
(2) concrete measurements, that is, specific values expressed in specific and well-accepted quantitative terms;
(3) concrete upper and lower limits of accepted safety levels; and finally
(4) concrete enforcement procedures and specific penalties for unacceptable safety levels.

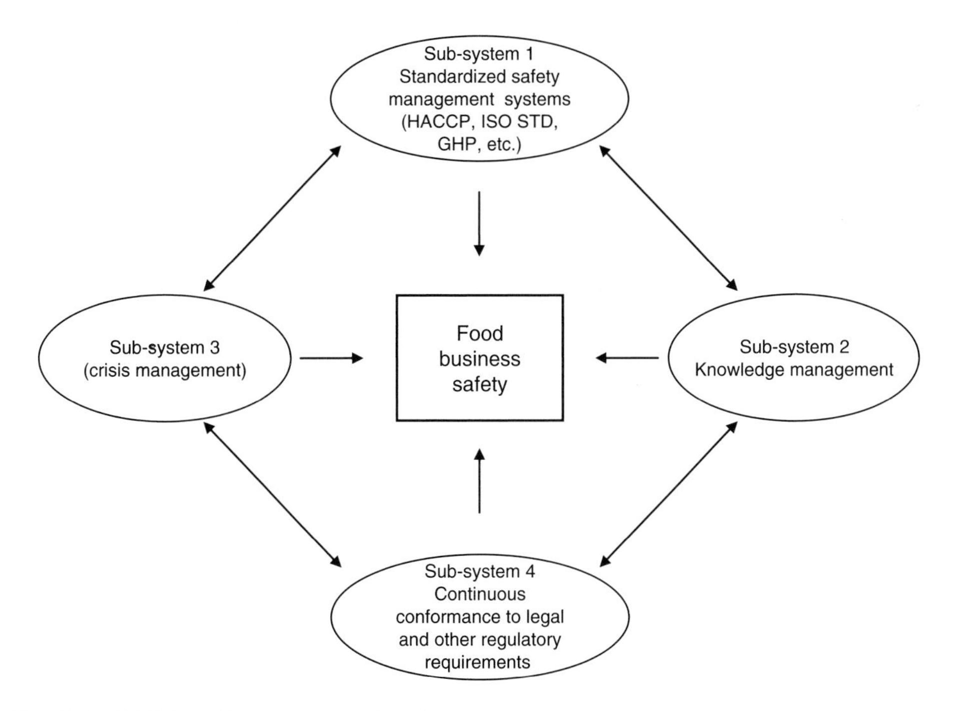

Fig. 4.10 LILFSS and its component sub-systems.

4.6.3.4.2 Sub-system 2: Knowledge Management System (KMS). Frequently, failures in safety matters are the result of omissions and/or wrong decisions and actions taken by the firm's human resources. Cognizant of the key role that the human plays in the quest for safety on the entire company level, LIFSS acknowledges that the continuous improvement of human resources knowledge is a necessary condition. For a company, keeping its people always knowledgeable, alert and quick action oriented is a must. No system can perform its function unless:

(1) people know what the system is designed to do;
(2) continually keep an eye on the system's functions; and
(3) quickly and effectively report the results to the proper company unit.

These ABCs are the soul of LIFSS's Sub-system 2: Knowledge Management System. Luckily, management educators, consultants and practitioners have over the last ten or so years developed a new discipline that deals with the creation, codification and dissemination of knowledge. Recognizing that knowledge is a firm's most valuable resource, organizations spend large sums of money daily in educating their human resources in the process of creating and using knowledge. In brief, knowledge creation and usage are not as automatic as most of us believe. This process must be designed, explained and 'sold' to all organizational participants. In addition, it must be made an integral part of the company's reward system. People will not share their knowledge unless they are rewarded for it. Otherwise, they will hoard it like gold.

LIFSS's Sub-system 2 KMS complies with the requirements explained above that are present in Sub-system 1; namely it has:

(1) concrete definitions;
(2) concrete measurements, that is, specific values expressed in specific and well-accepted quantitative terms;
(3) concrete upper and lower limits of accepted knowledge levels; and finally
(4) concrete enforcement procedures and specific penalties for unacceptable knowledge levels.

Granted, the measurements used in the KMS are not as 'concrete' as those used in Sub-system 1, and they are more of a 'qualitative' nature. However, they are accepted and 'bought' by human resources. They are part of the corporate culture. Therefore, those who do not 'obey' these 'standards' are expected to pay the price for it.

As in the Sub-system 1 activities, so in the Sub-system 2 there are Critical Control Points. These are expressed in the minimum level of knowledge an employee must possess so that they will not considered 'dangerous' for the overall safety of the company.

4.6.3.4.3 Sub-system 3: Crisis Management System (CMS). One of the basic assumptions made in designing LIFSS is that crises will occasionally occur. The history of the corporate world is replete with occasions of crises. Starting with the widely publicized Johnson & Johnson Tylenol crisis in the early 1980s and ending with the most recent Coca Cola dioxin case in Belgium in the 1990s, there are perfect examples of good vs. bad management of crisis. In the first case, the CEO was widely praised for the effective way in which he handled the company's crisis. Coca Cola's CEO was not as lucky. The clumsy way Dough Invester handled the case by refusing to accept the occurrence and take quick and effective corrective action damaged the company's reputation and its cash flow, and contributed towards the loss of his job.

Again, luckily, management educators, consultants and practitioners have over the last few decades developed a new discipline that deals with the prevention, early detection and swift and effective management of crises. Recognizing that dealing with a crisis in a disorganized manner can have severe consequences for the company's reputation and heavy financial burdens, Lotus has incorporated in its LIFSS a component Sub-system 4: Crisis Management System (CMS).

As in the case of Sub-system 2, the measurements used in the CMS are not as 'concrete' as those used in Sub-system 1: they are more of a 'qualitative' nature. However, they are accepted and 'bought' by the human resources. They are part of the corporate culture. Therefore, those who do not 'obey' these 'standards' are expected to pay the price for it. As in the Sub-system 1 activities, so in the Sub-system 3 there are Critical Control Points. These are expressed in the minimum level of crisis management skills an employee must possess so that they will not be considered 'dangerous' for the overall safety of the company.

4.6.3.4.4 Sub-system 4: Continuous Monitoring and Compliance System (CMCS). This sub-system monitors on a continuous basis the outcomes of the various sub-systems that are the domain of Sub-system 1. This sub-system is itself composed of three interconnected sub-systems. Sub-system 4.1, The Monitor, monitors on a continuous and automatic basis all activities of Sub-system 1. The results of this monitoring are the inputs to Sub-system 4.2, The Comparator, which are, in turn, compared with the legally set 'standards' that must be met. Finally, based on the results of these comparisons, Sub-system 4.3, The Corrector, recommends corrective actions which are fed back to the Sub-system 1. These corrective actions are the managerial activities that must be performed to bring the entire system back within the predetermined acceptable limits. These are the corrective actions that management must take to secure the functioning of the entire firm-wide system.

A reflective look at this system reveals that timely and accurate information is of utmost importance. Since dynamic systems have the tendency to overshoot and undershoot (i.e. to go out of control), LIFSS proposed here must be a 'metasystem', that is, a 'system of systems' that minimizes the information slack, i.e. the delays between detecting an error and the corrective action. The more delays there are, that is the more time we take to measure, compare and take corrective action, the higher the probability that overshooting and undershooting will occur with consequences that approach 'crisis levels'.

To secure the minimization of information delays, LIFSS introduces two additional sub-systems. The third sub-system aims at securing a well-informed and action-oriented human resource corporate body. The forth sub-system aims at securing fast and effective management of crises.

4.6.3.5 Track & Trace System – TRACER

An integrated track-and-trace systems belongs to the category of systems called Manufacturing Execution Systems (MES) and must have the capabilities of identification or coding, monitoring the production flow and the associated processes and quality controls as well as the collection and management of the necessary information. The system TRACER uses its own software, which can be used in many connected work stations and must cover the following general demands or specifications:

- Coverage of the specific demands and processes of the enterprise.
- Harmonious embedding and collaboration with the existing systems and processes of the business as well as with all information systems and automation systems such as enterprise resource planning (ERP), etc.
- Possibility of automatic connection with the existing coding systems such as various inkjet printers, thermal label printers and so on.
- Absolute collaboration with the various quality systems.
- Minimization of human intervention so as to avoid errors. Whenever it is necessary to have human intervention, the operations must be simple and easy to understand.
- Scaling possibility so that the system can satisfy all future needs.
- Capability of discrete management of the required information. For example, a company may choose to select a very detailed track-and-tracing system for its own use and at the same time provide 'outsiders', such as government agencies, only a portion of the available information, just enough to satisfy the regulatory requirements or other contractual obligations.
- Capability embodiment of the new technologies for automatic data capture, storage, transmission and in general data management via the INTERNET, Wi-Fi, RFID, DNA, and so on.
- Capability of communication with other information systems using XML < EDI and so on.

In general, the system TRACER (Fig. 4.11) follows closely the requirements of the European Union's FOODTRACE committee (www.eurofoodtrace.org).

4.6.3.6 Summary

Third Wave Automation depends on and capitalizes the cognitive capacities of the human being. During the last 50 or so years, a new science was created that aimed at enhancing human cognitive capabilities by offering a technology that assists humans in storing and recalling information. Thus, 21st-century organizations are essentially knowledge creating and using systems, which then develop capacities to learn, thus becoming learning organizations. It is this ability to learn that replaces the ability to automate the functions of mechanical devices that characterizes the automation quest of the 21st-century enterprises. Cybernation is the pinnacle of human accomplishments. Cybernation makes humans part of, not apart of or external to the system, thereby creating a suprasystem or a metasystem, a system of systems, that learns from its own mistakes and evolves to higher levels of complexity. Today's world requires a management system that has the requisite variety to deal with its enormous complexity.

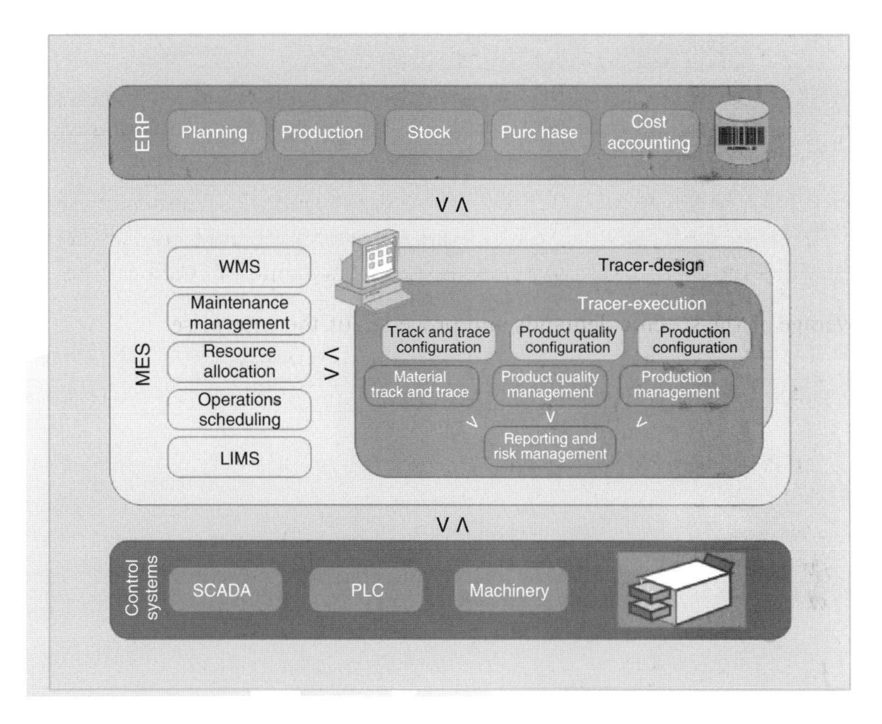

Fig. 4.11 Track-and-trace system (TRACER). Reproduced by kind permission of Theodorou Automation SA.

4.7 Lotus integrated safety system – a case study in the dairy industry

First of all, we will choose what to investigate: the *control domain*. In this example, 'ice cream production' is chosen.

The process of producing ice cream consists of several steps which can be examined separately. In this case, the step: 'Purchase of Milk for the Production of Ice Cream' is examined (Fig. 4.12).

There are several parameters contributing to the safety of the purchased milk and each one can also be examined separately. Thus, the appropriate 'control object' has to be chosen (Fig. 4.13).

Let us, for example, choose to investigate what the situation was, regarding the level of *Staphylococcus*, in the purchased milk. As sub-systems 1 and 4 are flashing, this means that our control object is related to some parameters contained in these two sub-systems (Fig. 4.14).

Thus, sub-system 1 highlights the process step, and sub-system 4 highlights the legal requirement for this measurement. It can be seen that, in the control measure column, these parameters relate to all management systems. For example, effective supplier assurance relates to ISO 9000, while the temperature and the time of pasteurization relate to HACCP and so on. In general, all the parameters related to each and every management system are represented and evaluated (Fig. 4.15).

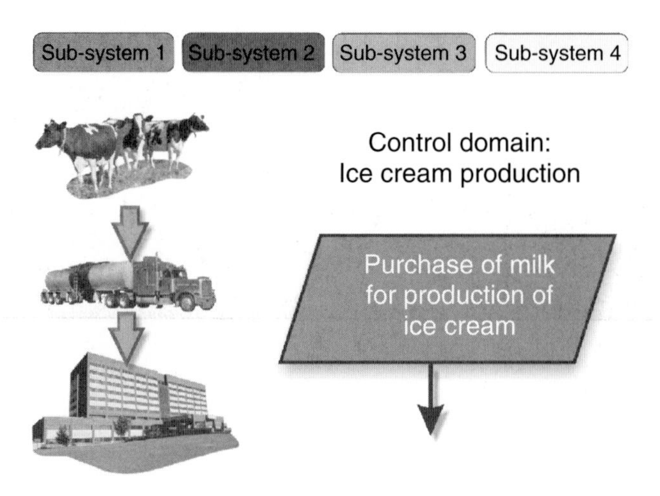

Fig. 4.12 'Purchase of milk for the production of ice cream' control step.

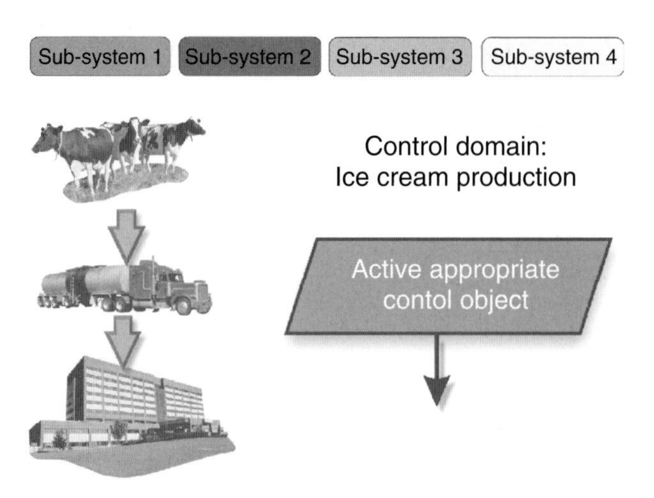

Fig. 4.13 Screenshot of Lotus Integrated Safety System.

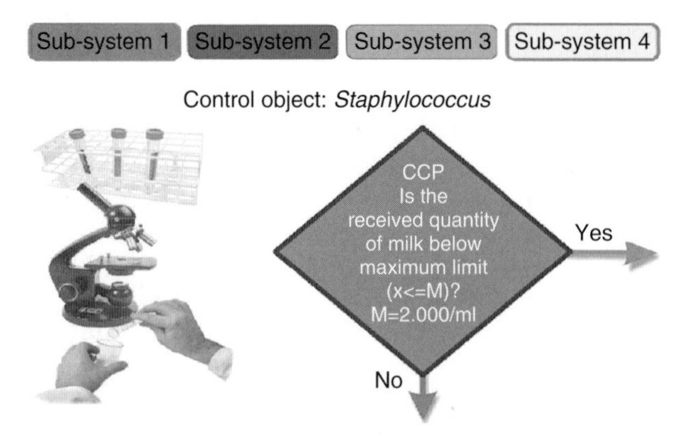

Fig. 4.14 Critical control point control for *Staphylococcus*.

Ice cream manufacture

Sub-system 1
Standardized safety management systems
(HACCP, ISO ST, GHP, etc.)

Process step	Control measure	Critical limits	Corrective action
1. Purchase of ingredients (e.g. fresh milk)	• Effective supplier assurance–audit • Certificate of analysis-agreed specification (maximum acceptable levels)	• Continued approved status (audit pass) • Legal limits	• Change supplier • Report to QA manager • Contact supplier • Reject consignment
2. Plastic tubs and film	• Correct choice of container and film (agree in specification) • Effective supplier assurance –audit	• Suitable for food use: - High-fat product - Compliance with legal migration limits • Continued approved status (audit pass)	• Change container/supplier
3. Pasteurization	Correct heat process, target level 82°C/15sec • Automatic direct	79.4°C–82°C Holding time 15 sec • Direct working	• Report to manager • Contact QA and discuss • Ensure divert working correctly • Repair thermograph • Hold product, until correct hear process verified; dump in not • START: Production does not start until corrected • END: Quarantine product; call engineer; notify QA and discuss

Sub-system 4
Continuous conformance to legal and other regulatory requirements

Legal requirements

1. EU DIRECTIVE:93/43/EEC(FOOD HYGIENE)
2. GREEK LAW :ΦΕΚ 1219/Β/04.10.2000(FOOD HYGIENE)
3. EU DIRECTIVE:98/93/EU(WATER QUALITY)
4. EU DIRECTIVE :2000/13/EU(FOOD LABELLING)
5. GREEK LAW :ΚΩΔΙΚΑΣ ΤΡΟΦΙΜΩΝ & ΠΟΤΩΝ
6. EU DIRECTIVE:92/46/EEC(MILK & MILK PRODUCTS)
7. GREEK LAW :ΦΕΚ 270/23.11.2001(ENVIRONMENTAL MANAGEMENT)
8. GREEK LAW:ΠΔ 56/95(MILK & MILK PRODUCTS)
9. EU DIRECTIVE:2001/114/EU(MILK)
10. GREEK LAW :LAW 1568/85(HEALTH & SAFETY AT WORK)

Fig. 4.15 Sub-systems 1 and 4 concerning the standardized safety-management systems and legal and regulatory requirements.

By measuring the level of *Staphylococcus* in the received milk, it can be established if it is below the maximum legal limit (M) or not (Figs 4.16 and 4.17). If yes, then proceed to the next step of production → click yes. If not: Reject the purchased milk and either

(1) ask for a report → CLICK Π.Α.Π.Π., which explains the Problems, the Causes, the Side Effects and the proposed solutions to this problem, or
(2) continue the process of examining various CONTROL OBJECTS by following the same procedure all over again.

The end result of this application is this printed report. By obtaining a report like this, and as much detail as requested, management can have a complete view of how

Fig. 4.16 Action taken in the case where the critical control point is below limits (i.e. the *Staphylococcus* count).

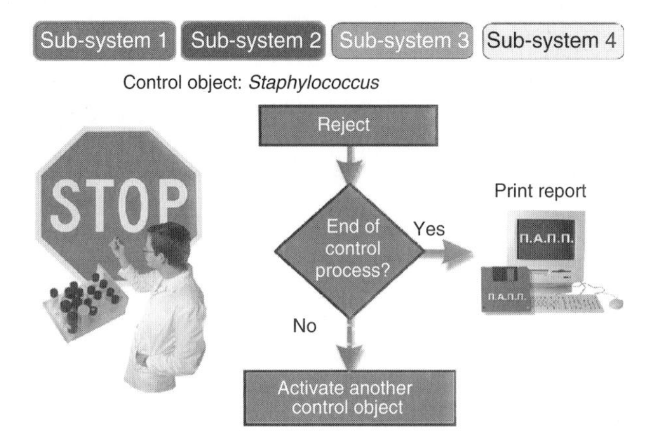

Fig. 4.17 Action taken in the case where the critical control point is above limits (i.e. the *Staphylococcus* count).

our safety systems work but can also have a closer look at the causes of the problems, side effects and possible solutions.

For the purposes of demonstration, only a limited number of parameters have been included. In fact, the real-time application of this system is advanced, but in any case it is user-friendly (Table 4.1).

By returning to each of the sub-systems and clicking on them, it is possible to determine the 'root' of the problem, understand it in a better way and propose *efficient* solutions for the benefit of the company (Fig. 4.18).

4.7.1 Summary

The aim of this case study was to demonstrate two things:

(1) to show that there is a need to integrate all of the most critical safety systems available to our industry, into a self-regulating (cybernetic) system that maximizes

Table 4.1 Problems, causes, side effects and solutions P.C.S.S. (Π.Α.Π.Π.) report table.

Problems (Προβλήματα)	Causes (Αιτίες)	Side Effects (Παρενέργειες)	Solutions (Προτάσεις)
Staphylococcus is above maximum acceptance limit on purchased milk	1. The supplier's premises do not conform to legal hygiene requirements 2. The milk tank was not cleaned properly 3. The supplier is not trained according to legal requirements	1. Smaller quantity of milk available for production (change of production plan) 2. Waste of resources (manhours, equipment, product) 3. Cost of re-auditing of the supplier	1. Re-auditing of supplier's premises 2. Purchase of milk from another supplier 3. Training of the supplier 4. Re-checking of cleaning procedures for the milk tank

safety by minimizing the errors (ε) between the standards (X) set by all these systems and the actual performance (Y);

(2) to show that using the tools provided by the new technology of Information and Telecommunications, a system can be designed that keeps management informed via the INTERNET. ANY TIME, ANY PLACE

4.8 Automation at the enterprise level

Automation at the whole enterprise level is heavily dependent on the process of gathering, processing and disseminating information up and down the organizational hierarchy and across departmental units. In order to be competitive, a company must have direct information. It is imperative that all systems provide immediate and reliable information. Information enables a company to make decisions which define its day-to-day direction and its long-term strategy. The collection of information is determined by its structure and systems.

Information is created in the three zones mentioned previously and is disseminated through the following levels of organizational hierarchy (Fig. 4.19):

- 1st level: managing directors;
- 2nd level: director;
- 3rd level: head of departments/foreman;
- 4th level: users.

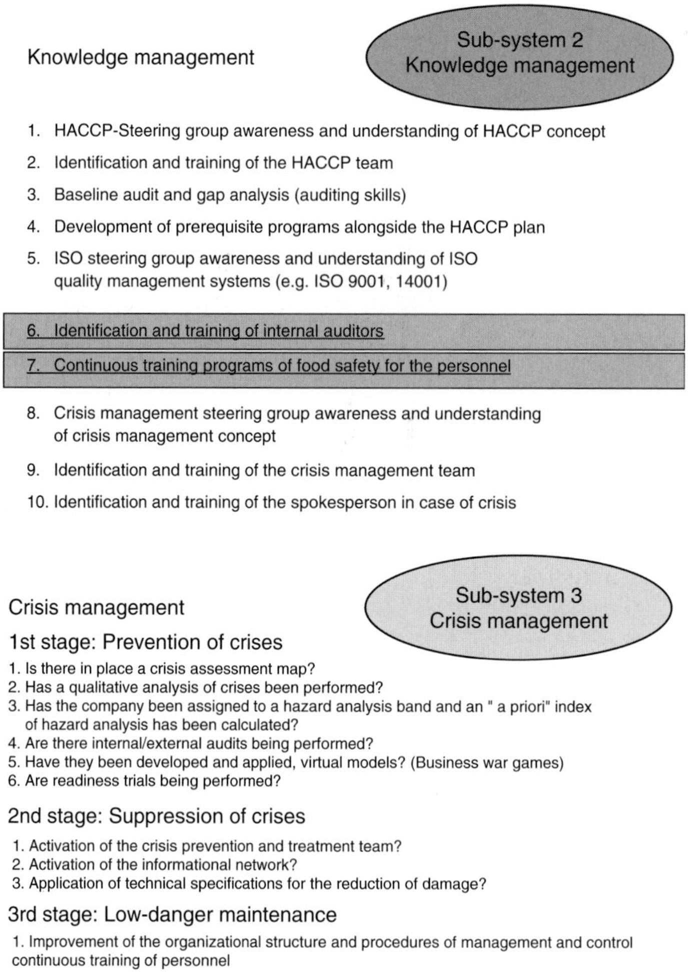

Knowledge management

Sub-system 2
Knowledge management

1. HACCP-Steering group awareness and understanding of HACCP concept
2. Identification and training of the HACCP team
3. Baseline audit and gap analysis (auditing skills)
4. Development of prerequisite programs alongside the HACCP plan
5. ISO steering group awareness and understanding of ISO quality management systems (e.g. ISO 9001, 14001)
6. Identification and training of internal auditors
7. Continuous training programs of food safety for the personnel
8. Crisis management steering group awareness and understanding of crisis management concept
9. Identification and training of the crisis management team
10. Identification and training of the spokesperson in case of crisis

Crisis management

Sub-system 3
Crisis management

1st stage: Prevention of crises

1. Is there in place a crisis assessment map?
2. Has a qualitative analysis of crises been performed?
3. Has the company been assigned to a hazard analysis band and an " a priori" index of hazard analysis has been calculated?
4. Are there internal/external audits being performed?
5. Have they been developed and applied, virtual models? (Business war games)
6. Are readiness trials being performed?

2nd stage: Suppression of crises

1. Activation of the crisis prevention and treatment team?
2. Activation of the informational network?
3. Application of technical specifications for the reduction of damage?

3rd stage: Low-danger maintenance

1. Improvement of the organizational structure and procedures of management and control continuous training of personnel

Fig. 4.18 Sub-systems 2 and 3 concerning knowledge management and crisis management.

Information at these four levels takes different forms. At the first level, the managing director's level, information is condensed and relates to some critical points that lead to decisions to bring about modifications by the managing directors.

Information at the second and third level is more analytical and relates to the responsibilities of each director and head of department. At these levels, the presentation of the information is possible, either in the way in which it is produced, or with modifications in the form and mode of presentation so as to be user-friendly and useful. Finally, at the fourth level, information must meet user requirements in a very satisfactory manner. The types and nature of information for each level are the responsibility of the managing director and are influenced by the tasks and the competences of the executives, as well as by the work flow dictated by the organizational pyramid.

Figure 4.20 presents a bird's eye view of a management information system for a dairy company. As can be seen from the figure, a total management information

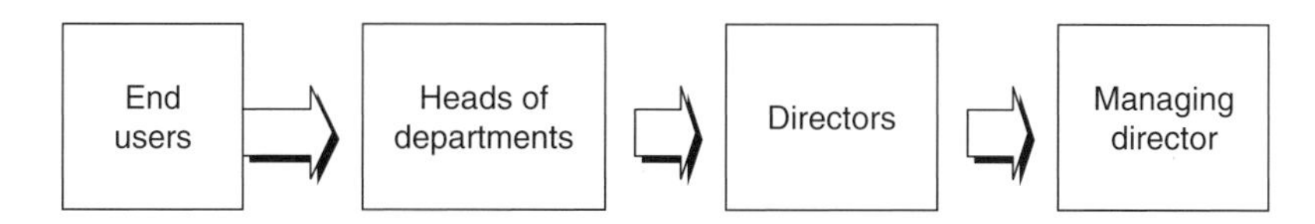

Fig. 4.19 Information flow through the hierarchy.

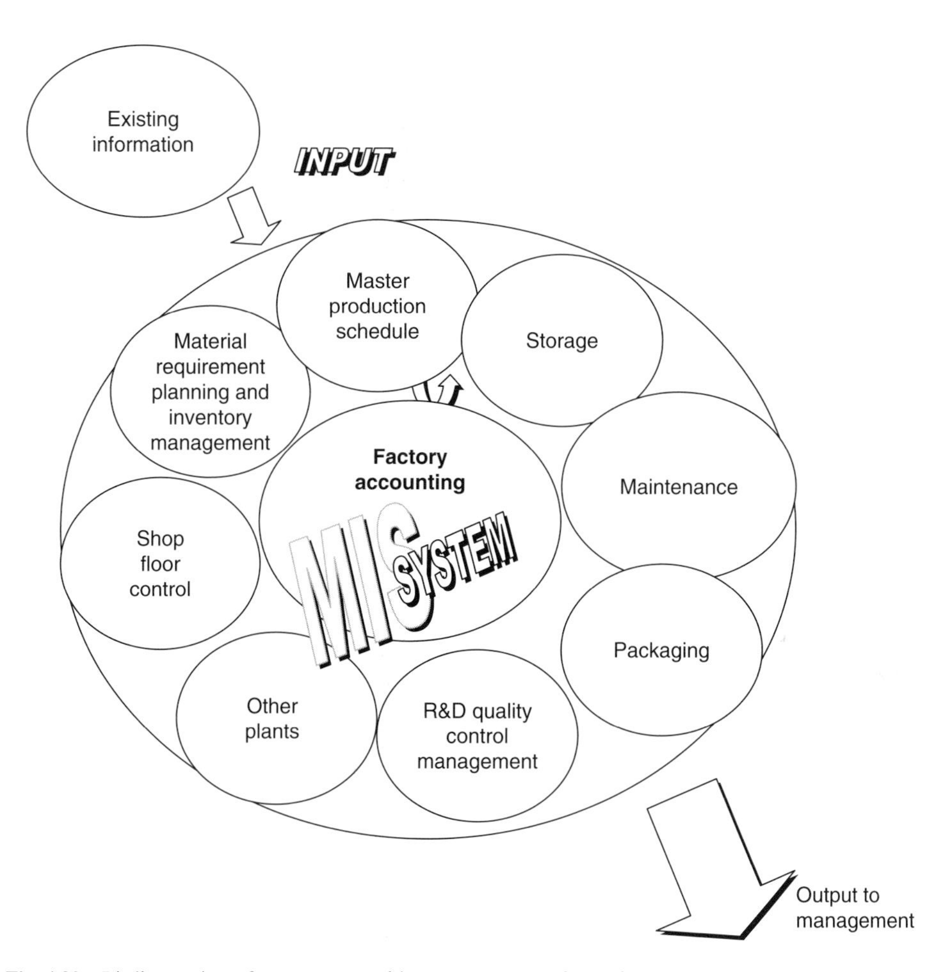

Fig. 4.20 Bird's eye view of an company wide Management Information System (MIS).

system integrates a number of special purpose systems. Systems designed to optimize the flow of raw material (ERP) and products throughout the company and between the company and its customers are all integrated into a total system that provides the company's decision-makers with timely and accurate information.

4.8.1 Logistics in dairy: how it helps

A real integrated, enterprise-wide, system is indispensable nowadays in the dairy. The very nature of the sector, regarding the highly sensitive and vital nature of its products, demands accurate, timely and fast information availability. The sector is characterized by:

- high sensitivity, not only of the raw material (milk), but also of the final product and its restricted shelf-life;
- a dairy's target group is a highly sensitive and fragile population, such as children, elderly and sick people;
- the use of a cold chain is critical in all stages from the raw material to the finished product.

The above particularities necessitate the management of an integrated automation system. In providing full support to the food processors, the chain management must be efficient. From the raw material (milk from the farm) to the end consumer, all necessary information and activities run vertically from the top management to the shop floor. Integrated logistics provides safety and security, lower labour costs, increased productivity, reduced scrap and waste, and finally decreased worker compensation claims. Automation on logistics, material handling and warehousing operation provides the real time trace and track functionality needed to support the food processors with safety and security and easy facilitation of any recalls that may be necessary.

The need for automation in the dairy is of great importance, as the number of different products is increasing every year. The need to control the warehouse operations, inventory and labour costs of a sophisticated system can be improved by automation. In other words, automation reduces the complexity of the entire process.

4.8.2 Enterprise Resource Planning

Programming business resources is a significant means allowing executives to have a complete picture of what is going on inside their company. ERP is a software system, which operates as a unit whose various parts communicate seamlessly. This particular characteristic differentiates ERP from traditional software operating as self-sufficient modules. In this way, management of business resources (namely collection, surveillance, processing and reporting) can be done automatically. ERP systems are tailored to dairy industries of all sizes at reasonable costs, thus making them accessible to even small companies. The main difficulty with ERP is the education that is required by dairy's executives, in order to understand and assimilate the new way of working. Furthermore, ERP's installation is a procedure that requires not only attention, but also a very good cooperation between the company that installs the software, the vendor and dairy's executives who will be called to run this system.

The purchase and the installation of an ERP system, like any software package – or more accurately software concept, as those working in Informatics wish to call it – is a 'cooperation' purchase, since work does not finish after ERP's installation, but continuous maintenance, improvements and changes of the system are required. In any case, ERP is a useful and essential software that enables a company to monitor its basic functions such as

- the monitoring of weight of products;
- the control of expiry date of materials;

- the check of the stock of products, of raw materials and of semi-ready products;
- the semi-ready products' content in materials;
- the recording of the quality-control results of both raw materials and ready-made products;
- the connection of materials with the ready-made product and with the traceability system;
- the pricing of the raw materials in the market;
- the pricing of the delivery of raw materials and of semi-ready and ready-made products;
- the pricing of the sales.

The above monitoring procedures through ERP constitute the body of the system. Each dairy has the opportunity to adjust the ERP system to its organizational demands, in order to achieve the best possible outcome. However, dairies have to educate their personnel so that their staff are acquainted with the system's potential. Thus, the software's purchase cost and the cost of the other systems that are necessary for its implementation are not the most crucial parts. On the contrary, installation, staff training and the maintenance of the system are more important with respect to both time and cost considerations.

4.8.2.1 ERP: benefits

In general, automation of the procedures and the guidance through ERP systems reduce the complexity and increase management control. In particular

- ERP increases efficiency and decreases expenses. Automation through ERP contributes to the reduction in personnel. Fewer job positions will be needed in order to support the new situation in the company. Overall, the total costs will be reduced.
- ERP contributes to the amelioration of the quality of the service that companies offer because it reduces bureaucracy.
- ERP improves programming and control.
- ERP helps monitoring the procedures in real time.
- ERP leads to an increase in competitiveness.
- ERP leads to improvements of productivity.
- ERP leads to better decision-making.
- ERP helps in cooperation between customers and suppliers.
- ERP minimizes mistakes and problems that result from production stoppages due to the lack of resources (raw materials, staff and packing).

4.8.2.2 Summary

The brief description of The Second Wave: Automation clearly shows the gradual but steady increase in the importance that machines play in the various stages of the milk-processing industry. Although the role of human beings has diminished considerably, humans are still the 'masters' telling the 'slaves' what to do.

4.9 Conclusions

Automation – with its integrated logistics, material handling and warehousing systems – can both proactively and reactively address the critical issues of product safety and security. Proactively, automated systems reduce the number of people with direct access to products. This alone can improve product reliability significantly while reducing the risk of catastrophe. Reactively, automated systems include up-to-the-minute track-and-trace functionality. This key feature of automation can easily facilitate any recalls that may be required. But beyond enhancing food safety and security, a fully integrated automation system can provide many compelling business benefits. Advanced automation systems increase visibility into food-manufacturing operations by improving the transparency and traceability of vital business information. The result is lower labour costs, increased productivity, reduced scrap and waste, and meeting, and even anticipating, the continually increasing societal and legal demands.

References

http://www.sspindia.com/profile.html
http://www.foodprocessing-technology.com/papers/
http://www.automationworld.com/articles/Features/1559.html
Sand, G. (2004). Perfect Packaging: Thinking Outside the Box. *A-B Journal* **11,** 12–13.
Wiener, N. (1948). *Cybernetics or Control and Communication in the Animal and the Machine.* MIT Press, Cambridge, MA.

Chapter 5

Safety and Quality of Dairy Products

Peter J. Jooste and Lucia E. C. M. Anelich

5.1 Introduction

Potential food safety hazards are undesirable substances or organisms that contaminate food and constitute a risk to the health of the consumer (Anon., 2003). Three classes of contamination represent hazards in food, and these include: (1) biological hazards, such as bacteria, fungi and other microbial pathogens; (2) chemical hazards such as residues of medication in the lactating animal, pesticides and a variety of industrial and environmental contaminants that might contaminate the feed of the lactating animal and finally land in the milk; and (3) physical hazards such as discarded hypodermic needles, fragments of metal or glass and any other foreign object that may have found its way into the food, e.g. hair, feed particles, somatic cells, etc.

It is ironic that at a time when food is safer than ever before, certain breaches of food safety in Europe have generated much concern and insecurity about the safety of food (Mathot, 2002). The incidents referred to here include the contamination of animal feed with polychlorinated biphenyls in Belgium (Van Renterghem and Daeseleire, 2000) and the outbreak of bovine spongiform encephalopathy (BSE) in the UK (Reilly *et al.*, 2002). In the United States, the public views chemical residues (particularly pesticides) as the leading health threat in terms of food safety. Similarly, the World Health organization (WHO) reports public perception as placing hazards such as pesticide residues, environmental contaminants and food additives at the top of the list (WHO, 2005). This is despite findings by the Centres for Disease Control and the FDA that food-related illnesses from mycotoxins, drug residues, agricultural chemicals and hormones are either non-existent or limited (Tybor and Gilson, 2003).

On the other hand, experience has shown that most outbreaks of food-borne ailments are associated with microbiological contamination. In fact, it has been estimated that people are 100,000 times more likely to become ill as a result of microorganisms in food than as a result of pesticide residues (WHO, 2005). More than 200 known diseases are transmitted through food by bacteria, fungi, viruses and parasites; and each year, millions of illnesses around the world can be traced to food-borne pathogens (Oliver *et al.*, 2005).

In the United States alone, it has been estimated by Mead *et al.* (1999) that 76 million people become ill, more than 325,000 are hospitalized, and 5000 die each year from food-borne disease. Extrapolating these figures to the rest of the world would mean that up to one-third of the population in developed countries are affected by microbiological food-borne disease each year (Käferstein and Abdussalam, 1999). The situation in developing countries is even worse, where under-reporting must be greater due to fewer resources and lack of food-borne disease surveillance systems. It is thought that less than 1% of all food-borne disease incidences is reported in developing countries (WHO, 2005). While the risks of chemical and physical hazards and the associated public perceptions can never be ignored, preventing illness and death associated with food-borne pathogens remains a major worldwide challenge.

Food-safety incidents often originate in the early stages of the food production chain, sometimes far outside the direct control of the food manufacturer. The hazards in these early stages are more numerous and difficult to control than those in the more confined manufacturing environment. To produce safe and wholesome food, it has become necessary to control all the factors that may have a negative effect on food safety and the purity of the final product in each step of the production chain. This procedure is referred to as 'Integrated Chain Control for Food Safety' (Mathot, 2002.) As a result, the responsibility for food safety should be shared among all the players along the food chain.

5.2 Pathogens of special relevance

5.2.1 Introduction

Milk is the ideal medium for the growth of both pathogenic and spoilage microorganisms. Given the fact that dairy products are consumed safely on a daily basis by most individuals in the world, it is a tribute to the efforts of all involved in ensuring the safety of dairy foods that milk, yoghurt, ice cream and cheese are still among the safest foods on the market. Recent publications have indicated that dairy products account for a very small percentage of all food-borne illness cases reported annually (Byrne, 2004). Although milk and milk products are among the safest foods worldwide, they have an inherent potential for spreading food-borne illness that is a major concern to producers, processors, regulators and consumers.

Normally, milk is collected from a lactating animal (most commonly a dairy cow). Such milk can harbour a variety of microorganisms and can therefore be an important source of food-borne pathogens. Key sources of microbial pathogens that can contaminate milk are endogenous sources, such as the cow itself, and exogenous sources, such as the environment (soil, water, manure or human contact: Tybor and Gilson, 2003), collection and processing equipment, and human milk handlers on the farm and in the factory.

When diseases are transmissible from animals to humans and vice versa, they are known as zoonoses (Marinšek, 2000). Endogenous zoonoses are caused by zoonotic agents that are excreted by infected dairy animals, regardless of whether they cause disease in the animal or not (Michel and McCrindle, 2004). In contrast, exogenous pathogens derive from the other sources referred to above and are involved in secondary contamination of raw milk of healthy animals (Marinšek, 2000) or in post-pasteurization contamination of the milk (Michel and McCrindle, 2004).

In general, zoonoses can be caused by bacteria, protozoa or viruses. The significance of zoonotic agents has been categorized by Michel and McCrindle (2004) as high, moderate and low. The highly significant group includes the following organisms: *Brucella abortus*, *Yersinia enterocolitica*, *Salmonella* spp., *E. coli* 0157, *Campylobacter jejuni*, *Listeria monocytogenes*, *Bacillus* spp. and *Staphylococcus aureus*. All of these organisms can infect humans from both endogenous and exogenous sources, with the exception of *B. abortus*, which is solely endogenous. Zoonoses of moderate significance are listed by Michel and McCrindle (2004) as *Brucella melitensis, Mycobacterium bovis, Coxiella burnetti, Clostridium perfringens* and *Cryptosporidium*. Those zoonoses of low significance (Michel and McCrindle, 2004) include *Leptospira interrogans, Clostridium botulinum,* Rift Valley Fever and Foot and Mouth disease viruses, *Chlamydia psitacci* and *Toxoplasma gondii.* Other predominant microorganisms of cattle involved in milk-borne zoonoses according to Noordhuizen (2003) include *Shigella* spp. and the Norwalk virus.

Some of these infections cause disease in the dairy animal, while others do not, the latter situation making control more difficult. Apart from the microorganisms above that have highly specific symptoms in infected humans (e.g. haemolytic uremic syndrome caused by the enterohaemorrhagic *E. coli* 0157, Q-fever by *C. burnetti* and tuberculosis by *Mycobacterium bovis*), most of the organisms cause gastrointestinal disorders (Noordhuizen, 2003). Not every infectious agent will cause disease in humans either. Persons falling in the so-called YOPI group (the Young, the Old, the Pregnant and the Immunocompromised; Noordhuizen, 2003), however, stand a greater risk of often severe and frequently lethal infections by certain of the more significant zoonotic agents (Michel and McCrindle, 2004).

Most milk is pasteurized or thermally treated in some or other manner, thereby eliminating food-borne pathogens that may be present. The question could then be posed as to why there should be concern about the safety of the raw product. One of the reasons is the direct consumption of raw or unpasteurized milk, and outbreaks of disease in humans have been traced to this practice. Unpasteurized milk is often consumed directly by dairy producers, farm employees and their families, neighbours in the region and raw-milk advocates. It is also consumed directly by segments of the population in certain regions of the world through the consumption of raw-milk cheeses (Oliver *et al.*, 2005).

Entry of food-borne pathogens via contaminated raw milk into dairy-food processing plants can lead to the persistence and establishment of these pathogens in the form of biofilms; subsequent contamination of processed milk products and exposure of consumers to the pathogens. Inadequate or faulty processing may result in the survival of specific pathogens, resulting in such contaminants becoming a public-health threat (Oliver *et al.*, 2005). The survival of these organisms and the risk they pose will depend on many factors such as microbial numbers, survival capacity of individual microbial types and the opportunities for contamination presented during the harvesting and processing of the milk. In the ensuing sections, food-borne infectious agents will be dealt with under the following headings, namely prions, viruses, rickettsiae, protozoa and bacteria.

5.2.2 Prions

A prion is a proteinaceous infectious agent claimed by a large number of groups to be the infectious particle that transmits a disease from one cell to another and from one animal to another (Dealler, 2002). Prions entered the public's consciousness during the mad cow epidemic that hit England in 1986 (Guyer, 2004). Prions enter brain cells and there convert the normal cell protein PrPC to the prion form of the protein, called PrPSC. More and more PrPC molecules transform into PrPSC molecules, until eventually prions completely clog the infected brain cells. Ultimately, infected prion-bloated brain cells die and release prions into the tissue. These prions then enter, infect and destroy other brain cells (Guyer, 2004). As clusters of cells die, holes form in the brain, which takes on the appearance of a sponge, hence the medical term for the prion diseases namely "spongiform encephalopathies." Mad cow disease is also known as bovine spongiform encephalitis (BSE). The original source of the BSE agent remains unknown, but the disease was subsequently spread through contaminated meat and bone meal fed to cattle.

While a direct link between BSE and the human disease Creutzfeldt-Jakob disease (CJD) in humans has not been proven unequivocally, a general geographical association exists whereby the majority of BSE cases occurred in the UK, and the majority of cases of a CJD variant (vCVD) were also reported there. The emergence of BSE preceded vCJD, indicating a temporal association. It is now widely accepted that vCJD was transmitted to humans through the consumption of contaminated food (Reilly *et al.*, 2002). According to the Food Safety and Inspection Service of the USDA (2005), however, there is no scientific evidence to suggest that milk and dairy products carry the agent that causes BSE.

5.2.3 Viruses

Viruses have emerged as significant causes of food-borne and water-borne diseases in recent years (Svensson, 2000). From 1983 to 1987 in the United States, the Norwalk virus was fifth on the list of leading causes of food-borne disease, with Hepatitis A in the sixth place and other viruses (mainly rotaviruses) in tenth position (Bean *et al.*, 1990 as quoted by Cliver, 1997). In the period 1988–1992, Hepatitis A rose to fourth place, while Norwalk-like viruses dropped to ninth position (Bean *et al.*, 1996 as quoted by Cliver, 1997).

Viruses are obligate intracellular parasites. Having no intrinsic metabolism, they are dependent on a living host in order to multiply. They can consequently only be transmitted by food and water and do not replicate in these substrates (Svensson, 2000; Rzezutka and Cook, 2004). Virtually all food-borne viruses are transmitted via the faecal–oral route, with man as the only known reservoir for calicivirus and Hepatitis A, the two most important viruses at present associated with food-borne outbreaks (Svensson, 2000).

Human enteric viruses can be transmitted by raw- and under-pasteurized milk (Cliver, 1993). Experiments have been conducted where various dairy products and pasteurized milk were artificially inoculated with a variety of viruses to assess their survival in these products. In inoculated pasteurized and boiled milk, poliovirus and coxsackievirus B5 could survive for at least 90 days at 4°C and for 15 and 30 days at 25°C for each virus type, respectively. Echovirus could survive for 120 days in raw milk at 4°C. Yoghurt stored at 4°C supported the survival of poliovirus and coxsackievirus B5 for 90 days and echovirus for 120 days. In cottage cheese, each of the latter viruses could survive for 120 days (Tiron, 1992 as quoted by Rzezutka and Cook, 2004). Cliver (1993) evaluated cheddar cheese as a vehicle for viruses and found that poliovirus inoculated into cheese could persist during maturation but was inactivated through 6 log cycles by thermization of the cheese milk. Cliver (1993) concluded that subsequent pasteurization treatment was sufficient to control any viruses that may have contaminated the raw milk.

Regarding milk-borne outbreaks, the first enteric virus associated with such an incident was the poliovirus in the period before the Second World War. The virus was transmitted through water and unpasteurized milk (Cliver, 1988). Poliovirus is host-restricted to humans and therefore cannot infect cows, but inadequate milk-handling practices by infected workers and lack of pasteurization sometimes allow infection to occur.

In 1993, seven people were infected with the tick-borne encephalitis (TBE) virus after drinking unboiled goat's milk. Other cases of alimentary TBE were recorded in 1984 (four cases) and in 1989 (two cases). Both outbreaks were associated with the consumption of unpasteurized goat's milk (WHO, 1994). TBE belongs to the flavivirus family and is the only enveloped virus known to be associated with food-borne infections. The virus infects dairy animals via the tick vector, and infected animals shed the virus in their milk, which, if ingested without pasteurization, may infect humans (Svensson, 2000).

5.2.4 Rickettsiae

Coxiella burnetti, the causative agent of Q-fever, is a zoonotic organism that may infect the udder, probably by the hematogenous route. Consumption of, or contact with, the infected milk can lead to human infection (Chambers, 2002). This organism was found to be more heat-resistant than *Mycobacterium tuberculosis* and in 1956, the vat pasteurization temperature was raised from 61.7 to 63°C (holding time 30 min) to ensure destruction of the organism (Boor and Murphy, 2002). The current high-temperature short-time (HTST) pasteurization temperature of 72°C for 15 s is also effective in this regard. Milk alkaline phosphatase which serves as an indicator of pasteurization efficacy for bovine milk is more heat-resistant than *Coxiella burnetti* and is destroyed by the above pasteurization treatments (Boor and Murphy, 2002).

5.2.5 Protozoa

Over recent decades, parasitic protozoa have been recognized as having great potential to cause water-borne and food-borne disease. The organisms of greatest concern in food production worldwide are *Cryptosporidium, Giardia, Cyclospora* and *Toxoplasma*. Although other parasitic protozoa can be spread by food or water, current epidemiological evidence suggests that these four present the largest risks (Dawson, 2005). Very little documented evidence is available in which milk or dairy products are implicated in cases of food-borne disease caused by these organisms. Of the four genera referred to above, *Cryptosporidium* appears to be the most significant in terms of milk and dairy products. Even in the case of *Cryptosporidium*, however, the public health risk of food-borne spread from the point of view of milk and dairy products (such as butter, cheese and yoghurt), appears to be negligible (Dawson, 2005). Consequently, only *Cryptosporidium* will be discussed in more detail in this section.

It nevertheless remains important to prevent protozoan parasites from entering the food- (and also the milk-production) chain. Three main routes exist through which these parasites can enter the food-production process (Dawson, 2005):

- through contamination of food ingredients or raw materials on the farm;
- through contaminated water included in the final product for product processing or washing, or used for cleaning processing equipment;
- through transfer or spread via infected food handlers or food preparers in production, food service or domestic settings.

Preventive or control methods should therefore be devised to cover these three routes whenever they could be of significance to the final product consumed. It is also important that analytical laboratories tasked with examining outbreaks of infectious intestinal disease should have the necessary expertise in diagnosing the presence of parasitic protozoa in patients.

5.2.5.1 Cryptosporidium

Human cryptosporidiosis emerged as an important gastro-intestinal infection in the 1990s, due to the ingestion of contaminated water and foodstuffs containing the protozoan parasite, *Cryptosporidium parvum*. When ingested, it is capable of causing a high degree of morbidity in healthy populations. It results in mortality in vulnerable populations such as immunocompromised persons infected with HIV/AIDS or cancer patients receiving chemotherapy. There is no effective antimicrobial treatment to eradicate this agent from the gastrointestinal tract of symptomatic individuals.

Generally, cryptosporidiosis is a self-limiting disease presenting typically with diarrhoea, dehydration, abdominal cramps, vomiting, weight loss and electrolyte imbalance as symptoms, in immunocompetent individuals. Cryptosporidial infections can however be persistent in persons infected with HIV/AIDS to the point of being life-threatening (Deng and Cliver, 1999; Hunter and Nichols, 2002). Although there are ten recognized species of *Cryptosporidium* (Fayer *et al.*, 2000 as quoted by Millar *et al.*, 2002), human infection is mainly caused by *C. parvum* (Kosek *et al.*, 2001). *C. parvum* is an obligate intracellular parasite that infects the microvillus border of the epithelium in the gastrointestinal tract of humans and various animal hosts.

Water has been found to be the most important source of *C. parvum*, and because of the organism's chlorine resistance, it has been a particular threat in otherwise safe drinking-water supplies (Olson *et al.*, 2004; Dawson, 2005). Person-to-person contact is also well described, particularly in secondary cases in outbreak settings and in day-care facilities and hospitals (Guerrant, 1997 and Glaberman *et al.*, 2002, both quoted by Millar *et al.*, 2002).

Cryptosporidium oocysts have been isolated from several foodstuffs, and these have included fruit, vegetables and shellfish (Millar *et al.*, 2002). Several outbreaks of cryptosporidiosis have been associated with milk. In 1985, an outbreak of cryptosporidiosis occurred in Mexico in which 22 cases were confirmed. Contaminated milk is suspected to have been the cause (Elsser *et al.*, 1986). In 1995, 50 cases of cryptosporidiosis were confirmed in the United Kingdom. Junior schoolchildren were contaminated after drinking milk that was distributed to the school by a small-scale local producer. The on-farm pasteurizer was found to be faulty, and hence, the milk was not adequately pasteurized (Gelletlie *et al.*, 1997). In 1984, a mother and her one-year-old child were infected with *Cryptosporidium* after drinking unpasteurized goat's milk that had been purchased locally in Australia (WHO, 1984).

Cryptosporidium parvum cannot survive pasteurization of milk, and 100% inactivation was achieved by heating milk to 71.7°C for 5 s (Harp *et al.*, 1996). Deng and Cliver (1999) reported that *C. parvum* showed 0–5% viability after 48 h in ice cream stored at –20°C. These authors also found that prolonged storage of contaminated yoghurt for up to 240 h was not sufficient to destroy *C. parvum*. A decrease in viability of the organisms from 83% at time 0 to 61% after 240 h was noted, however.

5.2.6 Bacteria

5.2.6.1 *Brucella*

Brucella abortus and *B. melitensis* are zoonoses that can cause milk-borne disease outbreaks in humans and are the causative agents of brucellosis in cattle. Brucellosis eradication programmes in many countries have resulted in a decrease in outbreaks (Byrne, 2004). In many parts of the world, this eradication has been so successful that it no longer poses a hazard to human health (Chambers, 2002). Where the disease still occurs, pasteurization of milk has minimized the number of human outbreaks by these microorganisms (Byrne, 2004)

5.2.6.2 *Mycobacterium*

Mycobacterium bovis is a microorganism of concern to the dairy industry because it causes tuberculosis in cattle. It can also result in milk-borne *M. bovis* infections of humans. Cattle immunization and pasteurization of the milk has limited the number of tuberculosis cases associated with the consumption of milk or dairy products (Byrne, 2004).

Another *Mycobacterium* species that has generated increasing interest in latter times is *Mycobacterium avium* subspecies *paratuberculosis* (MAP). This organism is the aetiological agent of paratuberculosis or Johne's disease, a global disease of ruminants that was first described in 1895 (Johnes and Frothingham 1895 as quoted by Hillerton, 2003). The disease itself is a slow developing colitis in which the intestinal macrophages are infected. In the process, inflammatory reactions are induced in the host gut. This affects the ability of the gut to absorb protein from the diet resulting in clinical features that include diarrhoea and chronic weight loss (Hillerton, 2003).

Similarities between paratuberculosis and Crohn's disease in man were first reported in 1913 (Dalziel, 1913 as quoted by Hillerton, 2003). Although the homology of the two diseases is still under dispute, sufficient evidence exists to indicate that MAP should be prevented from entering the human food supply and that it should be regarded as a zoonosis until proven otherwise (Hillerton, 2003).

Milk may be contaminated by the natural shedding of infected macrophages or by faecal contamination (Hillerton, 2003). While the infectious dose is reported to be as low as 1000 organisms, clinically affected animals may shed up to 5×10^{12} mycobacterial cells per day (Chiodini *et al.*, 1984 as quoted by Hillerton, 2003). Occurrence of the disease in milk producing animals is consequently a challenge to animal health, milk quality and the safety of the milk supply (Hillerton, 2003).

The most accurate strategy for controlling MAP at present involves an approach based on three levels (Gallmann and Eberhard, 2004):

- *Control of MAP at the herd level* – Action to reduce the incidence of MAP infection in cattle appears to be worth while from the point of view of herd health as well as ensuring a reduction in the excretion of the organisms in the raw milk;
- *Control during milking* – The implementation of hygienic practices during milking is clearly necessary. While improved hygienic measures can reduce the risk of milk contamination with pathogenic bacteria, the risk to the consumer cannot be completely eliminated;
- *Control by pasteurization of the milk* – Although Grant *et al.* (2002) have reported that MAP can survive the normal pasteurization process, this work according to Gallmann and Eberhard (2004) has been shown on several occasions to contain certain significant flaws and should be disregarded. Several groups of workers have reported that heating at 72°C for 15 s reduced the number of viable MAP cells in milk by a factor of 10^4–10^5 (Lund *et al.*, 2002 as quoted by Gallmann and Eberhard, 2004). It is believed, therefore, that the presence of MAP in pasteurized milk should be ascribed to either the failure in applying the correct heat treatment or post-pasteurization contamination, rather than to the survival by the organisms of the proper heating process. This problem can be overcome by applying pasteurization guidance as recommended in EU directive 92/46 (Gallmann and Eberhard, 2004).

5.2.6.3 Enterobacteriaceae

The presence of any member of the *Enterobacteriaceae* family is undesirable in pasteurized dairy products. This is due to: (1) the inherent spoilage capacity of many genera in this family, (2) the fact that the presence of certain genera in water and foods may be indicative of faecal contamination and (3) the serious food safety implications that the presence in food or water of the many pathogens in this family may have.

5.2.6.3.1 Salmonella. The gastro-enteritic form of non-typhoid salmonellosis was not clearly linked to raw-milk consumption until the mid-1940s. Interest in milk-borne salmonellosis has peaked twice since the 1940s, first in 1966 when several large outbreaks were traced to non-fat milk powder and again in 1985 when one of the largest recorded outbreaks of food-borne salmonellosis involving more than 180,000 cases was traced to consumption of a particular brand of pasteurized milk in the Chicago area (Ryser, 1998 as quoted by Mostert and Jooste, 2002). Today, *Salmonella* and *Campylobacter* are generally recognized as the two leading causes of dairy-related food-borne illness in the United States and Western Europe, with rates of infection being particularly high in regions where raw milk is neither pasteurized nor boiled.

Raw milk can be a source of salmonellae, and 32 of 678 (4.7%) of raw bulk milk samples were reported to test positive in the United States (Byrne, 2004). Standard pasteurization destroys expected levels of salmonellae (<100 cfu ml^{-1}), with a wide margin of safety. Inadequate pasteurization and post-process contamination have occasionally resulted in milk and cream that test positive for *Salmonella*, as evidenced by the reported outbreaks. *Salmonella* is not particularly tolerant of heat, refrigeration, or salt, so the typical hurdles used in the dairy industry are effective in controlling this organism (Byrne, 2004).

5.2.6.3.2 Escherichia. *E. coli* is currently the best-known indicator of faecal contamination, primarily of water but also of raw food products. Its recovery from fresh dairy products consequently suggests that other organisms of faecal origin, including pathogens, may be present (Mostert and Jooste, 2002). *E. coli* strains are commonly associated with the normal facultative anaerobic microflora of the intestinal tracts of humans and animals (Olsvic *et al.*, 1991 as quoted by Greyling, 1998). Although many of these strains are harmless commensals, various *E. coli* strains have acquired genetic determinants (virulence genes rendering them pathogenic for both humans and animals). According to Holko *et al.* (2006), these pathogens are responsible for three main clinical syndromes, namely enteric and diarrhoeal diseases, urinary tract infections and sepsis/meningitis. On the basis of their distinct virulence properties and the clinical symptoms of the host, pathogenic *E. coli* strains can be divided into numerous categories or pathotypes (Holko *et al.*, 2006). The extra-intestinal infections are caused by three separate *E. coli* pathotypes, namely uropathogenic strains (UPEC), neonatal meningitis strains or MENEC (Willert, 1978 quoted by Holko *et al.*, 2006) and strains that cause septicaemia in humans and animals (Harel *et al.*, 1993; Dozois *et al.*, 1997; Martin *et al.*, 1997, all quoted by Holko *et al.*, 2006).

The diarrhoeagenic *E. coli* strains (Holko *et al.*, 2006) include enteroaggregative *E. coli* (EAEC), enteroinvasive *E. coli* (EIEC), enteropathogenic *E. coli* (EPEC) and enterotoxigenic *E. coli* (ETEC). Members of the latter group adhere to the mucosa of the small intestine and produce heat-labile and/or heat-stable enterotoxins (Cohen and Gianella, 1995 as quoted by Holko *et al.*, 2006). A group of ETEC strains produce cytotoxins called verotoxins or shiga-toxins (Calderwood *et al.*, 1996, quoted by Holko *et al.*, 2006), hence the acronyms STEC or VTEC, and these colonize the intestinal tract of healthy domestic animals, chiefly cattle. The STEC serogroups, 0157, 026 and 0111, are also designated enterohaemorrhagic *E. coli* or EHEC (Holko *et al.*, 2006). Only this latter group of organisms will be subjected to further discussion in this chapter.

Several outbreaks of *E. coli* gastro-enteritis have been traced to raw milk and dairy products (Anon., 1993 as quoted by Greyling, 1998). Infections with EHEC of the serotype 0157:H7 were first described in 1982 (Riley *et al.*, 1983 as quoted by Tsegaye and Ashenafi, 2005). *E. coli* 0157:H7 has become a pathogen of major concern for the food and dairy industries because of its ability to cause severe ailments such as haemorrhagic colitis (HC), haemolytic uremic syndrome (HUS), and thrombocytopenic purpura (TTP). The ailments affect all human age groups, and the pathogen is exceptional in the severe consequences of infection, its low infectious dose and unusual acid resistance (Buchanan and Doyle, 1997 as quoted by Tsegaye and Ashenafi, 2005). Cattle are the main reservoir of *E. coli* 0157:H7, while the most likely mode of transmission in foods is faecal contamination. The organism has an oral infectious dose within the range of 10–100 cells, or even lower in the case of susceptible groups (Lekkas *et al.*, 2006)

The survival of *E. coli* 0157:H7 in acidic foods such as yoghurt and its survival in fermented as well as fermenting dairy foods for long periods of time makes its hardy nature apparent (Tsegaye and Ashenafi, 2005). Such foods have been associated with *E. coli* 0157:H7 outbreaks (Lekkas *et al.*, 2006). Arocha *et al.* (1992; as quoted by Lekkas *et al.*, 2006) first reported that pH and acidity did not halt the growth of *E. coli* 0157:H7 during the manufacture of cottage cheese, the most well-known and popular acid-curd cheese.

Prompt action taken in response to recent haemorrhagic colitis outbreaks associated with acidic foods has resulted in a large number of validation studies in many countries. Particularly in the United States, mandates that food processors should guarantee a 5-log reduction of *E. coli* 0157:H7 during processing of fermented sausages and fruit juices have been adopted, but this does not seem to have been mandated for fermented milk products yet (Lekkas *et al.*, 2006).

The behaviour of *E. coli* 0157:H7 was studied during the manufacture and ripening of raw goat's milk lactic cheeses by Vernozy-Rozand *et al.* (2005). When cheese was manufactured from raw milk in the laboratory and inoculated with *E. coli* 0157:H7 to a final concentration of 10, 100 and 1000 cfu ml^{-1}, counts decreased to less than 1 $\log_{10} g^{-1}$ in curds just prior to moulding. However, viable *E. coli* 0157:H7 were found in cheeses throughout processing and even after 42 days of ripening. The conclusion reached by Vernozy-Rozand *et al.* (2005) was that the presence of small numbers of *E. coli* 0157:H7 in milk destined for the production of raw milk lactic cheeses can constitute a threat to the consumer. They regard it as imperative to ensure that milk used in the manufacture of such cheeses is of the highest microbiological quality.

Given the potential for contamination of milk by *E. coli* during milking, consumption of raw milk should be avoided. *E. coli* 0157:H7 is not heat-resistant and, like most *E. coli* strains, is readily destroyed by the pasteurization process. If good manufacturing practices are followed, consumption of pasteurized milk consequently poses little or no risk of containing this microorganism (Byrne, 2004). While the organism is unable to grow at less than 10°C, substantial growth can occur in temperature-abused milk.

5.2.6.3.3 Yersinia. *Y. enterocolitica* and other yersiniae falling within the Family Enterobacteriaceae are often referred to as 'environmental strains' (Stern and Pierson,

1979 as quoted by Yucel and Ulusoy, 2006). *Yersinia enterocolitica* is, nevertheless, a well-established pathogen of humans, causing acute gastroenteritis, enterocolitis and mesenteric lymphadenitis, as well as a variety of extra-intestinal disorders (Robins-Browne, 1997 and Bottone, 1999 as quoted by Yucel and Ulusoy, 2006). The symptoms of these ailments may be especially severe in children and individuals with underlying disease (Cornelis *et al.*, 1987 as quoted by Yucel and Ulusoy, 2006).

Yersinia enterocolitica is regarded as an unusual cause of milk-borne illness because of the low incidence of human pathogenic strains in the raw milk supply and the high susceptibility of the organism to pasteurization. Yersiniae may be contaminants of raw milk, and most contamination is thought to occur through contact with faeces or polluted water supplies (Byrne, 2004). Raw milk and inadequately pasteurized milk and dairy products have, nevertheless, been implicated in the transmission of *Y. enterocolitica* infections to humans. According to Yucel and Ulusoy (2006), the first recorded food-associated outbreak of yersiniosis occurred in New York, where more than 220 individuals were stricken with acute intestinal illness after consumption of contaminated milk. In addition, epidemiological studies have revealed that refrigerated food stored over prolonged periods of time poses an additional risk, since *Y. enterocolitica,* as a psychrotrophic microbe, is able to grow at temperatures as low as 0°C (Hanna *et al.*, 1976 and Black *et al.*, 1978 as quoted by Yucel and Ulusoy, 2006). *Y. enterocolitica* and atypical *Yersinia* spp. were also isolated from cheese samples in Turkey, and according to Yucel and Ulusoy (2006), the results indicated that *Yersinia* spp. are more likely to be isolated from foods with a high level of coliforms than from foods with low coliform counts. This lends impetus to the feasibility of the faecal origin of *Yersinia* contamination of raw milk.

5.2.6.4 Campylobacter jejuni

Campylobacter jejuni has been recognized since 1909 as an important cause of abortion in cattle and sheep. Improved isolation strategies have also implicated this organism as a causative agent of human diarrhoea. Altogether, 45 food-borne campylobacteriosis outbreaks (1308 cases) were reported in the United States between 1978 and 1986, over half of which involved ingestion of raw milk (Ryser, 1998 as quoted by Mostert and Jooste, 2002). Similar reports linking raw- and inadequately pasteurized milk to 13 outbreaks in Great Britain from 1978 to 1980 helped to further substantiate *C. jejuni* as an important milk-borne pathogen that has come to rival or even surpass *Salmonella* as an aetiological agent of human gastroenteritis worldwide (Ryser, 1998 as quoted by Mostert and Jooste, 2002).

Campylobacter jejuni can be isolated from the faeces of cattle infected or colonized with the organism and has been shown to cause asymptomatic bovine mastitis in which the organism is excreted directly into the milk of an infected cow (More O'Ferral-Berndt, 2000). In most outbreaks, however, *Campylobacter* could not be isolated from the milk after an outbreak. Nevertheless, a survey done in England in 1988 showed that 5.9% of milk samples tested were positive for *C. jejuni* (Humphrey and Hart, 1988 as quoted by More O'Ferral-Berndt, 2000). It was also found, in the latter survey that there was a significant association between the presence of *E. coli* in milk and that of *C. jejuni*.

More O'Ferral-Berndt (2000) reported that *C. jejuni* infections have been shown to be hyper-endemic in developing countries, including South Africa, with an age-related decrease in the incidence of infection in humans. Acquired immunity could be important in preventing infection or preventing illness after infection. Nevertheless, immunocompromised people are at risk of contracting the infection (Johnson *et al.*, 1984 as quoted by More O'Ferral-Berndt, 2000)

Campylobacter jejuni is killed by proper pasteurization, but outbreaks involving inadequately pasteurized milk have been described in England (Porter and Reid, 1980 and Fahey *et al.*, 1995 as quoted by More O'Ferral-Berndt, 2000). Failure in the public electricity supply and a faulty pasteurizer were identified as the causes of the problem. *Campylobacter* is also acid-sensitive, and this suggests that the genus will not survive a normal fermentation in a product such as yoghurt (Cuk *et al.*, 1987 as quoted by Robinson *et al.*, 2002).

5.2.6.5 Staphylococcus aureus

Staphylococcus aureus is a significant cause of mastitis in dairy cows throughout the world. The bovine mammary gland can be a significant source of enterotoxigenic strains of *S. aureus*. Enterotoxins produced by enterotoxigenic strains of this organism are classified according to serotypes into A–H groups and the so-called 'toxic shock syndrome toxin' or TSST (Oliver *et al.*, 2005). Takeuchi *et al.* (1998), as quoted by Oliver *et al.*, (2005) detected TSST in *S. aureus* strains isolated from milk of cows with clinical and sub-clinical mastitis and also from farm bulk tank milk. Various other studies referred to by Oliver *et al.* (2005) also implicated bulk tank milk as a potential source of enterotoxigenic *S. aureus* in milk and milk products, which may constitute a health hazard to consumers. Consumption of raw milk generally increases the chances of direct contact with food pathogens such as *S. aureus* and its toxins.

Workers involved in processing the milk and manufacturing dairy products can also be a source of *S. aureus* in the product. Asperger (1994) refers to various surveys that have demonstrated that 4–60% of humans are nasal carriers of *S. aureus* and that 5–20% of people carry the organism as part of the normal skin microflora.

Staphylococcus aureus was found to be the most frequent pathogen associated with cheese made from raw or 'unspecified' milk in *toxi-infections alimentaires collectives* (TIACs) reported in France. *Toxi-infections alimentaires collectives* (TIACs) are defined as the occurrence of two or more cases of a similar illness, usually gastro-enteritis, caused by the same food (De Buyser *et al.*, 2001). Although the majority of patients from the various outbreaks were hospitalized, no fatalities were reported. Other outbreaks of *S. aureus* enterotoxicoses are referred to by Boor and Murphy (2002). In 1985, 860 children were affected by drinking chocolate milk in Kentucky, USA. More than 14,000 people in Osaka, Japan, were affected by drinking contaminated pasteurized milk in the year 2000.

Staphylococcal enterotoxin production can be prevented by keeping the raw milk refrigerated until the milk can be effectively pasteurized. While pasteurization procedures will kill the cells of *S. aureus* in the milk, they will not destroy enterotoxin already present (Asperger, 1994). Identification of human carriers of toxigenic *S. aureus* in the processing and manufacturing line, as well as effective procedures for preventing post-pasteurization contamination, should further assist in preventing the incidence of this pathogen in the final product.

5.2.6.6 Listeria monocytogenes

Listeria monocytogenes has only recently emerged as a serious food-borne pathogen that can cause abortion in pregnant women and meningitis, encephalitis and septicaemia in newborn infants and immunocompromised adults (Ryser, 1998 as quoted by Mostert and Jooste, 2002). Although the disease is rare (i.e. 1–9 cases per 1,000,000 people per year) and accounts for only about 0.02% of all cases of food-borne illness, listeriosis accounts for about 28% of the deaths resulting from food-borne illness (Tompkin, 2002). This severe mortality rate emphasizes the necessity to minimize the exposure of the high-risk individuals referred to above. According to a number of studies referred to by Tompkin (2002), variability in virulence of *L. monocytogenes* strains is slowly gaining recognition and acceptance, showing that some strains have a greater potential to cause disease than others.

Throughout the world, three serotypes (4b, 1/2a and 1/2b) account for 89–96% of cases of human listeriosis, providing additional evidence that certain strains are more likely to cause illness (Farber and Peterkin, 2000 as quoted by Tompkin, 2002). Serovar 4b, for example, has caused several major outbreaks. One outbreak in Switzerland originated from soft cheese (Vacherin Mont d'Or) and involved 122 cases, resulting in 34 deaths during the period 1985–1987 (Tompkin, 2002).

It has been suggested that the infective dose of a virulent strain of *L. monocytogenes,* is in the order of 100–1000 cells (ADASC, 1999). In general, foods that have been implicated in listeriosis have contained >1000 cfu ml^{-1} or >1000 cfu g^{-1} (Tompkin, 2002). The most important property of *L. monocytogenes* is its ability to multiply in foods at refrigeration temperatures, and most foods incriminated in listeriosis have been held under refrigeration (ADASC, 1999). Any ready-to-eat food contaminated with *L. monocytogenes* and kept refrigerated may consequently yield population numbers that can present a threat to susceptible consumers. Milk and dairy products have been found to be sources of food-borne listeriosis, and after outbreaks involving a large number of cases (Arqués *et al.*, 2005), *L. monocytogenes* has become a pathogen of great concern to the dairy industry.

Listeria monocytogenes is commonly encountered in the dairy farm environment (Fedio and Jackson, 1992). The most likely sources of the organisms in raw bulk-tank milk are therefore environmental in nature with faeces/manure playing a major role. Contamination from within the udder as a result of shedding due to bovine listeriosis or listerial mastitis is rare (Fedio and Jackson, 1992). The organism can be transmitted to cows via feeds such as improperly fermented silage and other feedstuffs, causing infection in the animal (Faber and Peterkin, 2000). Incidences of *L. monocytogenes* of 4.2%, 2.2% and 2.6%, respectively, have been reported in farm milk samples of bovine, ovine and caprine origin (Arqués *et al.*, 2005).

Listeria has been isolated from many dairy products, including ice cream and various types of cheese (ADASC, 1999). In soft ripened cheese, contamination is limited to the first few millimetres under the rind. Hard cheese such as Parmesan does not favour growth, and other cheeses such as Colby, Swiss, Provolone, Munster, Feta and Limburger show gradual die-off of the bacteria. In Mozzarella cheese, the bacteria will survive the manufacturing process, but not the stretching temperatures. *L. monocytogenes* will not grow in Cottage cheese but can survive. Although the organism has been found to gain access to yoghurt as a post-pasteurization contaminant, it will not survive at pH levels below 4.6 (ADASC, 1999).

Listeria monocytogenes is quite well adapted to dairy factory environments. It is generally more common in factories where conditions tend to be wet and cool with areas of pooled water or liquid (ADASC, 1999). The organic load on the floors and in the drains, if high, can also contribute to the growth and survival of *Listeria*. The primary source of *Listeria* spp. in processing plants is probably floors and floor drains (Coleman, 1986 as quoted by Griffiths, 1989), especially areas around coolers or places subject to outside contamination. Due to the ubiquitous nature of the organism, all raw materials carried into the packaging area must be suspect. Cooling waters should also be considered as a possible source of contamination (Griffiths, 1989). Work practices of factory personnel and the dispersal of the organisms in water sprays and aerosols have also been identified as significant carriers of the organism throughout the factory (ADASC, 1999).

Many researchers have established that strains of *L. monocytogenes* can become established in a food-processing facility and remain members of the resident microbial flora for many years (Tompkin, 2002). Examples of niches in which the organisms can become established are hollow rollers on conveyors, cracked tubular support rods on equipment, the space between close-fitting metal-to-metal or metal-to-plastic parts, worn or cracked rubber seals around doors, on–off valves and switches for equipment as well as saturated insulation (Tompkin, 2002).

Listeria monocytogenes can become established on steel and rubber surfaces in the form of biofilms. Lee Wong (1998) reports that the organism was found to survive for prolonged periods on stainless steel and Buna-*N* (acrylonitrile butadiene) rubber. Under favourable conditions, it even multiplied on stainless steel. Temperature, relative humidity, soiling and the surface type affected the behaviour of surface-associated *L. monocytogenes*. In addition, the attachment surface affected the efficacy of sanitizers.

In terms of prevention or control procedures, *Listeria* spp. can contaminate raw milk from many environmental sources, and it is difficult to entirely prevent the presence of *Listeria* in raw milk. In most cases, control of raw-milk contamination should, however, be readily achieved by good sanitation and milking practices (Fedio and Jackson, 1992).

In controlling the presence of *L. monocytogenes* in the processing area and the product itself, the following strategies can be followed:

- In the first place, steps should be taken to prevent entry of *L. monocytogenes* to plant areas: These steps basically include the isolation of the milk reception area, and associated personnel, from the processing and packaging area and prevention of any raw product from entering the processing area (ADASC, 1999). Any product and container returns should similarly not be permitted back into the manufacturing or packaging area. There should be no unsealed openings to the manufacturing area from other areas.

- Listeriae should be prevented from establishing and growing in niches or other sites that can lead to contamination of the processed food (Tompkin, 2002). This should include a thorough check of sweetwater and glycol cooling systems, cracks and crevices in storage tanks and other equipment components and effective cleaning and sanitizing. All drains must be properly constructed, and cleaned and sanitized each day. Floor drains should not be located under or near filling and packaging equipment (ADASC, 1999).
- Effective pasteurization is the key step to control *Listeria* in the processing area. In the case of HTST pasteurization of milk, a minimum of 72°C for 15 s is essential. Products containing higher fat or sugar levels, require higher temperatures to ensure effective destruction of *Listeria* spp. It is recommended that all such products be heated to 75°C for 15 s to be safe (ADASC, 1999).
- Cleaning and sanitizing programmes are vital in ensuring that post-pasteurization contamination does not occur. Absorbent items such as rags and sponges should be eliminated to reduce potential harbourage and spreading of the organisms. Separate brushes should be used for product contact and non-product contact surfaces (ADASC, 1999).
- A sampling programme should be implemented that can assess in a timely manner whether the food-processing environment is under control. There should be a rapid and effective response to each positive sample obtained.
- Finally, there should be verification by follow-up sampling that the source has been detected and corrected (Tompkin, 2002).

5.2.6.7 Bacillus cereus

Bacillus cereus is a well-known food-poisoning organism that may cause illness through the production of either an emetic (vomit-inducing) toxin or at least three diarrhoeal toxins or enterotoxins (Granum, 2001 as quoted by Rossland *et al.*, 2005). Raw milk appears to be the major source of *B. cereus* in the pasteurized milk, and post-pasteurization contamination along the milk processing line is possibly only a minor source. Nevertheless, these organisms readily form biofilms on food contact surfaces that can cause serious problems in cleaning and disinfection operations. Peng *et al.* (2002 as quoted by Salo *et al.*, 2006) found that *B. cereus* cells attach to stainless steel surfaces and are capable of forming a biofilm, a situation that can present a major problem for the food industry.

Endospore formers such as *Bacillus* spp. which have survived the pasteurization process can cause a variety of spoilage problems in dairy products, and *B. cereus* has been detected in ice creams, milk powders, fermented milks and pasteurized milks (Rossland *et al.*, 2005). Although *B. cereus* is able to grow at low temperatures, pH is a definite hurdle, and pH 5 is critical for the growth of this organism (Rossland *et al.*, 2005). It is expected therefore that growth in fermented dairy products with a pH lower than 5 will be minimal. Even where population levels of a toxigenic *B. cereus* strain in milk did attain levels of up to 9×10^7 cfu g^{-1} (Agata *et al.*, 2002), the emetic toxin (cereulide) production was very low. Milk and milk products are consequently not likely to be sources of food poisoning caused by emetic toxin- producing *B. cereus* strains.

5.3 Chemical hazards

5.3.1 Mycotoxins

Mycotoxins are secondary metabolites produced by fungi of various genera when growing on agricultural products before or after harvest or during transportation or storage (Briedenhann and Ekermans, 2004). Contamination by such fungi can occur at many stages during feed production, e.g. during plant growth, harvesting, storage and processing (McEvoy, 2002). Mycotoxins are found mostly on feed ingredients such as maize, sorghum, wheat and groundnuts. Mycotoxins pose a risk to both animal and human health, and depending on the chemical structure of the mycotoxin, pathology may result due to the carcinogenic, oestrogenic, neurotoxic, dermonecrotic or immunosuppressive activity of the toxin (Briedenhann and Ekermans, 2004).

Fungal species of greatest concern in the dairy industry are *Aspergillus flavus*, *A. parasiticus* and *A. nomius* (Heggum, 2003). These species produce Aflatoxin B_1 under optimum conditions of temperature, water activity and nutrient availability (Tsaknis and Lalas, 2004). In recent years, concern has been expressed about the presence of Aflatoxin M_1 in milk and milk products, and a direct link exists between the growth of the above mycotoxic fungi in feed and feed ingredients and the occurrence of Aflatoxin M_1 in milk. Following consumption of Aflatoxin B_1 contaminated feeds, the metabolite, Aflatoxin M_1, is secreted in the milk (McEvoy, 2002).

The aflatoxins have a very high toxigenicity, and Aflatoxin M_1 levels in dairy products are regulated in at least 22 countries (Tsaknis and Lalas, 2004). The EU maximum limits (MLs) for different feed commodities vary between 0.05 and 0.005 mg kg^{-1} (Heggum, 2003). Provided that these MLs for Aflatoxin B_1 (and other mycotoxins) in feeds are observed, there should be no problem with harmful residues in milk (Jonker *et al.*, 1999 as quoted by McEvoy, 2002). The maximum Codex limit in milk for Aflatoxin M_1 is 0.5 $\mu g\ kg^{-1}$.

Mycotoxins produced by fungal species other than *Aspergillus* and possibly *Penicillium* are of minor concern for dairy products (Tsaknis and Lalas, 2004). Nevertheless, contamination of feed and forage with zearalenone (a *Fusarium* spp. mycotoxin) has been shown to result in residues of zeranol in forage-fed cattle (Kennedy *et al.*, 1995 and Kennedy *et al.*, 1998 quoted by McEvoy, 2002). Zeranol, a hormonal growth promoter, is specifically prohibited from use in food animals in the EU. Work by Kennedy *et al.* in 1998 (quoted by McEvoy, 2002) has demonstrated that hydrogenation of alfa-zearelenol, probably in the rumen, is responsible for the formation of zeranol. This finding of a 'natural' zeranol in cattle has complicated control measures. This makes it necessary to differentiate zeranol arising from feed and forage contamination from deliberate growth promoter abuse.

Mycotoxins cannot be totally eliminated from feed once the feed has been contaminated, and processing the feed by means of heat treatment has little effect in eliminating the mycotoxins. Continuous monitoring of levels needs to take place to ensure that contamination does not exceed acceptable tolerance levels. Preventing contamination at source is one of the most effective methods of reducing the risk of mycotoxin contamination. Suitable measures need to be applied during crop production, handling, storage and processing (Briedenhann and Ekermans, 2004). Practical steps that can be taken to reduce mycotoxin contamination of grains and feed ingredients (McEvoy, 2002) include the pre-harvest selection of resistant seed varieties, prevention of physical damage to crops by insects and the use of appropriate crop rotation. At harvest, precautions to be taken include proper handling to avoid physical damage and crop cleaning to remove field soil. Storage practices include keeping crops dry and clean and proper labelling of crops (dates, etc.) to ensure that if problems do occur, traceability is ensured (Marquardt, 1996 quoted by McEvoy, 2002).

5.3.2 Antimicrobials

The most frequently and commonly used antimicrobials associated with milk are antibiotics, employed to combat mastitis-causing pathogens in the dairy cow. National surveys in developed countries show that between 0.1 and 0.5% of tanker milk samples test positive for antibiotic residues (Tsaknis and Lalas, 2004).

The occurrence of antibiotic residues in milk may have economic, technological and even human health implications. In the first place, such residues can lead to partial or complete inhibition of acid production by cheese starter cultures. This can lead to inadequate ripening and maturation of the cheese, resulting in flavour or texture defects and substantial financial loss for the dairy industry (Tsaknis and Lalas, 2004). There has also been increasing public concern over the possible links between veterinary drug residues in milk and the transfer of antibiotic resistant organisms and resistance genes to humans as a result of veterinary and zootechnical use of antibiotics in food animals (McEvoy, 2002). A third concern raised was that sensitive individuals could exhibit allergic reactions to drug residues or their metabolites, especially in the case of β-lactam antibiotics. According to Tsaknis and Lalas (2004), the allergy risk, however, is very low.

As from 1990, maximum residue limits (MRLs) have been set in Europe for veterinary drugs in foodstuffs of animal origin like milk (Reybroeck, 2003). Most dairy companies also use rapid tests to monitor all incoming milk for the presence of β-lactam antibiotics. Some of these companies are claiming compensation for the costs of disposing of the milk of a contaminated tanker load from the responsible farmer.

The solution to the problem of drug residues in milk lies in the application of the general principles of 'Good Farming Practice'. These include the following principles (Reybroeck, 2003):

- Good farm management should in the first place be directed towards the prevention of infectious diseases, such as (sub)clinical mastitis, in order to limit the use of veterinary drugs.
- In the process, farmers must keep their animals in sound physical condition by ensuring hygiene and good housekeeping practices and implementing sound farm management.

- In preventing mastitis, the use of properly functioning milking machines is of primary importance. The use of veterinary drugs, nevertheless, remains necessary, but this option should only be exercised after a correct diagnosis by a veterinarian. Only registered pharmaceutical products with a known depletion pattern should be used.
- Correctly administering the drugs is also very important in terms of prescribed dose, frequency and route of administration.
- The keeping of reliable records of such drug use is also essential. This remains the responsibility of the milk producer to respect the prescribed withdrawal period. In the process, the treated animals need to be marked clearly to allow for correct identification (e.g. by taping a hind leg).
- Treated cows need to be milked last, and during the withholding period the milk must be discarded in the proper way. The milking equipment should also be cleaned properly after contact with the contaminated milk.
- Special care should be taken with milk of cows that have been treated with long-acting dry cow products or with milk from cows that have been recently purchased.
- Good communication is also important. Everyone on the dairy farm should be informed of any treatment, although the number of people authorized to administer antibiotic drugs should be limited. If in any doubt, the milk should be tested.

5.3.3 Allergens

Cows' milk allergy or hypersensitivity is commonly encountered with a prevalence of 2–3% in infants and 0.5–3% in adults. This allergy in infants and children is in most cases a transient condition lasting from several months to a few years, after which tolerance is inclined to develop (Bindels and Hoijer, 2000). The dietary management of this ailment includes strict avoidance of the allergenic proteins.

Detailed information on the allergenic epitopes of major bovine milk proteins (α_{s1}-casein, β-lactoglobulin and α-lactalbumin) has recently been published (Bindels and Hoijer, 2000). It can be inferred from these new insights that in individual patients, several proteins and per protein, several epitopes (allergenic segments or polypeptides), are often involved. Furthermore, multiple combinations of allergenic proteins and epitopes may occur. Allergenic epitopes are located in specific regions of the proteins, both on the hydrophilic surface and in hydrophobic regions which become exposed upon denaturation and/or digestion.

Because present knowledge of the exact mechanisms of sensitization and tolerance induction is still far from complete, it cannot be excluded that specifically fractionated and/or modified cows' milk proteins, in combination with other bioactive components, will in future be useful in preventing sensitization and/or promoting the induction of tolerance.

5.3.4 Industrial and environmental contaminants

5.3.4.1 Pesticide residues

Pesticides include insecticides, herbicides and fungicides. The most common insecticides in turn include organochlorines, organophosphates and carbamates. Organochlorine pesticides enter the food chain as a result of their lipophilic properties, thereby causing a potential health risk for consumers. Milk is considered as one of the more convenient indicators for measuring the extent of persistent residues that have originated in contaminated animal feed. The main route of human exposure to many organochlorine pesticides is through food of animal origin of which milk is the most important product (Kodba, 2000).

Typical contaminants of milk are persistent fat-soluble organochlorine pesticides such as hexachlorobenzene (HCB), dichloro-diphenyl-trichloro-ethane (DDT) and, to a lesser extent, the cyclodiene compounds (Kodba, 2000).

5.3.4.2 Dioxins and polychlorinated biphenyls

The term 'dioxins' covers a group of 75 polychlorinated dibenzo-*p*-dioxin (PCDD) and 135 polychlorinated dibenzofuran (PCDF) congeners, of which 17 are of toxicological concern. The most toxic congener is 2,3,7,8-tetrachloro dibenzo-*p*-dioxin (TCDD), which is a known human carcinogen. It has been concluded that the carcinogenic effect of dioxins does not occur at levels below a certain threshold (Council Regulations, 2001). The maximum level set in the latter Regulations for dioxins and dioxin-like PCBs in milk and milk products is 3 pg World Health Organization Toxic Equivalents (WHO-TEQ) per gram of fat.

The polychlorinated biphenyls (PCBs) in turn are a group of 209 congeners which can be divided into two groups according to their toxicological properties. Twelve congeners exhibit toxicological properties similar to dioxins and are therefore termed 'dioxin-like PCBs'. The other PCBs have a different toxicological profile (Council Regulations, 2001).

Dioxins and PCBs are extremely resistant to chemical and biological degradation and therefore persist in the environment and accumulate in the feed and food chain. Dioxins and PCBs arise during the production of chloro-organics and in emissions of industrial and municipal incineration- and pyrolysis processes (Fiedler, 1999 as quoted by Parzefall, 2002). Contamination of animal feed occurs via particle-bound distribution on grass and other fodder plants (Parzefall, 2002). The accumulation of dioxins in animals is mainly from these contaminated feeding stuffs. Human foods of animal origin in turn contribute to approximately 80% of the overall human exposure to dioxins (Council Regulations, 2001).

For these reasons, feeding stuffs and, in some cases soil, raise concerns as potential sources of dioxins. Like the organochlorine pesticides, the dioxins and PCBs are fat-soluble. Case studies that have involved these contaminants include the Belgian PCB incident in 1999 (McEvoy, 2002). Feedstuffs produced from a contaminated source were sent to 2500 farms, and nearly every category of agrifood (pork, milk, chicken and eggs) was affected. In another incident, dairy product dioxin levels increased from the low level of 0.6 pg International Toxic Equivalents (I-TEQ) per gram of fat (summer of 1997) to 4.9 pg I-TEQ per gram of fat in February 1998 (Malisch, 2000). The source was citrus pulp from Brazil used as feed material for ruminants. The pulp was contaminated with about 5–10 ng I-TEQ per kilogram of dioxin, a 20–100-fold higher contamination than the normal feed background.

In 2000, as part of a survey programme, the German authorities detected high levels of dioxins in a choline chloride (CC) premix used as an animal food component.

Analysis of a large number of samples of pure CC, pine sawdust, almond shell and other substances used in the premix confirmed significant amounts of dioxins in pentachlorophenol (PCP)-contaminated sawdust (Llerena *et al.*, 2003).

5.3.4.3 Heavy metals

The term 'heavy metals' is a general term that applies to a group of metals and metalloids with an atomic density greater than 6 g cm^{-3}. Although it is only a loosely defined term, it is widely recognized and usually applied to elements such as Cr, Ni, Cu, Zn, Hg and Pb that are commonly associated with pollution and toxicity problems. Although these elements differ widely in their chemical properties, they are used extensively in electronics, machines and artefacts of everyday life as well as in 'high-tech' applications (Okonkwo, 2002).

The widespread distribution, contamination and multiple effects of heavy metals in the environment has become a global problem. Pb is among the most common metal pollutants. Major sources of lead exposure include lead in paint, gasoline, water distribution systems, food, industrial emissions and lead used in hobby activities and (Kinder, 1997; Malykh, et al. 2003). Lead exposure attributable to automobile air emissions was a major exposure source prior to 1976. Between 1976 and 1990, lead used in gasoline declined by 99.8% in the United States, but not in some other countries where lead is permitted in gasoline, according to the National Association of Physicians for the Environment (NAPE, 1993 as quoted by Kinder, 1997).

The adverse effects of Pb are now well recognized, and these include effects on the development of cognitive brain function in children, as well as on the kidney and haematopoietic system (Okonkwo, 2002). In a study by Nicholson *et al.* (1999), the highest metal concentrations in dairy cattle feeds were for Zn and Cu. Mineral supplements contained higher concentrations of Ni, Pb, Cd, As and Cr than did other feed components.

5.3.5 Procedures to minimize risk of feed and milk contamination

(1) The production, processing, storage, transport and distribution of safe, suitable feed and feed ingredients is the responsibility of all participants in the food chain. These participants include farmers, feed ingredient manufacturers, feed compounders, transport contractors, etc. Each participant is responsible for all the activities under their direct control, including compliance with applicable statutory requirements (Jooste and Siebrits, 2003). The Dutch Animal feed sector has opted for a quality-assurance system based on the HACCP approach applied by the European food industry. This emphasizes that the animal feed industry and the preceding ingredient suppliers are part of the food chain. Hence, their slogan of 'Feed for Food' (Den Hartog, 2003).
(2) Production of feed and feed ingredients on the farm: Adherence to good agricultural practices (GAP) is encouraged in the production of natural, improved and cultivated pastures, and forage and cereal grain crops used as feed or feed ingredients for food-producing animals. Following GAP, prescriptions will minimize the risk of biological, chemical or physical contaminants entering the feed chain.

Crop residuals and stubbles used for grazing after harvest should also be considered as livestock feed. The same applies to livestock bedding, since most livestock will consume a portion of their bedding. Straw or wood shavings should therefore be managed in the same manner as animal feed ingredients. Rational grazing and dispersion of manure droppings should be applied to reduce cross-contamination between animals. Other factors should also be taken into consideration such as the proximity of the agricultural land to industrial operations where effluent or gas emissions can lead to feed contamination. Similarly, chemical fertilizers, manure, pesticides and other agricultural chemicals should be stored, managed and disposed of correctly.

(3) Monitoring and identification of health hazards: When purchasing feed ingredients from suppliers, such suppliers should be able to demonstrably guarantee product safety (Den Hartog, 2003). Audit procedures can include inspection, sampling and analysis for undesirable substances. Feed ingredients should meet acceptable and, if applicable, statutory standards for levels of pathogens, mycotoxins, pesticides and other undesirable substances that may constitute a health hazard for the consumer. Any contaminated feed or feed ingredient is unsuitable for animal feed and should be discarded. Traceability of feed and feed ingredients, including additives, should be enabled by proper labelling and record keeping at all stages of production and distribution.

(4) Processing, storage and distribution of feeds and feed ingredients. The effective implementation of GMPs and, where applicable, HACCP-based approaches should ensure that the following areas are addressed:
   - *Premises* – Buildings and equipment must be constructed to permit ease of operation, maintenance and cleaning. Water should be of a suitable quality, and effluent should be adequately disposed of.
   - *Receiving, storage and transportation* – Feed and feed ingredients should be stored separately from fertilizers, pesticides, etc. Processed material should also be stored separately from unprocessed ingredients. The presence of undesirable substances should be monitored for and controlled. The finished products should be delivered and used as quickly as possible. During storage, special precautions should be taken to restrict microbial growth in feedstuffs and ingredients.
   - *Personnel training* – Personnel should be adequately trained and aware of their role and responsibility in protecting food safety.

5.4 Physical hazards

Physical hazards, like microbiological and chemical hazards, can enter a food product at any stage in its production. There is a huge variety of physical items that can enter food as foreign material, and some of these are hazards to food safety (Teagasc, 2004). The main physical food safety hazards include items such as glass, metal (including broken needles), stones, wood, plastic, dust and hair. The main risk area for physical contamination of milk at the farm level is the stored milk in the bulk tank. Producers should, therefore, assess potential physical hazard risks in storage areas (e.g. breakable glass, loose material, etc.). Corrective action to be taken in the area housing the bulk milk storage tank is, for example, to use shatterproof light-covers to prevent glass contamination from taking place (Teagasc, 2004).

During the processing of milk, it is invariably subjected to procedures that will remove any physical contaminant. Examples of such procedures may include filtration, while centrifugal clarifiers are standard equipment in any commercial milk-processing operation. However, similar to food beverages such as fruit juice (FDA, 2004), consideration of potential hazards associated with glass breakage should be part of an HACCP plan in any processing plant that may be packaging liquid milk products in glass. Glass fragments caused by glass bottle breakage may result in serious injury and can be caused in a number of ways, including damage to the bottles in transit to the processing plant, or damage to bottles during mechanized handling (cleaning, filling or capping) of the bottles (FDA, 2004).

If control measures for glass fragments in the product are deemed necessary, one way is the use of online glass detection equipment such as X-ray detection (FDA, 2004). In this method, the product itself is continuously monitored after the last step at which glass inclusion is reasonably likely to occur (e.g. after bottling and sealing of the product; FDA, 2004). The HACCP critical limit may then be designated as 'no glass in the finished product'.

Another way to control glass fragments, applicable in operations where the containers are manually (not mechanically) handled and sealed (FDA, 2004), involves inspecting glass containers visually before they are filled to ensure that glass fragments are not present in the containers. An appropriately trained individual at a container inspection step in the process may do this.

A third way to control glass fragments is visual inspection at steps in the process where glass breakage can result in glass entering the product (FDA, 2004). This can take place upon reception of the glass containers, during glass container storage, mechanical conveying, mechanical filling and mechanical capping. The inspection looks for any evidence of glass breakage in those areas.

The consideration of potential hazards associated with metal fragments may be less relevant in dairy products such as milk. In dairy products containing fruit pulp or in the processing of fruit juice, metal fragments may become a part of the hazard analysis if operations such as the grinding of fruit or cutting operations are involved (FDA, 2004). In such cases, it is possible that metal fatigue or metal-to-metal contact can occur in the processing operation. If hazard analysis shows that it is reasonably likely that metal fragments can be a problem, controls for metal fragments should be established in the HACCP plan of the operation.

The following ways to establish control measures for metal fragments in the product are recommended by the FDA (2004). One way involves the use of online metal-detection equipment. With this method, the equipment continuously monitors the product after the last step at which metal inclusion is reasonably likely to occur (e.g. after bottling and sealing of the product) at a process step designated for metal detection. A second way to control metal fragments involves the use of a separation device such as a screen after the last step at which metal inclusion is reasonably likely to occur, at a process step designated for screening. A third way to control metal fragments involves visually inspecting equipment for damage or missing parts at process steps such as extracting and grinding, where such damage or loss of parts could lead to metal fragments in the fruit pulp or juice. This approach may only be feasible for relatively simple equipment that can be fully inspected visually in a reasonable time period.

5.5 Traceability of ingredients

Traceability is an important criterion for the evaluation of a food supply chain. The root cause of a safety or quality problem, for example, can only be identified if the raw material can be traced back to its source (Hamprecht *et al.*, 2004). The International Standards Organization (ISO) defines traceability as 'the ability to trace the history, application or location of that which is under consideration' (Golan *et al.*, 2004). This definition of traceability is necessarily broad because food is a complex product, and traceability is a tool for achieving a number of different objectives.

Such objectives (Golan *et al.*, 2004) may include the improvement of supply management, the facilitation of trace back for food safety and quality and/or the differentiation and marketing of foods with subtle or undetectable quality attributes. No traceability system is complete, however, since a system for tracking every input and process to satisfy every objective would be enormous and very costly. The extent to which a traceability system is applied will consequently depend on the characteristics of a production process and the specific traceability objectives (Golan *et al.*, 2004).

Current best-practice examples (Hamprecht *et al.*, 2004) show that in the case of a meat supply chain, for example, the agricultural producer can be retraced very rapidly. In other supply chains, such as the raw-milk supply chain, traceability is more complicated, since supplies from producers are mixed before entering the next processing stage. In a raw-milk supply chain, it is necessary to take and maintain product samples at the handover points (Hamprecht *et al.*, 2004). In the event of contamination, the source can then be identified by testing these stored samples.

While traceability is very important *per se* and yields definite financial and other benefits, it still remains only one element of any supply management or quality/safety control system (Golan *et al.*, 2004). Traceability, which entails a documentation of processes, should slot in with Food Safety and Quality activities at the level of HACCP critical control points and Standard Operating Procedures (Hamprecht *et al.*, 2004).

In Tables 5.1 and 5.2, a selection of the records that should be maintained in the milk supply chain to achieve traceability beyond the process of milking the cows is presented (Hamprecht *et al.*, 2004).

The interaction between traceability and food safety/quality assurance activities throughout production, transport and storage of the raw milk can be illustrated with reference to HACCP critical control points (CCPs) described by Hamprecht *et al.* (2004). Controls of the milk producer, in this case study, concentrate on the following four CCPs:

- CCP1: use of livestock medicines and other chemicals;
- CCP2: cooling and storage of milk;
- CCP3: equipment sanitation (cleanliness);
- CCP4: use of water for cleaning of milk contact surfaces.

Table 5.1 Records for ensuring traceability throughout the production process of raw milk (Cocucci *et al.* 2004 as quoted by Hamprecht *et al.* 2004).

Process phases	Traceability record
Herd management	Livestock treatment records (medication administered to the cattle)
Plot management	Records of soil treatment (nature and sources of fertilizers and pesticide treatments)
Feeding management	Feed registration records (nature of feed fed to cattle and documentation of batch numbers)
Herd management	Livestock movement registration (registration of cows bought or sold as well as registration of cows exchanged with other producers)
Feeds management	List of approved suppliers and definition of externally procured feedstuffs (list of all feed ingredients included in the feeding regime)

Table 5.2 Selection of records for ensuring traceability during transport of the raw milk (Cocucci *et al.* 2004 as quoted by Hamprecht *et al.* 2004).

Process phases	Traceability record
Tank storage	Records of milk batches collected from the farm storage tank by the transporting milk tankers and particulars about raw-milk samples and storage
Transport	Registration of the collection rounds organized by the logistics provider; mention of the sources of milk batches in the transport document

Table 5.3 Selection of records for controlling food safety at farm level (Dairy Farmers of Canada 2001 as quoted by Hamprecht *et al.* 2004).

CCP No.	CCP description	Hazard description	Record for controlling the hazard
CCP1a	Use of livestock medicines	Residues of medications in milk	Livestock treatment records
CCP1b	Use of pesticides	Pesticide residues in milk	Record of all pasture and fodder treatments with chemical substances
CCP2	Cooling and storage of milk	Growth of bacteria if milk temperatures are too high	Farm tank (or bulk tank) temperature log-book
CCP3	Equipment sanitation (cleanliness)	Growth of bacteria in unclean tanks; residues of cleaning agents	Milk equipment sanitation record sheet (record of cleaning operations)
CCP4	Use of water for cleaning of milk contact surfaces	Biological and chemical hazards from contaminated water	Water records (annual laboratory tests)

Two of the records maintained at these CCPs and referred to under traceability concern the feeding of the cattle and medication. A more detailed description and breakdown of these CCPs further illustrate the relevance to traceability and are shown in Table 5.3.

As mentioned earlier, traceability also slots in with Food Safety and Quality activities in terms of Standard Operating Procedures (Hamprecht *et al.*, 2004). Such procedures are described in greater detail in Table 5.4.

Table 5.4 Selection of prerequisite records for raw-milk production (Hamprecht *et al.* 2004).

Prerequisite Record	Document description
Standard operating procedure for pre-milking	List of steps compiled by the farmer; the steps describe the necessary actions to ensure that the milking equipment is properly setup by all employees milking the cows
Standard operating procedure for milking	List of steps compiled by the farmer; writing down the various actions helps ensure that all farm employees milk those cows in the same manner

The above records are maintained by the producers. Further down the supply chain, the logistics providers transporting the milk to the dairy as well as personnel of the processing dairy concentrate their controls around three control points (Hamprecht *et al.*, 2004), namely:

- CCP1: sanitation (cleanliness) of the bulk-tanker;
- CCP2: raw-milk delivery to the processing dairy and control prior to discharging the milk (e.g. milk temperature and antibiotic residue determination);
- CCP3: raw-milk delivery to the processing dairy and control following discharge of the milk (detailed laboratory analysis).

In the processing plant itself, traceability systems also help firms to isolate the source and extent of safety or quality-control problems. This helps reduce the production and distribution of unsafe or poor-quality products, which in turn reduces the potential for bad publicity, liability and recalls (Golan *et al.*, 2004). The better and more precise the tracing system, the faster the processor can identify and resolve food safety or food-quality problems. One milk processor surveyed by Golan *et al.* (2004) coded each item to identify time of processing, line of processing. place of processing and sequence. With such specific information, the processor can trace a faulty product to the minute of production and determine whether other products from the same batch are also defective.

In conclusion, a programme should be in place to ensure identification and traceability at all stages of manufacture and storage for raw materials through to the finished product. The programme must allow trace-back and trace-forward of all dairy products and ingredients, and must be validated (Dairy Food Safety Victoria, 2002). All dairy manufacturers are required by the Dairy Food Safety Victoria (2002), *Code of Practice for Dairy Food Safety,* to have a product-recall plan that is also validated to ensure its ongoing effectiveness. For this purpose, *The Food Industry Recall Protocol, a Guide to Conducting a Food Recall* (ANZFA, 2001 quoted by Dairy Food Safety Victoria, 2002) must be followed.

References

ADASC. (1999) *Australian Manual for Control of* Listeria *in the Dairy Industry*. Australian Dairy Authorities' Standards Committee, July 1999.

Agata, N., Ohta, M. and Yokoyama, K. (2002) Production of *Bacillus cereus* emetic toxin (cereulide) in various foods. *International Journal of Food Microbiology* **73**, 23–27.

Anon. (2003) Proposed draft Code of Practice on Good Animal Feeding. In: *Report of the 4th session of the ad hoc Intergovernmental Codex Task Force on Animal Feeding*; *Alinorm* 03/38A, Copenhagen, Denmark. 25–28 March 2003.
Arqués, J.L., Rodriguez, E., Gaya, P., Medina, M. and Nuñez, M. (2005) Effect of high pressure treatment and bacteriocin-producing lactic acid bacteria on the survival of *Listeria monocytogenes* in raw milk cheese. *International Dairy Journal* **15**, 893–900.
Asperger, H. (1994) *Staphylococcus aureus*. In: *Monograph on the Significance of Pathogenic Microorganisms in Raw Milk*. International Dairy Federation, Brussels. pp. 24–42.
Bindels, J.G. and Hoijer, M. (2000) Allergens: latest developments, newest techniques. In: *Safety in Dairy Products. Bulletin of the International Dairy Federation, Doc. No.* 351/2000. pp. 31–32.
Boor, K.J. and Murphy, S.C. (2002) Microbiology of market milks. In: Robinson, R.K. (ed.) *Dairy Microbiology Handbook*, 3rd edn. John Wiley and Sons, London. pp. 91–122.
Briedenhann, E. and Ekermans, L.G. (2004) Ensuring feed safety and quality during sourcing, processing and storage. In: *A Farm to Table Approach for Emerging and Developed Dairy Countries: Proc. IDF/FAO Int. Symp. Dairy Safety Hygiene,* Cape Town, 2–5 March, 2004, pp. 61–64.
Byrne, R. (2004) Microbiological hazards that need to be managed during and after processing (an overview). In: *A Farm to Table Approach for Emerging and Developed Dairy Countries: Proc. IDF/FAO Int. Symp. Dairy Safety Hygiene*, Cape Town, South Africa, 2–5 March 2004, pp. 127–129.
Chambers, J.V. (2002) Microbiology of raw milk. In: Robinson, R.K. (ed.) *Dairy Microbiology Handbook*, 3rd edn. John Wiley and Sons, London. pp. 39–90.
Cliver, D.O. (1988) Virus transmission via foods. *Food Technology* **42**, 241–248.
Cliver, D.O. (1993) Cheddar cheese as a vehicle for viruses. *Journal of Dairy Science*, **56**, 1329–1330.
Cliver, D.O. (1997) Virus transmission via food. *Food Technology* **51**, 71–78.
Council Regulations (2001) Council regulation (EC) no. 2375/2001 of 29 November 2001. *Official Journal of the European Community.*
Dairy Food Safety Victoria (2002) Code of Practice for Dairy Food Safety. http://www.dairysafe.vic.gov.au. Accessed December 2005.
Dawson, D. (2005) Food borne protozoan parasites. *International Journal of Food Microbiology*, **103**, 207–227.
Dealler, S. (2002) Information concerning transmissible spongiform encephalopathy for the scientific and business world. http://www.bse.airtime.co.uk/welcome.htm. Accessed December 2005.
De Buyser, M-L., Dufour, B., Maire, M. and Lafarge, V. (2001) Implication of milk and milk products in food-borne diseases in France and in different industrialised countries. *International Journal of Food Microbiology* **67**, 1–17.
Den Hartog, J. (2003) Feed for food: HACCP in the animal feed industry. *Food Control* **14**, 95–99.
Deng, M.Q. and Cliver, D.O. (1999) *Cryptosporidium parvum* studies in dairy products. *International Journal of Food Microbiology*, **46**, 113–121.
Elsser, K.A., Moricz, M. and Proctor, E.M. (1986) *Cryptosporidium* infections: a laboratory survey. *Canadian Medical Association Journal*, **135**, 211–213.
Faber, J. and Peterkin, P.I. (2000) *Listeria*. In: Lund, B.M., Baird-Parker, A.C. and Gould, G. (eds.) *The Microbiology of Food*, pp. 1178–1232. Chapman & Hall, London.
FDA (2004) *Juice HACCP Hazards and Controls Guidance*, 1st edn. US Department of Health and Human Services, Food and Drug Administration, Center for Food Safety and Applied Nutrition, February, 2004. http://www.cfsan.fda.gov/guidance.html (accessed December 2005).
Fedio, W.M. and Jackson, H. (1992) On the origin of *Listeria monocytogenes* in raw bulk-tank milk. *International Dairy Journal* **2**, 197–208.
Gallmann, P.U. and Eberhard, P. (2004) New developments in heating technology for food preservation and safety. In: *A Farm to Table Approach for Emerging and Developed Dairy*

Countries: Proceedings of IDF/FAO International Symposium on Dairy Safety Hygiene, Cape Town, South Africa, 2–5 March 2004, pp. 141–147.

Gelletlie, R., Stuart, J., Soltanpor, N., Armstrong, R. and Nichols, G. (1997) Cryptosporidiosis associated with school milk. *Lancet* **350**, 1005–1006.

Golan, E., Krissoff, B., Kuchler, F., Calvin, L., Nelson, K. and Price, G. (2004) Traceability in the US Food Supply: Economics, Theory and Industry Studies. AER-830, USDA/ERS, March 2004. http://www.ers.usda.gov/Amberwaves/April04/Features/FoodTraceability.htm (accessed December 2005).

Grant, I.R., Hitchings, E.I., McCartney, A., Ferguson, F. and Rowe, M.T. (2002) Effect of commercial-scale high-temperature, short-time pasteurisation on the viability of *Mycobacterium paratuberculosis* in naturally infected cow's milk. *Applied and Environmental Microbiology* **68**, 602–607.

Greyling, L. (1998) Hygienic and Compositional Quality of Milk in the Free State Province. MSc dissertation. Department of Food Science, Faculty of Natural Sciences, University of the Orange Free State, Bloemfontein, South Africa.

Griffiths, M.W. (1989) *Listeria monocytogenes*: Its Importance in the dairy industry. *Journal of the Science of Food and Agriculture* **47**, 133–158.

Guyer, R.L. (2004) Prions: puzzling infectious proteins. *NIH – Research in the News*. http://www.science-education.nih.gov/nihHTML/ose/snapshots/multimedia/ritn/prions1 (accessed November 2004).

Hamprecht, J., Noll, M., Corsten, D. and Fahrni, F. (2004) *Controlling Food Safety, Quality and Sustainability in Agricultural Supply Chains*. Kuehne-Institute for Logistics, University of St Gallen, St Gallen, Switzerland.

Harp, J.A., Fayer, R., Pesch, B.A. and Jackson, G.J. (1996) Effect of pasteurization on infectivity of *Cryptosporidium parvum* oocysts in water and milk. *Applied and Environmental Microbiology*, **62**, 2866–2868.

Heggum, C. (2003) Control of aflatoxin in cattle feed – the Danish system. In: *Proceedings of the Conference on Quality Management at Farm Level. IDF World Dairy Summit and Centenary,* 7–12 September, Bruges, Belgium, pp. 263–267.

Hillerton, J.E. (2003) Control of MAP in milk. In: *Proceedings of the Conference on Quality Management at Farm Level. IDF World Dairy Summit and Centenary*, 7–12 September 2003, Bruges, Belgium, pp. 251–253.

Holko, I., Bisova, T., Holkova, Z. and Kmet, V. (2006) Virulence markers of *E. coli* strains isolated from traditional cheeses made from unpasteurised sheep milk in Slovakia. *Food Control* **17**, 393–396.

Hunter, P.R. and Nichols, G. (2002) Epidemiology and clinical features of *Cryptosporidium* infection in immunocompromised patients. *Clinical Microbiology Reviews* **15**, 145–154.

Jooste, P.J. and Siebrits, F.K. (2003) Minimizing the risk of milk contamination by feed-borne residues. In: *Proceedings of the Conference on Quality Management at Farm Level. IDF World Dairy Summit and Centenary,* 7–12 September 2003, Bruges, Belgium, pp. 255–261.

Käferstein, F. and Abdussalam, M. (1999) Food safety in the 21st century. *Bulletin of the World Health Organization* **77**, 347–351.

Kinder, C. (1997) Lead contamination in our environment. www.yale.edu/ynhti/curriculum/units/1997/7/97.07.05.x.html

Kodba, Z.C. (2000) Organochlorine pesticide residues in dairy milk. In: *Bulletin of the International Dairy Federation, Doc. No.* 351/2000 'Safety in Dairy Products', IDF, Brussels, pp. 34–35.

Kosek, M., Alcantara, C., Lima, A.A.M. and Guerrant, R.L. (2001) Cryptosporidiosis: an update. *Lancet* **1**, 262–269.

Lee Wong, A.C. (1998) Biofilms in food processing environments. *Journal of Dairy Science* **81**, 2765–2770.

Lekkas, C., Kakouri, A., Paleologos, E., Voutsinas, L.P., Kontominas, M.G. and Samelis, J. (2006) Survival of *E. coli* 0157:H7 in Galotyri cheese stored at 4 and 12°C. *Food Microbiology* **23**, 268–276.

Llerena, J.J., Abad E., Caizach, J. and Rivera, J. (2003) An episode of dioxin contamination in feeding stuff: the choline chloride case. *Chemosphere* **53**, 679–683.

Malisch, R. (2000) Increase of the PCDD/F-contamination of milk, butter and meat samples by use of contaminated citrus pulp. *Chemosphere* **40**, 1041–1053.

Malykh. O., Nikonov, B., Gurvich, V., Kuzmin, S., Privalova, L., Kosheleva, A., Katsnelson, B., Marshalkin, A., Prokopyev, A. and Busyrev, S. (2003) Lead contamination of the environment and its health effects in children dwelling in the vicinity of copper smelters in the Sverdlovsk region. olgam@ocsen.ru.info@urcee.ru Accessed 27 Jul. 2007.

Marinšek, J. (2000) Hygienic irrevocability of milk and milk products – guarantee of safety. In: *Safety in Dairy Products. Bulletin of the International Dairy Federation, Doc. No.* 351/2000, pp. 24–26.

Mathot, P.J. (2002) Presentation on 'Integrated Food Chain Management' at the World Dairy Congress 'Congrilait', Paris, September 2002.

McEvoy, J.D.G. (2002) Contamination of animal feedingstuffs as a cause of residues in food: a review of regulatory aspects, incidence and control. *Analytica Chimica Acta* **473**, 3–26.

Mead, P.S., Slutsker, L., Dietz, V., McCaig, L.F., Bresee, J.S., Shapiro, C., Griffin, P.M., and Tauxe, R.V. (1999) Food-related illness and death in the United States. *Emerging Infectious Diseases* **5**, 607–625.

Michel, A.L. and McCrindle, C.M. (2004) An overview of zoonoses with importance to the dairy industry. In: *A Farm to Table Approach for Emerging and Developed Dairy Countries: Proceedings of the IDF/FAO International Symposium on Dairy Safety and Hygiene*, 2–5 March 2004 Cape Town, South Africa, pp. 47–51.

Millar, B.C., Finn, M., Xiao, L., Lowery, C.J., Dooley, J.S.G. and Moore, J.E. (2002) *Cryptosporidium* in foodstuffs – an emerging aetiological route of human food borne illness. *Trends in Food Science & Technology* **13**, 168–187.

More O'Ferral-Berndt, M. (2000) A comparison of selected public health criteria in milk from milk-shops and from a national distributor. M.Med.Vet.(Hyg.) dissertation, Faculty of Veterinary Science, University of Pretoria, Pretoria, South Africa.

Mostert, J.F. and Jooste, P.J. (2002) Quality control in the dairy industry. In: Robinson, R.K. (ed.), pp. 655–736 *Dairy Microbiology Handbook,* 3rd edn. John Wiley, London.

Nicholson, F.A., Chambers, B.J., Williams, J.R. and Unwin, R.J. (1999) Heavy metal contents of livestock feeds and animal manures in England and Wales. *Bioresource Technology* **70**, 23–31.

Noordhuizen, J.P. (2003) Microbiological contaminants (zoonoses). In: *Proceedings of the Conference on Quality Management at Dairy Farm Level. IDF World Dairy Summit and Centenary,* 7–12 September 2003, Bruges, Belgium, pp. 241–250.

Okonkwo, J. (2002) *Chemistry in industry and the environment: the search for happier and healthier lives.* Professorial Inaugural Address. Faculty of Sciences, Tshwane University of Technology, Pretoria, South Africa.

Oliver, S.P., Jayarao, B.M. and Almeida, R.A. (2005) Food borne pathogens in milk and the dairy farm environment: food safety and public health implications. *Foodborne Pathogens and Disease* **2**, 115–129.

Olson, M.E., O'Handley, R.M., Ralston, B.U.J., McAllister, T.E.A. and Thompson, O.K. (2004) Update on *Cryptosporidium* and *Giardia* infections in cattle. *Trends in Parasitology*, **20**, 185–191.

Parzefall, W. (2002) Risk assessment of dioxin contamination in human food. *Food and Chemical Toxicology* **40**, 1185–1189.

Reilly, A., Tlustos, C., Anderson, W., O'Connor, L., Foley, B. and Wall, P.G. (2002) Food Safety: a public health issue of growing importance. In: Gibney, M.J., Vorster, H.H. and Kok, F.J. (eds.) *Introduction to Human Nutrition.* Blackwell Science, Oxford pp. 292–317.

Reybroeck, W. (2003) Role of the farmer in preventing residues of antibiotics in farm milk. In: *Conference of Quality Management at Farm Level. IDF World Dairy Summit and Centenary*, 7–12 September 2003, Bruges, Belgium, pp. 239–240.

Robinson, R.K., Tamime, A.Y. and Wsolek, M. (2002). Microbiology of fermented milks. In: Robinson, R.K. (ed.) *Dairy Microbiology Handbook*, 3rd ed., pp. 367–430. John Wiley, London.

Rossland, E., Langsrud, T. and Sorhaug, T. (2005) Influence of controlled lactic fermentation on growth and sporulation of *Bacillus cereus* in milk. *International Journal of Food Microbiology* **103**, 69–77.

Rzezutka, A. and Cook, N. (2004) Survival of human enteric viruses in the environment and food. *FEMS Microbiology Reviews*, **28**, 441–453.

Salo, S., Ehavald, H., Raaska, L., Vokk, R. and Wirtanen, G. (2006) Microbial surveys in Estonian dairies. LWT *Food Science and Technology* **39**, (5), 460–471.

Svensson, L. (2000) Diagnosis of food borne viral infections in patients. *International Journal of Food Microbiology*, **59**, 117–126.

Teagasc (2004) Physical food safety hazards. Irish Agricultural and Food Development Authority. http://www.foodassurance.teagasc.ie/faol (accessed December 2005).

Tompkin, R.B. (2002) Control of *Listeria monocytogenes* in the food processing environment. *Journal of Food Protection* **65**, 709–725.

Tsaknis, J. and Lalas, S. (2004) Chemical hazards – an overview. In: *A Farm to Table Approach for Emerging and Developed Dairy Countries: Proceedings of the IDF/FAO International Symposium on Dairy Safety and Hygiene*, 2–5 March 2004, Cape Town. pp. 151–155.

Tsegaye, M. and Ashenafi, M. (2005) Fate of *E. coli* 0157:H7 during the processing and storage of Ergo and Ayib, traditional Ethiopian dairy products. *International Journal of Food Microbiology* **103**, 11–21.

Tybor, P.T. and Gilson, W. (2003) *Dairy Producer's Guide to Food Safety in Milk Production.* Cooperative Extension Service, College of Agriculture and Environmental Science, University of Georgia, Athens, Georgia, USA. http://www.inform.umd.edu/EdRes (accessed June 2003).

USDA (2005) Food Safety and Inspection Service, United States Department of Agriculture. http://www.fsis.usda.gov/Fact_Sheets/Bovine_Spongiform_Encephalopathy_Mad_Cow_Disease (accessed October 2005).

Van Renterghem, R. and Daeseleire, E. (2000) Polychlorinated biphenyls. In: *Bulletin of the International Dairy Federation,* Doc. No. 351/2000, International Dairy Federation, Brussels, Belgium. pp. 11–17.

Vernozy-Rozand, C., Mazuy-Cruchaudet, C., Bavai, C., Montet, M.P., Bonin, V., Dernburg, A. and Richard, Y. (2005) Growth and survival of *E. coli* 0157:H7 during the manufacture and ripening of raw goat milk lactic cheeses. *International Journal of Food Microbiology* **105**, 83–88.

WHO (1984) Cryptosporidiosis surveillance. World Health Organization. *Weekly Epidemiological Record* **59**, 72–73.

WHO (1994) Outbreak of tick-borne encephalitis, presumably milk-borne. World Health Organization. *Weekly Epidemiological Record* **19**, 140–141.

WHO (2005) Basic food safety for health workers. Chapter 1: Food borne Illness; 5–16. World Health Organization. http://www.who.int/entity/foodsafety/publications/capacity/en (accessed April 2005).

Yucel, N. and Ulusoy, H. (2006) A Turkey survey of hygienic indicator bacteria and *Yersinia enterocolitica* in raw milk and cheese samples. *Food Control* **17**, 383–388.

Chapter 6

Modern Laboratory Practices – Analysis of Dairy Products

Thomas Bintsis, Apostolos S. Angelidis and Lefki Psoni

6.1 Introduction

Consumers of dairy products need to be assured that the products they purchase and consume are wholesome, safe and accurately labelled. Although the majority of the milk supply produced worldwide comes from cows, the production of milk from other species and in particular ovine, caprine and water-buffalo milk plays an important socio-economic role in several countries mainly due to its use for the manufacture of traditional cheeses such as Feta, Halloumi, Mozzarella and Roquefort.

Dairy-product wholesomeness, appearance and cost are factors driving consumer acceptance. Government regulatory agencies in conjunction with the dairy industry and individuals involved in primary production are assigned with the responsibility of delivering foods that are of high quality, unadulterated and biologically and chemically safe. Food safety nowadays is achieved through end-product analysis, but also via careful monitoring and control of potential risks that can occur in all steps of the food-production chain, beginning from primary production in the field until the finished product reaches the consumer's household, the so-called 'from farm-to-table' approach. Appropriately equipped and accredited laboratories, sufficiently trained personnel and analytical methods that are fast and reliable are among the key factors necessary for achieving product quality and safety.

Standards on milk and milk products have been developed through the collaboration of the International Dairy Federation (IDF) with the International Organization for Standardization (ISO), with its Technical Committee ISO/TC 34 and the Association of Official Analytical Chemists (AOAC International). This approach has included consideration of all necessary aspects of validation and related uncertainty, and has been done in order to facilitate the national, regional and worldwide trading of dairy products. More specifically, the tripartite has provided standard sampling methods as well as methods for analyses, that is, reference and routine methods for the evaluation of the physical, chemical and microbiological properties of dairy products.

The aim of this chapter is to present reference and/or modern methods of analysis used for the evaluation of the quality and safety of milk and milk products.

6.2 Laboratory quality assurance

A Quality Assurance (QA) system, focused on the key issues which determine quality results, costs and timeliness, is vital for obtaining reliable results for physical, chemical and microbiological analyses of dairy products. Appropriate QA enables a laboratory to document that it possesses adequate facilities and equipment for carrying out the required analyses, and that the work is carried out in a controlled manner by competent staff, following documented validated methods. A laboratory may decide to design its own QA system based on Good Laboratory Practices (GLP), or it may follow one of the available established protocols (Smittle and Okrend, 2001).

The principles of GLP regarding the control of chemicals, using common managerial and scientific practices, were developed in 1980 by an international group of experts from the Organization for Economic Cooperation and Development (OECD).

The latest edition of *OECD Principles of Good Laboratory Practice* (OECD, 1998) contains guidelines concerning the organizational processes and conditions under which laboratory studies related to certain regulatory work are carried out. In this guideline, GLP is defined as 'a quality system concerned with the organizational process and the conditions under which non-clinical health and environmental safety studies are planned, performed, monitored, recorded, archived and reported'. More recently, the scope of GLP has been extended as described in European Commission (EC) Directive 2004/10/EC (EC, 2004).

6.2.1 Accreditation of laboratories

Accreditation of laboratories to internationally accepted Standards (ISO, 1999a) is currently recognized as a prerequisite for the assurance of analyses and has been frequently used as the basis of contracts for analytical work. Standard ISO 17025:1999 (ISO, 1999a) addresses the technical competence required by laboratories to carry out specific tests and calibrations and is used by laboratory accreditation bodies worldwide as the basis of the requirements for accreditation, and the definitions used are shown on Table 6.1. The role and content of a quality-assurance manual (QAM) are discussed by Garfield *et al.* (2000), and this manual describes the requirements of the Standard.

More specifically, a QAM, covers issues pertaining to:

- laboratory management;
- procedures for management review;
- procedures for internal audits;
- the training and the academic and professional qualifications of the staff employed;
- the suitability of the laboratory equipment and its maintenance;
- methods (standard or in-house developed), calibration procedures and validated test methods;
- the calibration of equipment and test materials;
- traceability;
- proficiency testing;
- sample handling and identification;
- recording of analytical data;
- reporting of the results;
- procedures for dealing with complaints;
- subcontracts with other laboratories; and
- procedures for ensuring the quality of goods and services purchased from outside suppliers.

The ultimate goal of an accreditation system is to identify laboratories with a demonstrated ability to produce accurate, reliable and consistent results using validated methods. Accreditation is granted to a laboratory for a specified set of activities (i.e. microbiological tests, or tests for pesticides) and specific categories of products (i.e. dairy products) following an external audit from an accreditation body. Such an

Table 6.1 Definitions of terms used in quality-assurance laboratory systems[a].

Term	Definition
Accreditation	Procedure by which an authoritative body gives formal recognition that a body or a person is competent to carry out specific tasks.
Certification	Procedure by which a third party gives written assurance that a product, process or service conforms to specific requirements.
Reference material	Material or substance, one or more of whose property values are sufficiently homogenous and well established to be used for the calibration of an apparatus, the assessment of a measured method or assigning values to materials.
Certified reference material	Reference material, accompanied by a certificate, one or more of whose property values are certified by a procedure, which establishes its traceability to an accurate realization of the units in which the property values are expressed, and for which each certified value is accompanied by an uncertainty at a stated level of confidence.
Traceability	Property of the result of a measurement or the value of a standard whereby it can be related to stated references, usually national or international standards, through an unbroken chain of comparisons all having stated uncertainties.
Measurement uncertainty	A parameter associated with the result of a measurement that characterizes the dispersion of the values that could reasonably be attributed to the measurand.
Limit of detection of an analyte	Often determined by repeat analysis of a blank test portion and is the analyte concentration, the response of which is equivalent to the mean blank response plus 3 standard deviations.
Limit of quantitation	The lowest concentration of analyte that can be determined with an acceptable level of uncertainty.
Ruggedness or robustness	Where different laboratories use the same method, they inevitably introduce small variations in the procedure and may or may not have a significant influence on the performance of the method. The ruggedness of a method is tested by deliberately introducing small changes to the method and examining the consequences.
Sensitivity	The difference in analyte concentration corresponding to the smallest difference in the response of the method that can be detected.
Accuracy of a method	The closeness of the obtained analyte value to the true value; it can be established by analysing a suitable reference material.
Precision of a method	A statement of the closeness of agreement between mutually independent test results and usually stated in terms of standard deviation. It is generally dependent on the analyte concentration, and this dependence should be determined and documented. Precision is a component of measurement uncertainty.
Repeatability	The type of precision relating to the measurements made under repeatable conditions (i.e. same method, same material, same operator, same laboratory, narrow time period).
Reproducibility	The concept of precision relating to measurements made under reproducibility conditions (i.e. same method, different operator, different laboratories, different equipment, long time period).

[a]After EURACHEM/CITAC (2002).

audit will typically include an examination of the analytical procedures in use, the quality-management system and the relevant documentation. The analytical procedures will be examined to ensure that they are technically appropriate for the intended purpose and they have been validated. The performance of tests may be witnessed to ensure that documented procedures are being followed. The laboratory's participation in proficiency testing schemes may also be examined. Audits may additionally include a test where the laboratory is asked to analyse samples supplied by the accreditation body, and is expected to achieve acceptable levels of accuracy, namely a kind of a proficiency testing.

6.2.2 Validation of analytical methods

An essential part of any sound QA system is the use of validated methods. All methods should be validated as being fit for purpose before their use in the laboratory. It is important that the methods used are fully documented, the laboratory staff is trained in their use, and control measures are established to ensure that the procedures are under statistical control.

Validation of analytical methods and measurements is vital to ensure that the results obtained can be relied upon as being accurate and free from false-negative and false-positive outcomes. When standard methods are used, laboratories should verify their own ability to achieve satisfactory performance against the documented performance characteristics of the method, before any samples are analysed. For this purpose, it is essential to employ suitably qualified, well-trained personnel. Methods developed in-house shall be validated and authorized before use. Where available, certified reference materials should be used to determine any systematic bias, or, where this is not possible, results should be compared with other techniques/methods, preferably based on different principles of analysis. Determination of uncertainty must form a part of this validation process and is essential for ongoing quality assurance.

6.2.3 Quantifying uncertainty, calibration and traceability

Without knowledge of uncertainty, it is impossible to judge if the measurement will serve its intended purpose. The ISO 17025:1999 Standard requires that the uncertainty of measurements be estimated and included on the test report. A good way to evaluate the uncertainty of an analytical method is to compare the results of the method with a reference value, namely a value established by means of a reference method which is known to be free of significant bias and has been correctly applied by the laboratory. Alternatively, the method's results can be compared with values obtained using certified reference materials.

Table 6.2 shows the uncertainties (95% confidence limit expressed as a percentage of the certified value) that have been achieved by reference materials from the Community

Table 6.2 Uncertainties (95% confidence limit expressed as a percentage of the certified value) on selected certified reference values in milk powder (MP) and milk fat (MF) reference materials from the Community Bureau of Reference[a].

Component	Matrix	Certified content	Uncertainties (%)
Lactose	MP	4.50 g 100 g^{-1}	1.1
Total fat	MP	29.6 g 100 g^{-1}	1.1
Ash (550°C)	MP	6.07 g 100 g^{-1}	0.7
Kjeldahl N	MP	4.80 g 100 g^{-1}	0.7
Butyric acid	MF	3.49 g 100 g^{-1}	1.7
Ca	MP	12.6 mg 100 g^{-1}	2.1
OCl pesticides	MP	6–70 $\mu g\ g^{-1}$	5–15
Cd	MP	2.90 ng g^{-1}	41
Aflatoxin M1	MP	0.09–0.76 ng g^{-1}	6–30

[a]Reproduced with permission from Wagstaffe (1993), IDF Special Issue 9302, International Dairy Federation, 41, Square Vergote, B-1040 Brussels, Belgium.

Bureau of Reference (BCR) for some selected properties in milk powder and milk fat reference methods. The uncertainties tend to increase as the content level decreases (e.g. lactose vs. cadmium content). Of course, there are laboratories that may achieve results with smaller uncertainties. Although it is not always necessary to achieve the smallest uncertainty for a particular measurement, it is essential that the laboratory has estimated and recorded the magnitude of its uncertainty.

Calibration is the process of establishing the relationship between values shown by a measuring system or equipment and the values obtained by measurement standards. Calibration is the fundamental process for establishing traceability, as it is through calibration that traceability to appropriate reference standards is actually achieved in practice (EURACHEM/CITAC, 2003). The usual way to perform calibration is to use certified reference materials and monitor the measurement response. When such materials are not available, a material with suitable properties and stability should be selected or prepared by the laboratory and used as a measurement standard. Instruments such as chromatographs and spectrometers need to be frequently calibrated as part of their normal operation, whereas, devices such as balances and thermometers are calibrated less frequently.

In principle, at least, it would be possible for a laboratory to undertake a thorough study of all sources of error and to demonstrate that an unbroken chain of comparison has been followed and that its measurements were traceable. In practice, for most laboratories, this approach is neither economical nor technically feasible. As stated by Wagstaffe (1993), certified reference materials provide an easy way to establish traceability, since it is only necessary to show that the method gives a result which is in agreement with a certified value of a reference material of similar matrix composition, i.e. by establishing an unbroken chain of comparison with a standard having essentially the same characteristics as the sample.

6.2.4 Quality aspects of microbiological media

Asperger (1993) has reviewed the quality parameters of media in the microbiological laboratory and stated that the significance of media to the precision and accuracy of microbiological methods differs. Precision is influenced mainly by the distribution of microorganisms in the sample material, the expense on dilutions and parallel plates or tubes, and the adherence to the procedure of the method. Accuracy, on the contrary, largely depends on the type and quality of the medium. Since, nowadays, the microbiological media are delivered in dehydrated powdered or granulated form, it is important to follow the manufacturer's instructions with respect to storage conditions and shelf lives. More specifically, one should:

- note the date of the opening of the bottle;
- monitor the autoclave conditions;
- check and record the pH at 25°C;
- use aseptic technique to dispense; and
- label the containers including the name of the medium, the date and the autoclave run.

The required volumes of the media should be managed so that the minimum storage time for prepared media will be used. The media should be inspected visually for volume, closure, colour and label before use. It is advisable to plan the purchases for a turnover of not longer than one year and follow the first-in-first-out (FIFO) principle.

6.2.5 Laboratory safety

Laboratory safety should be an essential element of any laboratory procedure. In the laboratory, there are chemicals that are toxic, flammable, explosive, corrosive or carcinogenic. Laboratory safety procedures need to be specially focused on the protection of the personnel as suggested by Directive 2000/54/EC of the European Parliament and of the Council (EC, 2000). It is important for a laboratory to develop a procedure for discarding samples of dairy products that have been examined for chemical analysis, in order to prevent their consumption, since these samples are often left at room temperatures for a long time.

In a microbiological laboratory, infective microorganisms are classified into four risk groups:

- Risk Group 1 – very low or no individual and community risk (i.e. microorganisms that are unlikely to cause human or animal disease);
- Risk Group 2 – moderate individual, low community risk (i.e. pathogens that can cause human or animal disease, but are unlikely to be a serious hazard to laboratory workers, the community, livestock or the environment; laboratory exposure may cause serious infection, but effective treatment and preventative measures are available, and the risk of infection is limited);
- Risk Group 3 – high individual, low community risk (i.e. pathogens that usually cause serious human or animal disease, but do not ordinarily spread from one infected individual to another; effective treatment and preventative measures are available); and
- Risk Group 4 – high individual and community risk (i.e. pathogens that usually cause serious human or animal disease and can be readily transmitted from one individual to another, directly or indirectly; effective treatment and preventative measures are not usually available).

Laboratories, on the other hand, are designated according to their design features, construction and containment facilities as: Basic – Biosafety Level 1 (i.e. working with Risk Group 1 microorganisms); Basic – Biosafety Level 2 (i.e. working with Risk Group 2 microorganisms); Containment – Biosafety Level 3 (i.e. working with Risk Group 3 microorganisms); and Maximum Containment – Biosafety Level 4 (i.e. working with Risk Group 4 microorganisms).

A very useful safety checklist to assess the safety status of a laboratory has been outlined in the *Laboratory Biosafety Manual* (WHO, 2003). The list includes assessment of laboratory's premises, storage facilities, sanitation and staff facilities, heating and ventilation, lighting, services, security, fire prevention, flammable liquid storage, electrical hazards, handling of compressed and liquefied gases, personal protection, the safety of laboratory equipment, and the handling and disposal of infectious materials, chemicals and radioactive substances.

6.3 Sampling

A representative sample taken by a trained person, and appropriately transported to the laboratory is the basis for any chemical, microbiological and/or physical analysis, and consequently for any QA laboratory system. Dairy products require special handling due to their high perishability in order to prevent contamination and the subsequent growth of such contaminants during transportation and storage. The adequacy and condition of the sample received for examination are of primary importance. If samples are improperly collected, mishandled or not representative of the sampled lot, the laboratory results will be meaningless. The way sampling is conducted always depends on the scope of the analysis, and it should be carried out as described in the joint IDF/ISO Standard (IDF, 1995a).

The preparation of test samples, initial suspensions and decimal dilutions for microbiological examination of dairy products is described in the joint IDF/ISO standard (IDF, 2001a), and the diluents suggested are listed in Table 6.3. Because interpretations about a large consignment of food are based on a relatively small sample of the lot, established sampling procedures must be applied uniformly. A representative sample is essential, especially for microbiological examinations, where pathogens or toxins are sparsely distributed within the food. Thus, adequate agitation for liquid milk products is of special importance, and specifications for devices used for manual and mechanical mixing are described by IDF (1995a). For solid products, specially designed triers (e.g. cheese triers of shape and size appropriate for the cheese to be sampled) are also described.

The number of units that comprise a representative sample from a designated lot of a food product must be statistically significant, and the homogeneity of the samples must be validated. Sampling plans may be random, systematic or sequential,

Table 6.3 Diluents used for microbiological analyses of dairy products[a].

Diluent	Product
Diluents for general use	
Peptone-salt solution	General use
Quarter-strength Ringer's solution	General use
Peptone solution	General use
Phosphate buffer solution	General use
Diluents for special purposes[b]	
Pre-enrichment medium: buffered peptone water	
Sodium citrate solution	Cheese and dried milk
Dipotassium hydrogen phosphate solution	Cheese, dried milk, fermented milk, caseinates, dried acid whey and sour cream
Dipotassium hydrogen phosphate solution with antifoam agent	Acid casein, lactic casein and rennet caseins
Tripolyphosphate solution	Alternative solution for rennet casein
Diluent for general use with α-amylase solution	Infant foods with high starch contents

[a]After: IDF (1995a, 2001).
[b]These diluents should only be used for the preparation of initial suspensions.

and they may be undertaken to obtain quantitative or qualitative information or to determine conformance or non-conformance with a specification. The batch is accepted or rejected according to the sampling plan that is based on the size of the batch and the acceptable error; the latter usually varies from 1 to 10%. Dairy products may be examined by taking analytical data on their chemical or microbiological characteristics, namely, sampling by variables (IDF, 1992), or they may be examined and classified into different groups, namely, sampling by attributes (IDF, 2004a). A variable may be described as a characteristic that can be measured quantitatively and may have any value within certain limits (e.g. a bacterial count). Attribute sampling is used to qualitatively classify a unit to be acceptable or unacceptable, namely, only two outcomes are possible (e.g. a sterile product or not).

Whenever possible, samples should be submitted to the laboratory in the original unopened containers. If products are in bulk or in containers too large for submission to the laboratory, representative portions should be transferred to sterile containers under aseptic conditions. In a large vessel with a bottom discharge outlet, samples should preferably be taken through the manhole. If taken from the discharge outlet valve or the sampling cock, sufficient milk must be discharged to ensure that the sample is representative of the whole (Mostert and Jooste, 2002). Proportionate sampling is done by taking representative quantities from each container and mixing the portions in amounts that are proportional to the quantity in the container from which they were taken. There can be no compromise in the use of sterile sampling equipment and the use of aseptic technique. One-piece stainless steel spoons, forceps, spatulas, and scissors should be sterilized in an autoclave, or dry-heat oven.

6.3.1 Sample collection

Sampling should be performed by an authorized person, free from any infectious disease, properly trained in the appropriate technique. Samples should be taken in duplicate, or in plural in the case of a legal requirement or an agreement between the parties concerned. When collecting liquid samples, an additional sample should be taken, as a temperature control and the temperature of the control sample should be checked at the time of collection and upon receipt at the laboratory. Whatever strategy is used for the sampling, it is vital that the person carrying out the sampling keeps a clear record of the procedures followed, so that the sampling process may be repeated exactly. In addition, it may be useful to include a diagram as part of the documentation to show the pattern of sampling when more than one sample is taken from the original product (i.e. hard cheeses). Containers should be clean, dry, leak-proof, wide-mouthed, sterile and of a size suitable for samples of the product; the use of certified sample collectors is suggested.

For handling raw-milk samples, the following procedures are suggested:

- Crushed ice, ice-water, pre-frozen icepacks or dry-ice should be used to cool and keep samples at 0–4°C.
- The ice level in the sample case should be kept slightly above the milk level in plastic or rigid-walled sample containers; special care should be taken not to freeze the samples prior to analysis, because it can cause disruption of bacterial cells.

- Samples should be protected from possible contamination, with special care to keep the level of ice-water lower than the milk sample. Microbiological analysis should be started as soon as possible, but no later than 36 h after collection at the farm, preferably within 24 h. If the control temperature sample on arrival is >4°C, the samples should not be tested.

The sample label is another important aspect of the documentation and should unambiguously identify the sample; labelling must be firmly attached to the sample packaging, and care should be taken to keep the label unaffected from fading or spillages. Preservatives of a nature that do not interfere with subsequent analyses and provided that samples will not be tested for their sensory characteristics may be added to some milk products. Such preservatives include chemicals like potassium dichromate (0.03%), sodium azide (0.08%) and Bronopol (2-bromo-2-nitro-1,3-propanediol, 0.02%).

6.3.2 Sampling report

Samples should be accompanied by a report, signed by the authorized sampling personnel, which needs to include (IDF, 1995a):

- the place, date and time of sampling;
- the names and affiliations of the sampling personnel and of any witnesses;
- the precise method of sampling, if different from that specified by the respective Standards;
- the nature and number of units constituting the consignment, together with their batch code markings, when available;
- the identification number and any code markings of the batch of which the samples were taken;
- the place to which the samples are to be sent; and
- any other information required, e.g. the condition of the product container, the temperature of the environment, whether a preservative substance (nature and quantity) has been added to the samples.

The samples should be delivered to the laboratory promptly with the original storage conditions maintained as closely as possible. For all samples, a full record of the times and dates of collection and of arrival at the laboratory is essential. If possible, the samples should be examined immediately upon receipt; detailed information on the general requirements and guidelines of the sampling procedure can be found in ICMSF (1986), Grace *et al.* (1993), IDF (1995a), Andrews and Hammack (1998) and ICMSF (2002).

6.4 Chemical analyses

Chemical analysis of milk and milk products normally requires determination of some or all of the following components: fat, protein, lactose, ash, moisture, urea and salt.

Standard methods have been accepted as official reference methods and have been used for years to evaluate the composition of dairy products. These methods often tend to be relatively tedious and demanding in terms of human resources, while automated routine methods have been developed that allow faster, simpler and sometimes cheaper procedures. Detailed descriptions of chemical methods can be found in the literature (Kirk and Sawyer, 1991; Bradley *et al.*, 1993; Ardö and Polychroniadou, 1999; Horwitz, 2005).

6.4.1 Fat content

The determination of fat in foods and especially in dairy products is important for both regulatory and nutritional information purposes. Fat can be determined using a variety of official reference methods (Table 6.4), but most often the following three gravimetric methods are used (Kirk and Sawyer, 1991):

- *Röse–Gottlieb method*: Alcohol and ammonia are added to the sample. The alcohol causes precipitation of the protein, which then dissolves in the ammonia. The fat is then extracted into a mixture of diethyl ether and petroleum ether. The amount of fat extracted is determined gravimetrically after removal of the solvent for a variety of dairy products (IDF, 1987a, 1987b, 1987c, 1987d).
- *Werner–Schmid method*: A portion of the sample is digested with hydrochloric acid, and after cooling, the fat is extracted using a mixture of diethyl ether and light petroleum or alcohol. The solvent is then removed by evaporation, and the amount of fat remaining is determined by weighing.
- *Weibull–Berntrop method*: A portion of the sample is digested by boiling with dilute hydrochloric acid. The hot digest is then filtered through a wetted filter paper, and the fats present in the digest are retained on the filter. The filter is dried and the fat extracted from it using a refluxing solvent such as *n*-hexane or light petroleum in a Soxhlet extraction apparatus. The solvent is then removed by evaporation, and the weight of extracted substances is determined.

Ardö and Polychroniadou (1999) have described routine butyrometric methods, which are commonly used in dairy laboratories for cheese samples: cheese is digested in a Gerber butyrometer (Gerber Instruments, Efflueticon, CH) by heating with a strong mineral acid, and fat is then separated, after the addition of amyl alcohol to obtain better fat separation, by centrifugation forming a clear layer. The percentage fat content is directly read on the butyrometer scale to the nearest 0.05%. An excellent review for the methods used for the determination of fat in butter has been provided by Evers (1999).

6.4.2 Protein content

The traditional method for determining the protein content of dairy products is by determination of the nitrogen content by the Kjeldahl method and by multiplying the resulting value by the factor 6.38; this result is expressed as ‘crude protein’.

Table 6.4 Methods adopted by international organizations for the chemical analysis of different dairy products[a].

Component	Product	Principle	Standard	
Fat content	Milk	Determination of fat content – Gravimetric method		ISO Standard 1211:1999
	Cheese and processed cheese products	Determination of fat content – Gravimetric method	IDF Standard 005:2004	ISO Standard 1735:2004
	Dried milk and dried milk products	Determination of fat content – Gravimetric method		ISO Standard 1736:2000
	Evaporated and sweetened condensed milk	Determination of fat content – Gravimetric method		ISO Standard 1737:1999
	Whey cheese	Determination of fat content – Gravimetric method		ISO Standard 1854:1999
	Cream	Determination of fat content – Gravimetric method		ISO Standard 2450:1999
	Caseins and caseinates	Determination of fat content – Gravimetric method		ISO Standard 5543:2004
	Skimmed milk, whey and buttermilk	Determination of fat content – Gravimetric method		ISO Standard 7208:1999
	Milk-based edible ices and ice mixes	Determination of fat content – Gravimetric method		ISO Standard 7328:1999
	Milk-based infant foods	Determination of fat content – Gravimetric method		ISO Standard 8381:2000
	Dried milk, dried whey, dried buttermilk and dried butter serum	Determination of fat content (Röse Gottlieb method)	IDF Standard 009C:1987	
	Evaporated milk and sweetened condensed milk	Determination of fat content (Röse Gottlieb method)	IDF Standard 013C:1987	
	Cream	Determination of fat content (Röse Gottlieb method)	IDF Standard 016C:1987	
	Skimmed milk, whey and buttermilk	Determination of fat content (Röse Gottlieb method)	IDF Standard 022B:1987	
	Whey cheese	Determination of fat content (Röse Gottlieb method)	IDF Standard 059A:1986	
	Milk-based edible ices and ice mixes	Determination of fat content (Röse Gottlieb method)	IDF Standard 116A:1987	
	Infant foods	Determination of fat content by the Weibull–Bentrop gravimetric method	IDF Standard 124A:1988	ISO Standard 8262-1:1987
	Edible ices and ice-mixes	Determination of fat content by the Weibull–Bentrop gravimetric method	IDF Standard 125A:1988	ISO Standard 8262-2:1987
	Milk products and milk-based foods (special cases)	Determination of fat content by the Weibull–Bentrop gravimetric method	IDF Standard 126A:1988	
	Caseins and caseinates	Determination of fat content by the Schmid–Bondzynski–Ratzlaff gravimetric method	IDF Standard 127A:1988	
	Butter, edible oil emulsions and spreadable fats	Determination of fat content	IDF Standard 194:2003	ISO Standard 17189:2003

Contd.

Table 6.4 (Contd.)

Component	Product	Principle	Standard	
Nitrogen content	Milk	Determination of nitrogen content – Part 1: Kjeldahl method	IDF Standard 020-1:2001	ISO Standard 8968-1:2001
Nitrogen content	Milk	Determination of nitrogen content – Part 2: Block-digestion method (Macro method)	IDF Standard 020-2:2001	ISO Standard 8968-2:2001
Nitrogen content	Milk	Determination of nitrogen content – Part 3: Block-digestion method (Semi-micro rapid routine method)	IDF Standard 020-3:2004	ISO Standard 8968-3:2004
Nitrogen content	Milk	Determination of nitrogen content – Part 4: Determination of non-protein-nitrogen content	IDF Standard 020-4:2001	ISO Standard 8968-4:2001
	Milk	Determination of casein-nitrogen content – Part 1: Indirect method		ISO Standard 17997-1:2004
Nitrogen content	Milk and milk products	Determination of nitrogen content – Method using combustion according to the Dumas principle	IDF Standard 185:2002	ISO Standard 14891:2002
Total solids	Cheese and processed cheese–sweetened	Determination of the total solids content	IDF Standard 004:2004	ISO Standard 5534:2004
	condensed milk	Determination of total solids content	IDF Standard 015B:1991	ISO Standard 6734:1989
	Milk, cream and evaporated milk	Determination of total solids content	IDF Standard 021B:1987	ISO Standard 6731:1989
Moisture content	Milk-fat products	Determination of water content – Karl Fischer method	IDF Standard 023:2002	ISO Standard 5536:2002
	Dried milk	Determination of moisture content	IDF Standard 026:2004	ISO Standard 5537:2004
	Casein and caseinates	Determination of water content	IDF Standard 078C:1991	ISO Standard 5550:1978
	Butter	Determination of moisture, non-fat solids and fat contents – Part 1: Determination of moisture content	IDF Standard 080-1:2002	ISO Standard 3727-1:2001
	Butter	Determination of moisture, non-fat solids and fat contents – Part 2: Determination of non-fat solids content	IDF Standard 080-2:2002	ISO Standard 3727-2:2001
Urea	Milk	Determination of urea content – Enzymatic method using difference in pH		ISO Standard 14637:2004
Lactose	Dried milk, dried ice-mixes and processed cheese	Determination of lactose content – Part 1: Enzymatic method utilizing the glucose moiety of the lactose	IDF Standard 079-1:2002	ISO Standard 5765-1
	Dried milk, dried ice-mixes and processed cheese	Determination of lactose content – Part 2: Enzymatic method utilizing the galactose moiety of the lactose	IDF Standard 079-2:2002	ISO Standard 5765-2

[a]After Webber *et al.* (2000).

The Kjeldahl method involves digesting the sample with concentrated sulfuric acid, along with potassium sulfate and a catalyst. The oxidation results in conversion of the nitrogen present to ammonium sulfate. After digestion is complete, the digest is made alkaline by adding concentrated sodium hydroxide solution. This releases ammonia, which is determined by titration.

The Dumas method is also used for total nitrogen determination. The sample is dropped into the furnace of the LECO instrument (Elemental Microanalysis NA2100, St Joseph, MI) at 1000°C. The gases produced are removed, except nitrogen, which is then detected by a thermal conductivity cell. The Dumas method is suggested as a replacement of the Kjeldahl method for nitrogen determination in foods (Simonne *et al.*, 1997), since the latter is a rather tedious and time-consuming method.

The dye-binding methods rely on the ability of dyes (e.g. amido black) to combine with polar groups of proteins of opposite ionic charge at low pH and form an insoluble protein–dye complex. This complex is then removed by centrifugation or filtration, and the concentration of unbound dye, remaining in the supernatant, is assessed by measuring its absorbance in the range of 550–620 nm. The resulting value is used to calculate the amount of protein as compared with the nitrogen content determined by the Kjeldahl method.

6.4.3 *Total solids*

Total solids are determined by drying a proportion of the sample to constant weight at a specified temperature. For most products, this temperature is set to 102 ± 1°C. Reference methods for different dairy products have been published (Table 6.4). For routine analysis, infrared (IR) absorption may be used to determine the moisture content of a variety of milk products (Section 6.4.2).

6.4.4 *Ash content*

Ash is the inorganic residue that remains after combustion of the organic matter and is determined by heating at a temperature of 525 ± 25°C. It does not correspond exactly in composition to the mineral matter, since some constituents are lost by volatilization (Prentice and Landridge, 1992).

6.4.5 *Lactose content*

Lactose is the main carbohydrate present in milk and most milk products. Methods for the determination of lactose content involve a volumetric determination using chloramine T (Kirk and Sawyer, 1991). Lactose can also be determined by polarimetry (Biggs, 1978), as well as enzymatically, based on the determination of NADH formed by the oxidation of β-galactose at 340 nm, or the determination of NADPH formed by the oxidation of glucose. In addition to lactose, other carbohydrates such as galactose, glucose, lactulose and epilactose can be assayed using gas chromatography or high-performance liquid chromatography (HPLC) (Ardö and Polychroniadou, 1999).

6.4.6 *Urea determination*

Urea is a normal constituent of milk and comprises part of the non-protein nitrogen fraction (van den Bijgaart, 2003). The reference method for the determination of urea is by measurement of pH difference (Table 6.4). Other methods include (Lefier, 1996): enzymatic methods measuring a by-product of the urea degradation (e.g. ChemSpec Analyzer, Bentley Instruments, Chaska, MN), IR spectroscopy (e.g. MilkoScan 4000, Foss Electric, Hilleroed, Denmark) and Fourier-transform infrared (FTIR) spectroscopy (e.g. MilkoScan FT120, Foss Electric).

6.4.7 *Salt content*

Salt is added as an additive in certain dairy products, and the salt content can be determined either using a titrimetric method (IDF, 2000), that is titration with silver nitrate and potassium chromate as an indicator, also known as the Mohr method, or via a potentiometric method (IDF, 1988, 2004b), i.e. using a potentiometer provided with a measuring electrode suitable for the determination of chloride ion.

6.4.8 *Routine instrumental methods*

In a typical dairy laboratory, proximate analysis is carried out using automated milk analysers based on IR, Mid-Infrared (MIR) and, more recently, FTIR spectroscopy.

In 1975, Foss Electric introduced the first single-cell dual-wavelength IR milk analysers. The analysis was based on absorptions by functional groups in the protein molecule at 6.46 μm, in the lactose molecule at 9.60 μm and in the fat molecule at a wavelength of 3.48 μm, which is the absorption characteristic of the C–H bond in the fatty acid chains (fat B) and 5.73 μm, which is the absorption characteristic of the ester bond between glycerol and the fatty-acid chain (fat A). The IR beam is passed through an optical filter, which transmits energy at the wavelength of maximal absorption for the component being measured. The filtered beam then passes through the sample cell and then to a detector. Signals from the detector are then used to compare the absorption at the two wavelengths and hence determine the concentration of the component (Biggs, 1978). The success of the IR method in the analysis of milk has been due, in great part, to the relatively consistent composition of bovine milk. However, according to Abd El-Salam *et al.* (1986), the use of the 3.48 μm filter improved the repeatability and accuracy of the fat determinations in caprine and ovine milks.

MIR spectrometry is the most frequently applied methodology for the compositional analysis of milk and milk products. The application of MIR spectrometry offers several striking advantages compared with traditional methods used in milk analysis, as it offers the possibility of analysis of up to 500 samples per hour. A recently introduced parameter in MIR spectrometry is the determination of pH in raw-milk samples (Baumgartner *et al.*, 2003). This comes along with an increased need for rapid routine methods to monitor the quality of milk and milk products within the QA system of an analytical laboratory. The pH value is a good indicator for milk quality, as it is affected by the hygiene conditions during milking and sampling, as well as by the storage time and temperature.

With the introduction of the Fourier Transform technique, infrared spectrometry has gained in performance. Accuracy has improved, and precision has remained almost unaltered. In addition, the simultaneous determination of parameters such as urea, pH, lactic acid, free fatty acids and freezing point has been made possible.

Routine methods used for compositional analyses must be calibrated, adjusted or at least compared with reference methods. Calibration against a reference method using a sufficient number of representative samples or against reference materials is required to correlate the instrument signal at various wavelengths and the concentration of a particular component. Calibration methods like principal-component analysis (PCA), multiple linear regression and partial least-squares regression are commonly used, and with such calibration it is possible to predict certain amounts of the component of interest. These methods work very well for amounts of component quantities larger than about 1%. Sørensen (2004) described the procedure for the use of routine methods in the determination of chemical parameters in milk and milk products, starting with the selection of test samples and giving practical examples. When the routine method has been calibrated, it is good practice to validate it using an independent set of samples, preferably sampled after the calibration period.

The use of rapid methods has been further encouraged by the EC Regulation 213/2001 (EC, 2001) as long as certain criteria are fulfilled, and the procedure for calibration and regular checking is described. In this Regulation, the term 'critical difference' is introduced. The calculation of the critical difference, required for each component, is dependent on the reproducibility and repeatability from both the reference and the rapid method.

6.5 Detection of antibiotic residues

Inflammation of the mammary gland (mastitis) is the most prevalent disease in dairy cattle and the leading cause of financial losses for milk producers. A 1995 survey among US veterinarians revealed that antibiotics are the drugs most often used or prescribed to lactating dairy cows (Sundlof *et al.*, 1995), and among them penicillin G is the antibiotic most frequently used for the treatment of bovine mastitis followed by other β-lactam antibiotics and oxytetracycline (Mitchell *et al.*, 1998). In a recent survey of 99 conventional US dairy herds, 84.9, 9.1 and 0.9% of farmers reported treatment with administration of antibiotics to 1–10, 11–25 and >25% of their dairy cows, respectively, during the 2-month period that preceded the interview (Zwald *et al.*, 2004). Insert labels for all veterinary drugs state a withdrawal period that must be observed by dairy farmers before returning the treated animal to the milking string. Non-adherence to these guidelines or use of the drug in an extra-label fashion (different dose, different route or frequency of administration) can result in violative levels of antibiotics entering the milk supply. In the US, the Animal Medicinal Drug Use Clarification Act of 1994 allows veterinarians to prescribe extra-label uses of certain approved animal drugs and approved human drugs for animals under certain conditions. An extra-label drug use algorithm has been made available through the American Veterinary Medical Association website as a guide to veterinarians when extra-label drug use is contemplated (AVMA, 2005).

The presence of antibiotic residues (especially of β-lactam antibiotics) in the milk supply can cause adverse reactions to already-sensitized individuals, and exposure of consumers to sub-therapeutic antimicrobial concentrations may lead to the development of antibiotic-resistant bacteria (Dewdney *et al.*, 1991; Sundlof, 1994). From a dairy-industry viewpoint, the presence of antibiotic residues in the milk supply can lead to serious problems, as antibiotics restrict growth of LAB that are used in various dairy fermentations (Hunter, 1949; Cogan, 1972). Therefore, milk obtained from lactating animals that have been treated with veterinary medicinal drugs must not contain drug residues which may pose health hazards for consumers. Hence, testing of milk for antibiotic residues of β-lactam antibiotics has become mandatory (US FDA, 2003), and regulatory agencies have established maximum residue limits (MRLs) for antimicrobial substances in foods of animal origin. Sischo (1996) has summarized the history of the US antibiotic residue testing programme. In the European Union (EU), MRLs are listed in Annex I of Regulation 2377/90 of the EC (EC, 1990). A number of amendments have been subsequently made, mostly in order to extend the list of compounds whose MRLs have been established. Commission Decision 2002/657 establishes the requirements, procedures and criteria that must be used to define the performance of analytical methods used for screening or confirmatory purposes (EC, 2002).

According to the 2003 annual report of the National Milk Drug Residue Database (a voluntary industry reporting programme) of the US Food and Drug Administration, Center for Food Safety and Applied Nutrition, a total of 4,382,974 samples of milk (bulk raw milk from dairy farms, tanker trucks, pasteurized milk and milk products in bulk or in package form and other sources) were analysed for animal drug residues. Of these, 2945 samples (*ca.*0.07%) were found positive for a residue leading to disposal of *ca.*76 million pounds of milk (US FDA, 2004). In that report, a 'positive result' means that 'the sample was found positive for a drug residue by a test acceptable for taking regulatory action in a certified laboratory by a certified analyst or the milk was rejected on the basis of an initial test by the milk processor'.

The methods that are used to detect the presence of antibiotic residues in milk can be classified as qualitative and quantitative, depending upon whether they provide quantitative residue concentration output besides denoting the presence of antimicrobials in test samples. Methods are also categorized according to the scientific principle employed for the detection of antibiotics. The first methods that were introduced relied on the microbial inhibition principle in which the presence of antimicrobial compounds in milk is denoted via the inhibition of growth of a susceptible assay microorganism (typically *Bacillus stearothermophilus* var. *calidolactis*, but other organisms as well) incubated in the presence of a suspect milk sample (Katz and Siewierski, 1995; Nouws *et al.*, 1999; Gaudin *et al.*, 2004). In the absence of inhibitory levels of antimicrobials, germination and growth of the organism are detected visually via the opacity in an agar growth medium or via a colour change of a pH indicator, resulting from changes in pH due to microbial metabolic activity. A combination of thin-layer chromatography with the microbial inhibition principle has also been reported (Ramirez *et al.*, 2003).

Immunological methods (immunoassays) rely on the specific recognition and binding of the antibiotic by an antibody. The most popular immunological chemical residue detection format is the enzyme-linked immunosorbent assay (ELISA) (Martlbauer *et al.*, 1994; Loomans *et al.*, 2003; Gaudin *et al.*, 2005). Recently, Huth *et al.* (2002) reported the development of a capillary-based fluorescent immunoassay for the determination of β-lactams.

Enzymatic colorimetric assays are usually based on the detection of the degree of inactivation of an assay enzyme as a result of its specific binding with the antibiotic present in the milk sample. After appropriate incubation, an appropriate reagent (enzyme substrate) is added to the enzyme–milk mixture. Reduced- or no enzyme-product formation (as determined by a redox colour indicator) is indicative of the presence of antibiotic residues in the test sample.

Receptor binding assays do not employ antibodies to bind antibiotics, although the principles of these tests are analogous to those of immunoassays. The binding molecules, called receptors, are bacterial proteins that bind to the antibiotic present in the test sample. Binding proteins are usually conjugated with an enzyme (e.g. horseradish peroxidase). Following proper incubation, washing and addition of enzyme substrate, unbound conjugate in the milk–conjugate mixture is detected colorimetrically after transfer of the mixture to an antibiotic-coated chamber.

A great number of antibiotic residue assays (test kits) have emerged during the last 15 years. A detailed classification of antibiotic residue detection assays along with detailed descriptions of the different assay formats and detection principles can be found in several reviews (IDF, 1991b; Cullor, 1992, 1993; Boison and MacNeil, 1995; Mitchell *et al.*, 1998). Surface plasmon resonance-based biosensor assays constitute a more recent addition in the field of residue testing and have been applied for the detection of sulfamethazine (Sternesjo *et al.*, 1995; Gaudin and Pavy, 1999), streptomycin and dihydrostreptomycin (Baxter *et al.*, 2001; Ferguson *et al.*, 2002), chloramphenicol (Gaudin and Maris, 2001) and β-lactams (Gustavsson *et al.*, 2002, 2004; Cacciatore *et al.*, 2004; Gustavsson and Sternesjo, 2004). The design of a novel biosensor-based assay for the detection of tetracycline, streptogramin and macrolides has been recently reported by Weber *et al.* (2005). Spectroscopic (Sivakesava and Irudayaraj, 2002) and electrophoretic (Cutting *et al.*, 1995) residue-detection methods have also been reported.

Rapid screening assays and in particular microbial inhibition assays are typically used by the dairy industry for screening incoming milk for antibiotic residues. Microbiological assays have also been proposed for residue verification purposes following initial milk screening (Nouws *et al.*, 1999). Different assays are available for different classes of antibiotics (e.g. β-lactams, aminoglycosides.). Multi-plate or tube microbial inhibition tests have been developed and evaluated in terms of their ability to detect multiple antimicrobial residues at or below their MRLs. None of the reported schemes so far have been proven capable of detecting all possible families of antibiotics (and individual antibiotics within a family) that can be found in milk, at or below their MRLs (Gaudin *et al.*, 2004).

The evaluation of the performance of rapid screening tests that are intended for testing bovine milk has been the subject of numerous scientific reports (Macaulay and

Packard, 1981; Senyk *et al.*, 1990; Cullor and Chen, 1991; Cullor *et al.*, 1992, 1994; Tyler *et al.*, 1992; Sischo and Burns, 1993; van Eenennaam *et al.*, 1993; Gardner *et al.*, 1996; Halbert *et al.*, 1996; Andrew *et al.*, 1997; Nouws *et al.*, 1998; Angelidis *et al.*, 1999; Hillerton *et al.*, 1999; Andrew, 2000; Gibbons-Burgener *et al.*, 2001; Kang and Kondo, 2001; Gaudin *et al.*, 2004, 2005; Suhren and Knappstein, 2004; Kang *et al.*, 2005). Considerably less research has been devoted to evaluation of assays' performance when used to analyse ovine milk (Althaus *et al.*, 2001; Althaus *et al.*, 2003a, 2003b; Berruga *et al.*, 2003; Molina *et al.*, 2003a, 2003b) or caprine milk (Zeng *et al.*, 1996, 1998; Contreras *et al.*, 1997). A current list of AOAC performance tested kits can be found on the Internet (AOAC, 2005a).

Commercial rapid screening tests do not permit compound-specific identification or quantitation. Therefore, sensitive, robust and quantitative analytical methods are necessary when contradictory results are obtained by different rapid test methods in order to confirm the presence of the antimicrobial(s), identify them and precisely determine their concentration (Harik-Khan and Moats, 1995; Moats, 1999; Riediker *et al.*, 2001). Liquid chromatography (LC) alone or in combination with mass spectrometry (MS) provides a powerful analytical tool for verification and quantitative determination of antibiotic residues in presumptive positive milk samples. Such analytical techniques have been used for the verification and quantitative determination of aminoglycosides (Carson and Heller, 1998; Heller *et al.*, 2000; van Bruijnsvoort *et al.*, 2004; Bogialli *et al.*, 2005), β-lactams and cephalosporins (Heller and Ngoh, 1998; Moats and Romanowski, 1998; Takeba *et al.*, 1998; Daeseleire *et al.*, 2000; Bruno *et al.*, 2001; Riediker and Stadler, 2001; Holstege *et al.*, 2002; Becker *et al.*, 2004; Bogialli *et al.*, 2004), chloramphenicol (Sørensen *et al.*, 2003; Guy *et al.*, 2004), sulfonamides (van Rhijn *et al.*, 2002; de Zayas-Blanco *et al.*, 2004), fluoroquinolones (Marazuela and Moreno-Bondi, 2004), macrolides (Dudrikova *et al.*, 1999; Stoba-Wiley and Readnour, 2000) and nitrofurans (Perez *et al.*, 2002). Di Corcia and Nazzari (2002) and, more recently, Gentili *et al.* (2005) have provided extensive reviews on the use of LC-MS as a confirmatory analytical approach for documenting the presence of veterinary drugs in animal-food products with specific emphasis on the recent advances that have occurred in interface and analyser technology.

The initial impetus regarding antibiotic residue screening assays has been on enhancing their residue detection abilities in order to make assays capable of detecting trace amounts of residues in foods and, thereby, ensure protection of public health. However, technical modifications to improve an assay's sensitivity may result in reduction in the assay's specificity because of the inverse relationship between these two parameters. Numerous published scientific reports have emphasized that results of rapid screening tests should be interpreted with caution, and several reasons have been listed. Such tests are qualitative, do not distinguish among different antimicrobial compounds within the same class and often have different detection limits for different compounds. Rapid screening assays often yield positive results, even when the concentration of the antimicrobial in milk is lower than the specified MRL. Rapid screening assays must be used only for testing milk of specified type and species, because several reports have shown that such assays can occasionally lead to false-positive results unless appropriately used (Cullor *et al.*, 1992, 1994; Tyler *et al.*, 1992; van Eenennaam *et al.*, 1993;

Angelidis *et al.*, 1999). In other words, because such tests have been designed and evaluated for testing commingled milk that arrives in the dairy plant via tanker trucks, their use for testing milk from bulk tanks and, even more so, milk from individual cows has often proven to be problematic. Test kits should be used only for applications for which they have been designed. These applications should be clearly stated in the product's insert. For instance, a test kit that has been designed to evaluate antibiotic residues in tanker truck bovine milk should not be used to determine whether a mastitic cow that has been treated with antibiotics can return to the milking parlour, or used to check for antibiotic residues in ovine milk. Specificity problems (false-positive results) have been repeatedly reported, particularly for microbial inhibition assays that have been applied to test milk samples from cows with natural (van Eenennaam *et al.*, 1993; Gibbons-Burgener *et al.*, 2001) or experimentally induced (Tyler *et al.*, 1992) mastitis. In this case, false-positive samples have been attributed to, or correlated with, elevated milk somatic cell count (SCC), bovine serum and elevated concentrations of natural antimicrobial compounds, such as lactoferrin and lysozyme in milk from mastitic animals (Carlsson *et al.*, 1989; Cullor *et al.*, 1992; Angelidis *et al.*, 1999; Kang and Kondo, 2001). Other milk constituents that have been shown to interfere with assay performance include protein content and fat (Andrew, 2000). Cullor (1994) has provided a testing scheme and a list of parameters that should be considered upon assessing the appropriateness of using such assays under field conditions (i.e. use on individual animal milk). Such parameters/definitions are listed below:

- *False-positive* result, the identification of an untreated, or antibiotic-free animal (or an animal having an antibiotic residue in its milk at concentrations below the MRL) as test-positive for antibiotic residues.
- *False-negative* result, the identification of an animal as test-negative when it is excreting violative levels of antibiotic into its milk.
- *Biomedical (epidemiologic) specificity,* the probability of correctly identifying true-negative (untreated) animals. This definition should be differentiated from the chemical (assay-development) definition of specificity, i.e. the ability of an assay to differentiate between two different drugs.
- *Biomedical (epidemiologic) sensitivity*, the probability of correctly identifying true-positive (treated with antibiotics) animals. Again, this definition should be differentiated from the chemical definition of sensitivity, i.e. the degree, or level of the assay's response, the detection limit of the assay.
- *Reliability*, the assay's credibility to be consistent with its intended use (i.e. to identify antibiotic residues).
- *Prevalence*, the proportion of the population (e.g. of lactating cows) that actually has antibiotic in the milk.
- *Predictive value of a positive test (positive predictive value)*, the probability that a test-positive animal actually contains violative antibiotic residues in its milk. The positive predictive value of a test is a function of the test's sensitivity, the test's specificity and also a function of the prevalence in the target population.
- *Predictive value of a negative test (negative predictive value)*, the probability that a test-negative animal actually does not have violative antibiotic residues in its milk.

The negative predictive value of a test is also dependent upon the test's sensitivity, specificity and prevalence in the target population.

False-negative results are of concern to consumers because milk containing violative levels of antibiotic residues may enter the food chain. False-positive results are of concern to dairy farmers, who will suffer unnecessary economic loss and may be accused of failing to withhold milk for the appropriate time period after administration of antibiotics to lactating cows. Veterinarians can also be accused of extra-label use of antibiotics. Because of the low prevalence of antibiotics in the milk supply (Anonymous, 1995; US FDA, 2004), the predictive value of a positive test that is not 100% specific is rather low. Hence, on a large scale (i.e. bulk tank milk arriving in the milk factory), initial screening of milk should be conducted using a highly sensitive and inexpensive assay; positive results obtained with the first assay should be confirmed using a different assay that has a high specificity and the ability to quantify the antibiotic.

Standardized protocols pertaining to various aspects of analysis of antibiotic residues have been developed from standardization organizations and are available at a cost. Such Standards and protocols are available through IDF (1993, 1999), ISO (2003a) and AOAC (methods 962.14, 982.15, 982.16, 982.17, 982.18, 988.08 and 995.04) (Horwitz, 2005). For a detailed coverage of issues pertaining to the analysis of drug residues in foods (including compounds other than antimicrobials), the reader is referred to specialized textbooks (Botsoglou and Fletouris, 2001) and to several published reviews (Shaikh and Moats, 1993; Kennedy *et al.*, 1998; Niessen, 1998; Oka *et al.*, 1998; Schenck and Callery, 1998; Stead, 2000; Di Corcia and Nazzari, 2002; Hernandez *et al.*, 2003; Gentili *et al.*, 2005).

The problem of antibiotic residues is better addressed through improvement in farm management practices. Improvements that reduce the farm's milk SCC levels by improving the animals' health status (reducing the incidence of mastitis) lead to decreased antimicrobial administration frequencies and therefore reduce the risk of antibiotic residue violations (Ruegg and Tabone, 2000; Saville *et al.*, 2000).

6.6 Detection of adulteration in dairy products

Adulteration of milk and milk products constitutes a fraudulent practice in the dairy industry that stems from several reasons depending on the type of adulteration. One of the most common types of adulteration is the mixing or substitution of ovine, caprine or water-buffalo milk with bovine milk for the production of cheeses. This form of adulteration stems from the fact that bovine milk is cheaper and available year-round in sufficient quantities. Addition of water in milk is probably the oldest form of adulteration and yet another form of financially driven adulteration. Other types of adulteration such as the substitution of milk fat (MF) with vegetable- or animal fat or the substitution of milk proteins with vegetable proteins have also been known to occur. Besides fraudulent substitutions, the precise proportions of milk from different species in mixed-milk cheeses is very important for ensuring the authenticity of certain Protected Denomination of Origin (PDO) cheeses, as such products should be manufactured using defined amounts of each type of milk. For mixed-milk cheeses, therefore, besides the need for the qualitative determination of the 'foreign' milk species, it is also important for regulatory authorities to be able to precisely quantify the different milk types. Substitution of caprine with ovine milk and vice versa may also occur for technological reasons.

Adulteration of dairy products is of great importance for financial, ethnic, regulatory and potentially health reasons. Consumers are deceived into paying for a product of inferior value and consuming a product whose chemical composition is mislabelled and may therefore be against their religious beliefs, and finally may be at risk for food allergies especially during the first years of life (Lara-Villoslada *et al.*, 2005). The organoleptic properties of adulterated products are also affected. It is therefore the joint responsibility of the dairy industry and the regulatory and inspection authorities to prevent dairy product adulteration and mislabelling. Several EU regulations state the need for applying reference analytical methods for the detection of adulteration in foodstuffs. Therefore, the dairy industry and inspection authorities need fast and reliable methods for routine control of the incoming milk to detect possible adulteration. Thus, analytical techniques (based on different scientific principles) have been developed (and continue to be optimized), aiming at detecting different forms of adulteration in milk and milk products. Such analytical techniques can be classified as chromatographic, electrophoretic, immunological, spectroscopic and molecular.

Three main classes of milk constituents have been targeted as biomarkers for the detection of adulteration in milk and dairy products: milk proteins (caseins (Cn) and/or whey proteins), MF and DNA. Although milk proteins are probably the ones most extensively studied among other food proteins, several factors can complicate their detection, identification and quantification. Such factors include intra- and interspecies genetic polymorphism, compositional variation (the protein composition of milk varies within species depending on breed and stage of lactation) and chemical differences originating from post-translational modifications (phosphorylations, glycosylations). Furthermore, owing to the plethora and complexity of the different technological processes involved in the manufacture of the already diverse collection of dairy products, product-specific factors such as the level (temperature) and extent (time) of the applied heat treatment, as well as the practice and duration of ripening with its associated proteolytic consequences, should be taken into consideration. Whey proteins are more heat-sensitive, but less labile to proteolysis during cheese ripening than caseins. Choices, therefore, regarding the selection of protein markers and the method of analysis must be based on product characteristics.

Methods of analysis based on qualitative or quantitative differences in MF are faced with several difficulties owing to the variability in MF composition originating from differences in genetic background (breed), nutritional status and diet of the animals, stage of lactation and season of the year. Finally, for purposes of quantitation, DNA-based methods need to take into account the possibility of variable DNA content in milk, because the source of DNA (leukocytes from the milk somatic cells) can also vary, primarily as a function of the health status of the udder (the SCC in milk increases as a result of mastitis in lactating animals). Other than that, DNA from milk cells is quite robust and has been shown to endure even conditions of prolonged cheese ripening.

The larger volume of research to date has been devoted to detection of adulteration in cheeses, where bovine milk is fraudulently added as a substitute for caprine, ovine or water-buffalo milk during cheese manufacture. For this purpose, Cn as well as whey proteins have been used as markers for detection of adulteration using chromatographic, electrophoretic, spectroscopic and immunological approaches.

HPLC relies on differences in chemical affinity (and therefore time of interaction) among sample molecules towards a mobile and stationary phase which ultimately determine the analytes' retention time in the chromatographic column and hence their elution time. HPLC has been used for the identification of homologous proteins in binary milk mixtures and fresh or ripened cheeses made from milk of different species using whey proteins as markers (Torre *et al.*, 1996; Ferreira and Cacote, 2003). Volitaki and Kaminarides (2001) used HPLC to detect adulteration of Halloumi cheese with bovine milk down to the 1% level using plasmin-hydrolysed caseins as markers. Bordin *et al.* (2001) used reverse-phase HPLC (RP-HPLC) with a C_4 column and reported the separation and quantification of the seven major proteins in different types of commercially available milk in a single run.

Among the different analytical approaches, polyacrylamide gel electrophoresis (PAGE) (separation of different proteins is based on their charge to mass ratio) is probably one of the earliest to be used for detection of adulterations (Aschaffenburg and Dance, 1968). PAGE is nowadays usually performed using pre-cast, thin polyacrylamide-agarose gels immersed in buffer followed by Coomasie Blue staining, a procedure that can be highly automated and can allow for fast resolution, staining and destaining of milk proteins in a few hours (van Hekken and Thompson, 1992). Appropriate choice of the running buffer pH can enable determination of genetically variant forms of milk proteins with charged amino acid substitutions (Strange *et al.*, 1992). PAGE has been also used in combination with immunoblotting for detection of adulteration in cheeses (Molina *et al.*, 1996). Sodium dodecyl sulfate PAGE separates proteins based on their molecular weight and, in conjunction with scanning densitometry, can give the absolute amounts of the main proteins present in milk.

Isoelectric focusing (IEF) separates proteins based on their electrophoretic mobility (net charge) at isoelectric pH, whereas two-dimensional gel electrophoresis relies on both the isoelectric mobility and molecular weight. The EU has published a reference method for the detection of bovine milk and caseinate in cheeses manufactured from milk of other animal species (EC, 1996). The EU method is based on the isolation of Cn from the cheese followed by their solubilization and hydrolysis with plasmin to convert β-Cn to γ-Cn. The resulting protein fractions are then submitted to IEF, and the protein profiles are compared with those of known composition standards (containing 0 and 1% bovine milk, respectively). The method is qualitative and cannot quantify the amounts of added bovine milk, a characteristic that is of interest in certain PDO cheese varieties that should be manufactured using specified quantities of different milks. Mayer *et al.* (1997) used IEF and cation-exchange HPLC with the para-κ-Cn fraction of caseins as a marker (a fraction that is unaffected by the degree of cheese-ripening) to differentiate bovine, ovine and caprine milk in mixed-milk cheeses. Recently, Cerquaglia and Avellini (2004) proposed improvements to the EU reference method by using more rapid sample preparation and ready-to-use polyacrylamide gel plates that improve the visibility of the separated protein bands. Veloso *et al.* (2002) compared RP-HPLC and PAGE in terms of their ability to detect adulteration of caprine and ovine milks with bovine milk using α-Cn as a marker. The authors reported that electrophoresis was more sensitive and able to detect both types of adulteration to the 5% level (the addition of bovine to ovine milk could not be detected by the HPLC method used).

In recent years, considerable research has been conducted on the development of two powerful analytical techniques, namely capillary electrophoresis (CE) and mass spectrometry (MS). CE is a relatively new analytical tool in food analysis (Zeece, 1992; Lindeberg, 1996) and has been used successfully for the determination of milk proteins and the assessment of the quality of dairy products (de Jong *et al.*, 1993; Recio *et al.*, 1997a). CE is performed in ultra-thin (typically 25–75 μm internal diameter) capillaries that are resistant to high electric fields. One end of the buffer-filled capillary is immersed in a buffer reservoir containing the anode and the other end in a buffer reservoir containing the cathode. Separation of the analytes is based mainly on their charge, their mass and their Stoke's radius. CE can be performed in several separation modes such as capillary zone electrophoresis, capillary gel electrophoresis and capillary IEF (Sørensen *et al.*, 1998). CE has several advantages over other analytical electrophoretic techniques. It is quantitative, highly sensitive and automated, and requires minimal sample volumes; having high resolving power, it results in fast and highly efficient separations. However, limitations of CE are the high capital cost and the technical expertise required. The fact that only one sample can be examined at a time renders this analytical approach unsuitable for routine examinations of dairy samples. In the field of dairy science, CE has been applied for detection of adulterations in dairy products, for the analysis of genetic or non-genetic (post-translational phosphorylation or glycosylation) polymorphisms in milk proteins from different species, for the evaluation of the heat treatment applied to milk and for the study of proteolysis during cheese-ripening (Recio *et al.*, 1997a, 1997b). Ramos and Juarez (1986) and Strange *et al.* (1992) have provided extensive reviews of the chromatographic and electrophoretic methods used for analysis of milk proteins.

Matrix-assisted laser detection ionization-MS (MALDI-MS) enables the fast identification of most proteins present in a sample of diluted milk. The technique has been used successfully to determine thermal damage in milk proteins following milk pasteurization or sterilization and therefore has been applied for the detection of the addition of powdered milk in fresh milk (Catinella *et al.*, 1996; Siciliano *et al.*, 2000). Also, MALDI-MS has been used for the detection of fraudulent additions of bovine milk to ovine, or water-buffalo milk. Angeletti *et al.* (1998) tested several different milk and cheese samples using MALDI-MS, and demonstrated the ability of the technique to differentiate milk from different species based on the differences in the resulting MALDI mass spectra, whereas Schmidt *et al.* (1999) used pyrolysis-MS for the detection of addition of whey proteins in milk.

Immunoassays and, in particular, the ELISA can be used in various configurations to detect adulteration in milk products (Moatsou and Anifantakis, 2003; Bonwick and Smith, 2004; Hurley *et al.*, 2004a, 2004b). Polyclonal or monoclonal antibodies have been used in a variety of formats (direct, indirect, sandwich and competitive ELISA). Monoclonal antibodies can nowadays be produced in large quantities via hybridoma technology and are preferred over polyclonal antibodies, as the latter have to be affinity-purified in order to be species-specific. ELISA is a simple, sensitive, automated and rapid test that can allow for the simultaneous analysis of many samples and can be used as a routine method for detection of substitution of ovine and caprine milks by bovine milk or the substitution of caprine milk with ovine milk in raw and heat-treated milk and cheese.

MF has also been used as a marker for detection of adulterations. As milk from different species is characterized by differences in the fatty acid profile of MF (Prager, 1989), Iverson and Sheppard (1989) used the lauric:capric (12:10) fatty acid ratio, which was obtained from gas-chromatographic analysis of cheeses, to detect the presence of bovine milk in ovine- or caprine-milk cheeses.

Substitution of MF in milk and milk products (such as butter and ice cream) with vegetable fats or with non-milk animal fats (beef tallow, lard or chicken fat) is another fraudulent practice. Analysis of MF has been conducted by gas–liquid chromatography, and decisions have been based on the measurement of the resulting concentrations or ratios of selected fatty acid moieties in the suspect samples. Butyric acid (C4) has been proven to be a suitable marker for MF quantitation in mixed-fat foods (e.g. spreads), as it is exclusively found in MF (Fox *et al.*, 1988; Molkentin and Precht, 1998, 2000). Triglycerides have also been used as indices to detect adulteration of milk with non-milk fat substitutes (Timms, 1980; Battelli and Pellegrino, 1994), but such approaches may not be appropriate for application to mature cheeses due to the extensive lypolysis in such products and the concomitant hydrolysis of triglycerides. Because phytosterols are either undetectable or present at trace amounts in MF, detection and quantitative determination of phytosterols (β-sitosterol) are the approaches used to detect addition of vegetable oils in MF (IDF, 1970; Sheppard *et al.*, 1985; Kamm *et al.*, 2002). Substitution of MF with other animal fats (as opposed to vegetable fats) is more difficult to detect because of the lack of suitable markers. Spectroscopic and calorimetric approaches for the detection of MF adulteration have also been published (Sato *et al.*, 1990; Coni *et al.*, 1994).

Among DNA-based methods for food authentication (Lockley and Bardsley, 2000), PCR is the most widely used. Ruminant milk from mammary glands contains a high number of somatic cells (leucocytes and epithelial cells) which contain genomic DNA. This DNA has been shown to persist during the various heat-treatment regimes for milk as well as cheese-ripening processes and can therefore be isolated and used for PCR amplification. Appropriate design of primers, adequate protocols for extracting inhibitor-free DNA and appropriate target sequences with sufficient species-to-species variation are necessary for the valid use of PCR for detection of adulterations. Two main approaches have been used to differentiate closely related species: the use of universal primers and the use of species-specific primers. The use of universal primers allows amplification of the target sequence in all species, and such an approach must be followed by sequence- or restriction analysis of the amplified target. Alternatively, the use of species-specific primers allows the unique identification of the target species in the mixed sample because such primers generate product only in the presence of DNA from the 'suspect' species in the mixture. Recently, Bottero *et al.* (2003) were able to detect bovine, ovine and caprine milk in mixed-milk dairy products in a single step, using a multiplex PCR approach (different pairs of primers in the same reaction mix). In PCR, both nuclear and mitochondrial (Mt) DNA can been used as amplification substrates. Mt DNA is characterized by a much higher copy number, and its sequence can be found in published databases and is characterized by species-specific sequence variation. PCR is rapid, sensitive, specific and reliable for highly processed samples and can be used for routine examinations. However, because the amount of extracted DNA from a milk or cheese sample is proportional to the somatic cell content of the milk and can also be influenced by the extraction protocol, simple PCR applications may not be appropriate for quantitative interpretations. To the authors' knowledge, to date the use of quantitative real-time PCR for detection of adulteration in dairy products has not been reported.

Whey is produced in vast amounts as a by-product of the cheese-making process and has a low price. Therefore, addition of whey (solids) to liquid milk or milk powder is economically attractive, and so analytical methods are needed that are able to detect "rennet" "whey" total solids in these products. Van Riel and Olieman (1995) used CE and caseinomacropeptide as a marker to detect addition of "rennet" "whey" solids in milk and buttermilk powder. Recio *et al.* (2000) used genuine, adulterated and commercial ultra-high-temperature (UHT) milk samples and the CE peak-area ratios of three different caseinomacropeptide-related peptides to propose a method for detecting addition of rennet–whey solids in UHT milk.

Soya bean is a low-cost legume with good nutritional qualities, and its proteins are often added to different non-vegetable food products to enhance their nutritional quality. Nonetheless, the addition of soya or other vegetable proteins to dairy products may be illegal in certain countries, their undeclared addition is considered fraudulent, and this can have health implications (e.g. wheat proteins for coeliac disease patients). Among other analytical approaches, chromatographic (Krusa *et al.*, 2000; Espeja *et al.*, 2001), immunological (Sanchez *et al.*, 2002) and electrophoretic protocols (Manso *et al.*, 2002) have been published for the determination of vegetable proteins in milk products.

If water is added to milk deliberately or accidentally, the density of milk will decrease, and its freezing point will become closer to that of pure water (the freezing point of milk is lower than that of pure water due to the presence of dissolved components, mainly lactose and chloride ions). Detection of added water in milk is conducted through cryoscope analysis. The method involves supercooling and crystallizing milk samples via mechanical vibration; the heat of crystallization causes a rise in sample temperature, until the sample reaches its precise freezing point. The cryoscope values are given in degrees Hortvet (°H). The baseline freezing point of normal, raw bovine milk is considered to be 0.540°H (Hooi *et al.*, 2004), although, depending on certain factors, slightly different average values can occur (Harding, 1995). Because the cryoscope method is sensitive enough to detect even 1% extraneous water in milk, in order to interpret cryoscope data for milk quality purposes, prior documented and precise knowledge of the freezing point of the milk produced by animals of a given herd, or of the milk supply produced by animals in a specified geographical area is required. Recently, Mabrook and Petty (2003) proposed an approach for detecting watered full-fat milk using single-frequency admittance measurements of milk at 8°C.

Table 6.5 presents recent published literature on the detection of adulteration in dairy products. A thorough description of the applications of different analytical techniques for the detection, characterization and quantitation of milk proteins has been

Table 6.5 Detection of adulteration in dairy products using different analytical methods[a].

Method	Product type	Type of adulteration	Milk (protein) marker	Detection limit (%)	Reference
HPLC	Halloumi cheese	BM in OMC	Plasmin-hydrolysed Cn	1	Volitaki and Kaminarides (2001)
RP-HPLC,	Milk	BM in CM	Bovine α-Cn	5	Veloso *et al.* (2002)
Urea-PAGE		BM in CM		5	
		BM in OM		5	
RP-HPLC	Milk, Terrincho cheese	BM in OMC and CMC	β-Lg	2	Ferreira and Cacote (2003)
		OM in CMC			
RP-HPLC	Terrincho cheese	BM in OMC	Bovine α-Cn	20	Veloso *et al.* (2004)
Urea-PAGE				10	
Cation-exchange HPLC	Halloumi cheese	CM in OMC	Caprine para-κ-Cn	5	Moatsou *et al.* (2004)
HPLC, ESI-MS	Milk	BM in CM	β-Lg	5	Chen *et al.* (2004)
MALDI-TOF MS	Cheese	BMC in OMC	Ovine γ_3-Cn/Bovine γ_2-Cn	10	Fanton *et al.* (1998)
Pyrolysis-MS	Milk	Whey-protein addition	Milk mass spectrum	2	Schmidt *et al.* (1999)
		BM in WBM	α-La	≤5	
MALDI-TOF MS	Milk	BM in OM	β-Lg	10	Cozzolino *et al.* (2001)
		PM in FM	α-La and β-Lg		
MALDI-TOF MS	Mozzarella cheese	BM in WBMC	α-La	5	Cozzolino *et al.* (2002)
		OM in WBMC	α-La and β-Lg	2	
SDS-PAGE	Buttermilk powder	Non-fat dry milk for buttermilk powder	Milk-fat globule membrane proteins		Malin *et al.* (1994)
PAGE	Halloumi cheese	BM in OMC	Bovine α_{s1}-Cn	2.5	Kaminarides *et al.* (1995)
PAGE	Ovine yoghurt	BM in ovine yoghurt	Bovine para-κ-Cn	1	Kaminarides and Koukiassa (2002)
ELISA (competitive)	Milk	BM in OM	Bovine α_{s1}-Cn	0.125	Ronald *et al.* (1993)
	Roquefort cheese	BM in OMC		0.5	
ELISA (sandwich)	Milk	BM in CM and OM	Bovine β-Lg	0.001	Levieux and Venien (1994)
IEF (immunoblotting)	Cheese	BM in OMC	Bovine β-Cn	0.5	Addeo *et al.* (1995)
ELISA (immunostick)	Milk	BM in OM	Bovine β-Cn	1	Anguita *et al.* (1996)
	Cheese	BM in OMC		0.5	
ELISA (indirect)	Milk	CM in OM	Caprine α_{s2}-Cn	0.5	Haza *et al.* (1996)
ELISA (competitive, indirect)	Cheese	BM in OMC and CMC	Bovine γ_3-Cn	0.1	Richter *et al.* (1997)
ELISA (competitive)	Milk and cheese	BM in CM ,CMC, OM and OMC	Bovine β-Cn	0.5	Anguita *et al.* (1997)

Contd.

Table 6.5 (Contd.)

Method	Product type	Type of adulteration	Milk (protein) marker	Detection limit (%)	Reference
ELISA (indirect)				0.5	Haza *et al.* (1997)
ELISA (indirect, competitive)	Milk	CM in OM	Caprine α_{s2}-Cn	0.25	
ELISA (sandwich)	Milk	BM in CM	Bovine β-Lg	0.03	Negroni *et al.* (1998)
ELISA (competitive, indirect)	Cheese	CMC in OMC	Caprine α_{s2}-Cn	0.5	Haza *et al.* (1999)
ELISA (competitive, indirect)	Milk	BM in CM, OM and WBM	Bovine IgG	0.1	Hurley *et al.* (2004a)
CE	Milk- and buttermilk powder	Addition of RWS	CMP	0.4	van Riel and Olieman (1995)
CE	Milk	BM in OM and CM	Bovine α_{s1}-Cn	8	Cattaneo *et al.* (1996)
CE	Milk, cheese	BM in OM	Bovine β-LgB/caprine	0.5	Cartoni *et al.* (1998)
		BM in OMC	α-La	2	
CE	Milk, cheese	BM in CM	Bovine β-LgA/caprine	2	Cartoni *et al.* (1999)
		BM in CMC	α-La	4	
CE	Cheese	BM in OMC and CMC	α-La and β-Lg	1	Herrero-Martinez *et al.* (2000)
CE	Halloumi cheese	BM in OMC	Bovine αs_1-Cn	5	Recio *et al.* (2004)
		CM in OMC	Caprine para-κ-Cn	2	
CE	Hypoallergenic infant formula	Bovine β-Lg (presence)	Bovine β-Lg		Veledo *et al.* (2005)
PCR-REA-PAGE	Cheese	BM in OMC and CMC	β-Cn gene	0.5	Plath *et al.* (1997)
PCR-RFLP	Feta cheese	BM in OMC and CMC	Mt cytochrome b gene		Branciari *et al.* (2000)
	Mozzarella cheese	BM in WBMC			
PCR	Cheese	BM in CMC	Control region of Mt DNA	0.1	Maudet and Taberlet (2001)
PCR	Milk	BM in CM	Mt cytochrome b gene	0.1	Bania *et al.* (2001)
Duplex-PCR	Milk, Mozzarella cheese	BM in WBM and WBMC	Mt cytochrome b gene	1	Rea *et al.* (2001)
Duplex-PCR	Milk, Mozzarella cheese	BM in WBM and WBMC	Mt cytochrome b gene		Bottero *et al.* (2002)
Multiplex-PCR	Cheese	BM in CM	Mt 12s and 16s rRNA genes	0.5	Bottero *et al.* (2003)
Duplex-PCR	Cheese	BM in OMC	Mt 12s and 16s rRNA genes	0.1	Mafra *et al.* (2004)
PCR	Milk	BM in OM and CM	Mt 12s rRNA gene	0.1	Lopez-Calleja *et al.* (2004)
PCR	Milk	CM in OM	Mt 12s rRNA gene	0.1	Lopez-Calleja *et al.* (2005a)
PCR	Milk, Mozzarella cheese	BM in WBM and WBMC	Mt 12s rRNA gene	0.1	Lopez-Calleja *et al.* (2005b)
PCR	Mozzarella cheese	BM in WBMC	Bovine Mt cytochrome oxidase subunit I gene	0.5	Feligini *et al.* (2005)

[a]BM = bovine milk; CE = capillary electrophoresis; CM = caprine milk; CMC = caprine milk cheese; CMP = caseinomacropeptide; Cn = casein; ELISA = enzyme–linked immunosorbent assay; ESI = electrospray ionization; FM = fresh milk; HPLC = high-performance liquid chromatography; Ig = immunoglobulin; La = lactalbumin; Lg = lactoglobulin; MALDI = matrix-assisted laser desorption ionization; MS = mass spectrometry; Mt = mitochondrial; OM = ovine milk; OMC = ovine milk cheese; PAGE = polyacrylamide gel electrophoresis; PCR = polymerase chain reaction; PM = powdered milk; REA = restriction enzyme analysis; RFLP = restriction fragment length polymorphism; RP = reversed phase; RWS = rennet whey solids; TOF = time of flight; WBMC = water-buffalo milk cheese.

presented by Tremblay *et al.* (2003), whereas O'Donnell *et al.* (2004) have summarized the advantages and limitations of the main analytical approaches used in milk proteomics. An excellent chapter on the topic of dairy product adulteration has been put together by Ulberth (2003). Results from analyses conducted on the same set of samples with the application of techniques that are based on different scientific principles have been recently published by Mayer (2005). This is one of the few articles in the literature where many different analytical techniques were applied on the same set of samples as part of the same study. In this study, standard mixtures of either bovine and ovine, or bovine and caprine milk, containing 0–100% bovine milk were tested as milk mixtures and also used for the pilot-scale manufacture of Camembert, Tilsit and Kashkaval model cheeses. Cheeses were tested at various time points in their ripening period. Although only a defined set of operating conditions was applied for each analytical method, the study helps to point out some of the advantages and limitations of each analytical format in detecting adulteration.

International standardization organizations have developed and standardized analytical methods (Standards) pertaining to the detection of adulteration of milk and dairy products. Standards have either been developed individually or jointly by more than one organization. In the area of dairy products' adulteration, relevant methods include two IDF Standards (IDF, 1970, 1995b) and two joint IDF/ISO Standards (IDF, 2002a, 2002b).

Chromatography- and immunochemistry-based commercial assays (test kits) for detection of adulteration in dairy products have been made available through several manufacturers. Examples are the 'Casein-IEF Consumable Kit' (ETC Elektrophorese-Technik, GmbH), the 'Proteon Test' (Zeu-Immunotech, Spain), the 'Bovine IgG Indicator' (Midland BioProducts Corporation, IA), the 'IC-Bovine' and 'IC-Caprine' assays (Neogen Europe Ltd, Ayr, UK), the 'RIDASCREEN product adulteration kits' (R-Biopharm Rhône Ltd, Glasgow, UK) and possibly others as well.

6.7 Detection of abnormal milk

For the purposes of this chapter, the term 'abnormal' denotes the milk originating from cows suffering from udder inflammation (mastitis) or the milk containing the mammary gland secretion produced during the first post-partum days (colostrum). Screening and confirmatory methods used to detect mastitic milk usually rely on the fact that milk from a mastitic quarter has an elevated number of somatic cells (>300,000 per ml) (Nierman, 2004). At the farm level, the California Mastitis Test (CMT) (Schalm and Noorlander, 1957) is still used as a rapid screening method to test for the presence of mastitis in individual udder quarters. The test is based on the reaction of cellular DNA from the somatic cells in the milk sample with the CMT reagent that is added to an appropriate four-cup tray (one cup for each quarter of the udder). The thickness of any gel formed is correlated with the concentration of somatic cells in the milk sample. A five-point (objective) milk-scoring is done immediately after a 10-s mixing of the milk with the CMT reagent. The test has been shown to be reliable for diagnosing mastitis at the farm level and is fast, cheap and easy to perform.

The increase in concentration of salts in milk from an inflamed udder can be detected as elevated electrical conductivity by means of portable conductivity meters. On a scale of 0 to 9, conductivity readings greater than 5 are usually, but not always, indicative of milk abnormality. Finally, methods based on ATP bioluminescence use the milk ATP content (originating from milk somatic cells) as an indicator of mastitis (SomaLite test by Charm Sciences, Lawrence, MA).

In the laboratory, the SCC of milk can be assessed directly on microscope slides using the compound microscope (Fitts and Laird, 2004) or using electronic somatic cell counters (IDF, 1995b). For electronic counting, both the Coulter counter method (Beckman Coulter, Brea, CA) and the fluoro-opto-electronic method (Fossomatic, Foss Electric) can be used. The former is based on electrical conductivity measurements, whereas the latter is based on the fluorescence of the somatic cells stained with a fluorescent dye. As for the aetiologic agent of mastitis, this is usually identified based on the animals' clinical manifestations in conjunction with the herd's medical history and also based on the results of microbiological examinations of milk samples that have been aseptically taken from suspect quarters.

Detection of colostrum in the milk supply is of importance for the dairy industry as colostrum is rich in immunoglobulins that readily denature during heat treatment of milk in the dairy factory creating a protein deposit that reduces heat transfer in the heat exchanger. Test kits are available for use by the dairy industry to detect the addition of colostrum in milk. The assays detect the elevated IgG concentration in milk. Examples are the Kalokit (Zeu-Immunotek), an immunoenzymatic test for colostrum detection in cow's milk and the Bov IgG-Test (ID Biotech, France), a radial immunodiffusion method that measures the immunoglobulin content in milk by means of the size of a precipitation zone formed on an agar plate that contains immunoglobulin antiserum.

6.8 Microbiological methods

The aim of the microbiological examination of milk and milk products is to ensure that the products are both safe to eat (i.e. enumeration and detection of certain pathogens) and of high microbiological quality, thus having a reasonable shelf-life (i.e. enumeration of contaminating and/or spoilage microorganisms).

For the analysis of the microbiological quality of milk and milk products, a number of reference methods have been introduced by international organizations (i.e. IDF and ISO) as shown in Table 6.6. These cover all aspects of microbiology from detection and isolation of specific organisms to identification and confirmation of isolates, and are recommended for a wide range of products such as milk, liquid and dried products, fermented milks, cheese and processed cheese, butter, cream, ice cream and frozen desserts. If a standard method is not followed as described, but is rather modified, method validation is required (Section 6.2.2).

6.8.1 Standard plate count

The colony count at 30°C is a very useful method for assessing the microbiological quality of a product, and the reference method is described by IDF (1991a). The microbes

Table 6.6 Reference microbiological methods for milk and milk products[a].

Microorganism	Product	Method		
Total count	Milk and milk products	Colony-count technique at 30°C	IDF Standard 100B:1991	
Total count	Butter, fermented milks and fresh cheese	Colony-count technique at 30°C	IDF Standard 153:2002	ISO Standard 13559:2002
Total count	Milk	Plate loop technique at 30°C	IDF Standard 131:2004	ISO Standard 8553:2004
Contaminating microorganisms	Butter, fermented milks and fresh cheese	Colony-count technique at 30°C	IDF Standard 153:2002	ISO Standard 13559:2002
Coliforms	Milk and milk products	Most Probable Number technique	IDF Standard 73B:1998	
Enterobacteriaceae	Food and animal feeding stuffs	Part 1: Detection and enumeration by MPN technique with pre-enrichment		ISO Standard 21528-1:2004
	Food and animal feeding stuffs	Part 2: Colony-count method		ISO Standard 21528-2:2004
Escherichia coli	Milk and milk products	Part 1: Most Probable Number technique	IDF Standard 170A:1999	
Escherichia coli	Milk and milk products	Part 2: Most Probable Number technique using 4-methylumbelliferyl-ß-D-glucuronide (MUG)		
Escherichia coli	Milk and milk products	Part 3: Colony-count technique at 44°C using membranes		
Bacillus cereus	Dried milk products	Most Probable Number technique	IDF Standard 181:1998	
Clostridium perfringens	Food and animal feeding stuffs	Colony-count technique		ISO Standard 7937:2004
Staphylococci-coagulase-positive	Milk and milk-based products	Most probable number technique	IDF Standard 60:2002	ISO Standard 5944:2002
Staphylococci-thermonuclease produced by coagulase-positive	Milk and milk products		IDF Standard 83A:1998	
Staphylococci-coagulase-positive	Milk and milk products	Colony-count technique	IDF Standard 145A:1997	
Staphylococcus aureus	Dried milk	Colony-count technique at 37°C	IDF Standard 138:1986	
Salmonella spp.	Milk and milk products		IDF Standard 93:2001	ISO Standard 6785:2001
Sulfite-reducing bacteria growing under anaerobic conditions	Food and animal feeding stuffs			ISO Standard 15213:2003
Yeasts and moulds	Milk and milk products	Colony-count technique	IDF Standard 94:2004	ISO Standard 6611:2004
Yersinia enterocolitica	Food and animal feeding stuffs			ISO Standard 10273:2003
Yoghurt-characteristic microorganisms	Yoghurt	Colony-count technique at 37°C	IDF Standard 117:2003	ISO Standard 7889:2003
Yoghurt culture (*Lactobacillus delbrueckii* subsp. *bulgaricus* and *Streptococcus thermophilus*)	Yoghurt	Identification of characteristic microorganisms (*Lactobacillus delbrueckii* subsp. *bulgaricus* and *Streptococcus thermophilus*)	IDF Standard 146:2003	ISO Standard 9232:2003

[a]After Webber *et al.* (2000).

that are assessed are aerobic mesophiles that can grow on non-selective media and include LAB, psychrotrophic bacteria, thermoduric bacteria and sporeformers, including pathogenic bacteria.

Although a colony count at 30°C gives limited information for the potential health-risk of the product, it is a very useful measure of the hygiene conditions and the sanitation level during harvest and processing. The main disadvantage of the standard plate count is the long incubation time, since the results take 72 h. Thus, faster methods are often used for quality control purposes.

6.8.2 Direct microscopic count

The Direct Microscopic Count method is based on the technique developed by Breed in 1911 (Hill, 1991a). With this technique, a small amount of milk is spread evenly over a predefined area of a microscope slide. After drying, treatments to remove fat, and staining of the smear, individual cells or bacterial aggregates are counted in several fields using a compound microscope and conventional bright field illumination (Hill, 1991a). Several variations of the staining and counting procedures have been developed, and most methods specify that 0.01 ml of the sample is spread over 1 cm^2 of the slide. Major disadvantages of the staining method are that it does not distinguish between dead and viable cells and that the small sample volume renders the method insensitive and subject to considerable error (Mostert and Jooste, 2002).

6.8.3 Direct epifluorescent technique

The Direct Epifluorescent Technique (DEFT) is a method in which individual bacteria or groups of bacteria are counted, after appropriate staining, using a specialized microscope following a procedure in which the bacteria are filtered out of the milk sample and concentrated on the surface of a filter membrane (Pettipher *et al.*, 1980). The procedure is carried out as follows (Hill, 1991b). The milk sample is first pre-treated with a proteolytic enzyme and a surfactant which lyses somatic cells and modifies fat glodules sufficiently so that a 2 ml sample can be filtered through a 0.6-μm-pore-size polycarbonate filter. Filtration concentrates the bacteria on the surface of the membrane filter; the bacteria are then stained with Acridine Orange, and the mounted filter is examined through an epifluorescence microscope. Metabolically active bacteria fluoresce orange-red, while inactive bacteria fluoresce green. The clumps of orange-red-fluorescing bacteria are counted in the field of view and a DEFT count per millilitre of the milk sample is calculated from the count over several fields. By using DEFT, a bacterial count for a milk sample can be obtained within 25 min. The main disadvantage of the method is the limited number of samples that can be tested due to the fact that it is a microscopic count. A semi-automated system has been described by Hill (1991b), but the sample preparation remains the time-consuming step.

6.8.4 Spiral plate counting

Spiral plate counting (SPC) is a mechanical procedure, employing a 'loop' to deposit undiluted sample onto a surface of a prepared agar plate. This procedure allows the measurement of bacterial numbers in raw-milk samples containing 500–500,000 bacteria ml^{-1}. An operator can conduct the SPC procedure on approximately 50 samples per hour, and within the count range mentioned, dilution bottles and pipettes are not needed (Brazis, 1991).

6.8.5 Bactoscan

The principle of Bactoscan 8000 (Foss Electric) is direct microscopic counting. The Bactoscan 8000 is a fully automated instrument, in which the bacteria are centrifuged and separated from the milk, stained with a fluorescent dye (e.g. Acridine Orange) and counted electronically as light impulses in a continuous flow fluorescent microscope. The sample is treated with a lysing solution (Suhren *et al.*, 1991), and so bacterial clumps are dissolved, a fact that leads to more accurate counts. The main advantage is speed (approximately 80 samples per hour). Lachowsky *et al.* (1997) stated that automated Bactoscan 8000 can provide a reasonable alternative to the classical cultural methods for enumeration of bacteria in bulk-tank raw milk when bacterial counts lie within the instrument's specified applicable range of 40,000–80,000 cfu ml^{-1}. Samples can be preserved for no more than 7 days prior to analysis by addition of chemicals such as boric acid or sodium azide. A possible disadvantage is the fact that the relationship between the standard plate count and Bactoscan values at lower plate counts (i.e. $<10,000$ cfu ml^{-1}) is less consistent than at higher cfu levels (Mostert and Jooste, 2002).

A newer development for the counting of individual cells is the introduction of Bactoscan FC from Foss Electric, which is based on flow cytometry. In the Bactoscan FC, the DNA/RNA of the bacteria is stained with the fluorescent dye ethidium bromide. Certain buffers and enzymes are added during sample preparation to reduce the influence of other milk constituents, and the bacterial clusters are also separated into single bacteria. Fluorescence is induced by laser, and the light emitted is detected when the stained particles pass as a hydrodynamically focused stream through a fluorescence detector. The accuracy of the estimation of the microbial load measured with Bactoscan FC is superior to the Bactoscan 8000. However, the capital cost of the equipment is higher, and somatic cell counts, when exceeding 1 million ml^{-1}, might influence bacterial counts (Mostert and Jooste, 2002).

6.8.6 Dye reduction tests

Dye reduction tests for the determination of the bacteriological quality of milk are based on the ability of certain bacterial enzymes, such as the dehydrogenases, to transfer hydrogen atoms from a substrate (e.g. Methylene Blue or resazurin) to biological acceptors. During the reaction, the dye is reduced at a rate that depends on the enzyme activity, and this has been used as an index of the number of bacteria present. The dye is added to the milk, and the colour change after incubation is monitored. The period of time required to change or to decolorize the dye is an index of the bacteriological load of the milk. The incubation period of the Methylene Blue test carried out on bulk milk with plate counts of approximately 100,000 ml^{-1} was 6 h at 37°C, while that of the resazurin test was 3 h (Lück, 1991).

6.8.7 Determination of pyruvate or ammonia

The enzymatic breakdown of carbohydrates, fat and protein produces, besides other metabolites, lactate, free fatty acids, ammonia, urea and pyruvate. Pyruvate is the main metabolic product of the breakdown of carbohydrates, lipids (via glycerol) and proteins (via amino acids such as alanine, serine and glycine) (Suhren and Heeschen, 1991). In the presence of lactate dehydrogenase and simultaneous reduction of NAD, lactate is oxidized to pyruvate:

$$\text{Lactate} + \text{NAD} \rightarrow \text{Pyruvate} + \text{NADH}_2.$$

The amount of $NADH_2$ formed is proportional to the lactate content and is measured at 340 nm. The amount of pyruvate can be affected by the health status of the udder, feeding variations and differences in the level of metabolic activities in the course of lactation.

Ammonia is another metabolite used for the assessment of the bacteriological quality of raw milk. Ammonia determination can be carried out using an ion-sensitive electrode, enzymatically or colorimetrically in a continuous flow system (Suhren and Heeschen, 1991). Freshly drawn milk has an ammonia content between 3 and 6 mg kg^{-1} (Söderhjelm and Lindqvist, 1980). By the time a significant increase in ammonia content is observed, the bacterial count should be very high, namely, 10^7–10^8 cfu ml^{-1}, while an ammonia value greater than 8.5 mg kg^{-1} indicates inferior milk quality (Suhren and Heeschen, 1991).

6.8.8 Contaminating microorganisms

The count for contaminating microorganisms (IDF, 2002c) is very much related to the total count. It differs from the conventional total count in that the culture medium is carbohydrate-free. The medium consists of peptone from casein, peptone from gelatin, sodium chloride and agar. Starters, that is LAB which require fermentable carbohydrates as a carbon source, are not able to develop colonies in the medium or, at best, are only capable of developing pinpoint colonies (Mostert and Jooste, 2002). The rationale is that non-lactic microorganisms (e.g. psychrotrophic bacteria) can be selectively detected, as far as the pinpoint colonies are not counted as contaminants. Recently, Angelidis *et al.* (2006) proposed that contaminating flora counts can serve as a quality index for fermented (dairy) foods for which the use of total aerobic counts is not appropriate.

Psychrotrophic bacteria are those bacteria able to grow at 7°C or less, regardless of their optimal growth temperature (Frank *et al.*, 1993). Psychrotrophic bacteria can grow at refrigeration temperatures, and so they can multiply in raw milk during storage. They are heat-sensitive organisms but produce heat-stable proteolytic and lipolytic enzymes, which can withstand pasteurization. *Pseudomonas* spp. is the most common genus, and they can be enumerated using selective media containing antibiotics, such as the *Pseudomonas* agar base supplemented with glycerol, cetrimide, fucidin and cephaloridine (Baylis, 2003). Certain defects in dairy products can be caused by this group of bacteria, and for products intended to be kept at refrigeration temperatures, the presence of psychrotrophic microorganisms is the most important factor in determining their shelf-life.

6.8.9 Thermoduric bacteria

Thermoduric bacteria are defined as those which are resistant to pasteurization (i.e. heating at 63°C for 30 min). Their presence in raw milk implies that they may be present in the final product as well. However, since they are unable to grow at refrigeration temperatures, they are not of primary importance when the refrigeration chain is properly maintained. Thermoduric bacteria isolated from milk usually include spore-formers such as *Bacillus* spp. and *Clostridium* spp. and non-spore-forming cocci (e.g. *Micrococcus* spp. and *Streptococcus* spp.) and rods such as *Microbacterium* spp. and other members of the coryneform group (Mostert and Jooste, 2002).

Aerobic spore-formers are bacteria belonging to the genus *Bacillus* that form endospores, which are resistant to a variety of adverse conditions such as heating and drying. Most spores withstand heating at 80°C for 10 min and therefore survive pasteurization. They are usually thermophilic, i.e. they have an optimum temperature of growth at about 55°C. *Bacillus cereus* is an aerobic spore-forming bacterium that is involved in food intoxification and produces two types of enterotoxins: diarrheal and emetic. A standard Most Probable Number (MPN) technique is described for the enumeration of *B. cereus* in dried-milk-based infant foods (IDF, 1998a). The method employs tryptone soya polymyxin broth and, after inoculation, subculture onto polymyxin pyruvate egg-yolk mannitol Bromothymol Blue agar or mannitol egg-yolk polymyxin agar. Identification of colonies is confirmed by morphological and biochemical tests.

Anaerobic spore-formers are of little importance in the spoilage of milk and most of the milk products, but they do pose a risk in cheese-making. Certain defects in cheeses have been attributed to the presence of species of clostridia such as *Clostridium butyricum* and *Clostridium tyrobutyricum* (Bergere and Lenoir, 2000). In addition, the use of milk powder in the manufacture of processed cheeses may pose a risk for these products.

6.8.10 Coliforms and Enterobacteriaceae

Coliforms, according to the definition of the IDF (1998b), are bacteria which form characteristic colonies on violet red bile lactose (VRBL) agar at 30°C, namely, dark-red colonies, 0.5 mm in diameter, which show evidence of precipitation of bile salts in the medium surrounding the colonies, and which can ferment lactose with the production of acid and gas under the conditions described. Coliform is not a taxonomic classification but rather a working definition used to describe a particular group of Gram-negative, facultative anaerobic rod-shaped bacteria belonging to the family of *Enterobacteriaceae*. Furthermore, according to the definition of the International Commission on Microbial Specifications for Foods (ICMSF, 1978) 'faecal coliforms' comprise a group of organisms selected by incubating inocula derived from a coliform enrichment broth at a temperature of 44–45°C.

Coliforms, and in particular *Escherichia coli,* are regarded as indicators of recent faecal contamination if they are found in water, since they die out rapidly in water, but in most dairy products they do not die out, and the conditions are favourable to their growth. Coliforms are referred to as indicator microorganisms, since their presence is used to indicate the potential presence of pathogens in foods. Almost always, the presence of coliform organisms in dairy products is due to contamination from equipment that has not been properly cleaned and sanitized, or due to incorrect operation of the heat treatment process.

The test for *Enterobacteriaceae* instead of coliforms is a more sensitive test for post-pasteurization contamination, since the test detects all of the heat-sensitive, non-spore-forming Gram-negative rods and provides good evidence that contamination has occurred. In this case, the media used for the test must contain glucose instead of lactose (e.g. violet red bile glucose (VRBG) agar). For coliform counts, the direct plating of an ice-cream sample on media like VRBL agar may cause some false-positive results, since non-lactose fermenting bacteria may ferment sugars contained in the undiluted sample (Papademas and Bintsis, 2002). It should be remembered that contamination with *Enterobacteriaceae* shows that other additional serious pathogens may have contaminated the product as well. It is believed by some investigators that the higher the number of indicators, the greater the possibility that pathogenic microorganisms will be present. However, this indicator/pathogen relationship is scientifically debatable and not accepted unanimously by the scientific community.

6.8.11 Enterococcus *spp.*

Bacteria of the genus *Enterococcus* are an important group of lactic acid bacteria (LAB), which have a predominant habitat in the gastrointestinal tract of humans and animals. They also persist in the extraintestinal environment and can colonize diverse niches due to their high heat tolerance and ability to survive under adverse environmental conditions. Thus, enterococci occur in large numbers in foods, especially those of animal origin, such as dairy products. The presence of enterococci in milk and dairy products has been considered as an indicator of poor sanitation during production and processing. High levels of contaminating enterococci may deteriorate the sensory properties of dairy products (Litopoulou-Tzanetaki, 1990; López-Diaz *et al.*, 1995). On the other hand, many scientists believe that enterococci play an important and beneficial role in the development of organoleptic characteristics of cheese during ripening. Due to their proteolytic and lipolytic activities (Centeno *et al.*, 1996; Arizcum *et al.*, 1997), their ability to metabolize citrate (Tsakalidou *et al.*, 1993; Sarantinopoulos *et al.*, 2001) as well as their ability to produce bacteriocins (Torri Tarelli *et al.*, 1994; Giraffa, 1995; de Vuyst *et al.*, 2002; Foulquié Moreno *et al.*, 2003), many authors include them in certain starter cultures for cheese production (Litopoulou-Tzanetaki *et al.*, 1993; Garde *et al.*, 1997; Centeno *et al.*, 1999; Sarantinopoulos *et al.*, 2002). Direct sources of milk contamination are human and animal faeces, and indirect sources include contaminated water, animal hides and milk tanks. The presence of enterococci in cheeses made from pasteurized milk is due to recontamination after heat treatment of milk and also due to their heat resistance. For the enumeration of enterococci, selective media have been developed such as the KF streptococcal agar (Kenner *et al.*, 1961), the synthetic media Citrate Azide agar, the Kanamycin Aesculin Azide agar and Slanetz and Bartley medium (Baylis, 2003).

6.8.12 Yeasts and moulds

Yeasts and moulds are heat-sensitive microorganisms, and their enumeration is of little importance in milk and most milk products. They play an important role, however, in spoilage, causing certain defects, and possibly in the development of flavour of certain cheeses (Bergere and Lenoir, 2000; Bintsis and Papademas, 2002). Of special significance is the presence of yeasts and moulds in yoghurt and especially the presence of yeasts in flavoured yoghurt, where they ferment added sugar to produce gas. The enumeration of yeasts and moulds is carried out as described in IDF (2004c) using either yeast-extract dextrose oxytetracycline hydrochloride agar or yeast extract dextrose chloramphenicol agar. The antibiotics are added to these media to inhibit bacterial growth.

6.8.13 Specific pathogenic bacteria

In addition to the enumeration of contaminating and indicator microbes, dairy products need to be examined for the presence of certain pathogenic bacteria as suggested by Regulations and Standards. For these purposes, reference methods are used (Table 6.6). Methods for the enumeration of specific pathogenic bacteria are fully described in the FDA's Bacteriological Analytical Manual (BAM) (US FDA, 2005), the American Public Health Association's (APHA) Standard Methods for Examination of Dairy Products (Flowers *et al.*, 1993) and the Campden and Chorleywood Food Research Association Group manual (Baylis, 2003).

6.8.13.1 Listeria monocytogenes

In early studies, it was noted that *Listeria* spp. are able to grow at low temperatures, a phenomenon now known to be largely aided by the organism's ability to accumulate cryoprotectants from foods (Angelidis *et al.*, 2002), and this feature has been used to isolate these bacteria from foods by incubation of agar plates for prolonged periods at 4°C until the formation of visible colonies (Gasanov *et al.*, 2005). This method of isolation takes up to several weeks and usually does not allow for the isolation of injured cells, which will not grow at low temperatures after being stressed. Thus, enrichment methods were developed, and reference methods for the detection of *Listeria* spp. in foods are the ISO 11290 method (ISO, 1998a) and the FDA method (US FDA, 2005). Both methods require enrichment of a 25-g food sample in a selective broth (i.e. containing acriflavin, naladixic acid and cycloheximide), designated to retard or inhibit the growth of competing microorganisms, prior to plating onto selective agar (i.e. Oxford, PALCAM or LPM) and biochemical identification of typical colonies (Gasanov *et al.*, 2005). Rapid methods for the detection and enumeration of *Listeria* spp. and *L. monocytogenes* are used in modern dairy laboratories (Section 6.8).

6.8.13.2 Staphylococcus aureus

The presence of *S. aureus* in food is usually taken to indicate contamination from the skin, mouth or nose of workers in the food-processing area, but inadequately cleaned equipment or raw animal products may also constitute sources of contamination (ICMSF, 1978).

Foods must contain at least 10^6 cfu g^{-1} enterotoxigenic *S. aureus* to induce illness. Small numbers of *S. aureus* present in thermally processed foods represent the survivors of very large populations (Mostert and Jooste, 2002). Thus, the presence of *S. aureus* in dairy products suspected of causing staphylococcal poisoning should be interpreted with caution. A reference method for the enumeration of coagulase-positive staphylococci in milk and milk products has been described (IDF, 1997a). It is based on the inoculation of a selective medium (i.e. Baird–Parker or Rabbit plasma fibrinogen) with a serially diluted sample. Typical colonies after 24–48 h incubation are then confirmed by means of a coagulase test. For samples that are suspected of having low numbers of staphylococci, the MPN technique is used (IDF, 2002d).

For the detection of heat-stable DNase (thermonuclease) produced from coagulase-positive staphylococci, a reference method has been developed (IDF, 1998c). This enzyme is extracted using acidification, centrifugation, treatment with trichloroacetic acid and heating, and the sample is tested for thermonuclease activity in Toluidine Blue O-DNA agar. A positive thermonuclease test indicates that coagulase-positive staphylococci have grown to levels of 10^6 cfu g^{-1} or more, and therefore a test for detection of entrotoxin should follow. Such a detection assay, using a radioimmunoassay technique, has been described by Flowers *et al.* (1993).

6.8.13.3 Escherichia coli

E. coli is the type species of the genus *Escherichia*, a member of the *Enterobacteriaceae* family. It is a catalase-positive, oxidase-negative, fermentative, short, Gram-negative, non-spore-forming rod. Enterohaemorrhagic *E. coli* (EHEC), sometimes also known as verotoxin-producing *E. coli* (VTEC), is the most important group of virulent *E. coli*; the other three (based on the mechanisms of virulence that they possess) are: enterotoxigenic *E. coli* (ETEC), enteroinvasive *E. coli* (EIEC) and enteropathogenic *E. coli* (EPEC). *E. coli* O157:H7 is the serotype of EHEC that is most frequently isolated from humans (Adams and Moss, 1995). It has been reported (Foster, 1990) that dairy cattle are a natural reservoir of *E. coli* O157:H7. Strains of *E. coli* O157: H7 and other VTEC are classified as Hazard Group 3 pathogens, and so laboratories working with such toxin-producing strains must have Containment Level 3 facilities (Section 6.2.5).

6.8.13.4 Salmonella *spp.*

The reference method for the detection of *Salmonella* spp. in milk and milk products is based on pre-enrichment in a non-selective medium (i.e. buffered peptone water) to resuscitate injured cells, followed by enrichment in selective media (i.e. Rapaport–Vassiliadis and selenite-cysteine broths) and plating out on selective solid media (i.e. xylose lysine deoxycholate agar or brilliant green agar) (IDF, 2001b). Confirmation of presumptive *Salmonella* spp. colonies is carried out using appropriate biochemical and serological tests (Flowers *et al.*, 1993; Baylis, 2003).

6.8.13.5 Yersinia enterocolitica

Yersinia enterocolitica is an asporogenous, short Gram-negative, facultative anaerobic rod belonging to the *Enterobacteriaceae* family. It is catalase-positive and oxidase-negative, and has the ability to grow well at refrigeration temperatures. It is widely distributed in the food environment and has been recovered from a range of foods, including milk and milk products (Swaminathan *et al.*, 1982). Owing to the fact that *Y. enterocolitica* is unlikely to survive pasteurization (Hanna *et al.*, 1977), the organism can only be a problem with post-pasteurization contamination. Cold enrichment is the usual practice, with homogenates prepared in non-inhibitory Tris-buffered or phosphate-buffered saline and incubated for up to 21 days at 4°C, or for 14 days at 9°C. A selective solid medium has been developed, containing desoxychlorate, cefsulodin, irgasan and novobiocin (CIN agar) (Swaminathan *et al.*, 1982; Harrigan, 1998).

6.8.13.6 Enterobacter sakazakii

Enterobacter sakazakii belongs to the genus *Enterobacter* of the family *Enterobacteriaceae* and is a Gram-negative rod. The organism was known as 'yellow-pigmented *Enterobacter cloacae*', until 1980, when it was renamed *Enterobacter sakazakii*. Urmenyi and Franklin (1961) reported the first two known cases of meningitis caused by *E. sakazakii* in 1961. Subsequently, cases of meningitis, septicemia and necrotizing enterocolitis due to *E. sakazakii* have been reported worldwide. Although most documented cases involve infants, infections in adults have been reported as well (Hawkins *et al.*, 1991). Overall, case-fatality rates have varied considerably, with rates as high as 80% in some instances. While a reservoir for *E. sakazakii* is unknown, a growing number of reports suggest a role for powdered milk-based infant formulas as a vehicle for infection (Biering *et al.*, 1989; van Acker *et al.*, 2001). Enumeration of *E. sakazakii* is carried out using enterobacteriaceae enrichment broth and after incubation at 37°C overnight and subsequent plating on VRBG agar; on this agar, *E. sakazakii* forms purple colonies surrounded by purple halos of precipitated bile acids (Baylis, 2003).

6.9 Rapid microbiological methods

Conventional methods are labour-intensive, require prolonged incubation periods and rely heavily on the ability of the microorganisms to replicate and also their growth rate (Gracias and McKillip, 2004). During the last decade, many rapid microbiological tests have been developed, including antibody and nucleic-acid-based methods, miniaturized biochemical kits, modified conventional methods and selective membranes. Automation in the enumeration or detection of bacteria in dairy products has been introduced in dairy testing, and this field is constantly evolving (Vasavada, 1993; Karwoski, 1996). These innovations are essentially modifications of the classical methods in terms of more rapid detection of cells using more advanced methods based on immunology or gene technology. A faster detection of microorganisms ensures the safety of the consumers. Fermented dairy products are compositionally complex and contain high levels of background microflora. Such factors often interfere with the detection assays, resulting in less-than-optimal detection limits. A prior sample-processing step is often required for achieving a higher sensitivity and specificity (Stevens and Jaykus, 2004).

6.9.1 Antibody-based methods

The basic principle for an antibody-based detection method (or immunoassay) is the specific binding of the antibody (animal-derived protein) to a target antigen, followed by the detection of the antigen–antibody complex. This reaction depends primarily upon the structurally complementary binding sites of the two molecules. The most significant feature of these methods is that this binding occurs in the presence of other organisms and interfering food components (Huis in't Veld and Hofstra, 1991). These antibodies can be polyclonal (mixture of several antibodies), reacting with different sites of the antigen, or monoclonal (one pure antibody) reacting with only one epitope of the antigen. However, the effectual use of antibodies to detect microorganisms depends on the stable expression of target antigens in a microorganism, which are often influenced by many parameters such as temperature, preservatives, acids, salts or other chemicals found in food. Antibody-based assays have been developed for microbial detection using different labels to generate the signal. Radioisotopes were the first to be utilized, but the use of enzymes became more attractive. Enzymes or other chemically active biomolecules can be linked to antibodies in a fashion that enables the resulting complex to retain its immunological and chemical activities. For an immunoassay, the stability and activity of these enzyme complexes are very important (Gerhardt *et al.*, 1994). Immunological methods are promising because of their sensitivity, rapidity and simplicity, and also because testing can be carried out directly from enrichment media without tedious sample preparation.

There are many different commercially available antibody-based assays for several microorganisms and their toxins. Some of these are presented in Table 6.7, and detailed lists are given in the official site of AOAC (2005b) and in BAM (US FDA, 2005). There are six basic formats of antibody-based assays. The simplest is the Latex Agglutination Test (LAT), a format that is increasingly being used. LAT detects the presence of an antibody or an antigen using latex beads coated by an antibody (or antigen). If the corresponding antigen (or antibody) is present in the sample, it is attached to the latex beads, and they agglutinate (clump together into visible particles) due to the formation of molecular cross-bridges (Huis in't Veld and Hofstra, 1991). In the LAT, an isolated pure colony is usually used as the sample (D'Aoust *et al.*, 1991; Feng, 1997); however, the use of a centrifuged sample from an enrichment culture has also been reported. The sensitivity of the method is 10^7–10^8 cfu ml^{-1} of sample. There are commercially available LATs for several food-borne pathogens such as *Campylobacter*, *E. coli*

Table 6.7 Some commercially available immunological test kits used for the microbiological analysis of dairy products[a].

Test kit	**Manufacturer**[b]	**Microbe or toxin**	**Method**	**Matrix**	**Approval**
E. coli O157:H7 VIA™	TECRA (TECRA International Pty Ltd, Australia)	*E. coli* O157:H7	ELISA	Food and environmental samples	AOAC 001101
Listeria VIA™	TECRA	*Listeria* spp.	ELISA	Food and environmental samples	AOAC 998.22
Salmonella VIA	TECRA	*Salmonella* spp.	ELISA	Food	AOAC 989.14
Salmonella ULTIMA	TECRA	*Salmonella* spp.	ELISA	Food	AOAC, AFNOR
S. aureus VIA™	TECRA	*S. aureus*	ELISA	Food, water, environmental samples, raw materials, cosmetics and pharmaceutical products	New Zealand Dairy Board Evalutest, Australia
Pseudomonas VIA™	TECRA	*Pseudomonas* spp.	ELISA	Food, water and environmental samples	
Assurance EHEC	BioControl (Biocontrol Systems SARL, France)	*E. coli* O157:H7	EIA	Food and environmental samples	AOAC 996.10
Assurance *Listeria* spp.	BioControl	*L. monocytogenes* + related species	EIA	Food and environmental samples	AOAC 996.14
Assurance Gold *Salmonella* spp.	BioControl	Motile + non-motile *Salmonella*	EIA	Processed foods and raw foods	AOAC 999.08
Assurance Gold *Campylobacter* spp.	BioControl	*C. jejuni* + thermophilic *Campylobacter*	EIA	Food and environmental samples	
VIP	BioControl	*E. coli* O157:H7	IMP	Food	AOAC 996.09
VIP	BioControl	*Listeria* spp.	IMP	Food	AOAC 997.03
VIP	BioControl	*Salmonella* spp.	IMP	Food	AOAC 999.09
1–2 Test	BioControl	*Salmonella* spp.	IMF	Food and environmental samples	AOAC 989.13
VIDAS[R] LSX	BioMerieux (BioMerieux Industry Inc., USA)	*Listeria* spp.	ELFA	Meat and dairy products	AOAC 100501

Contd.

Table 6.7 (Contd.)

Test kit	Manufacturer[b]	Microbe or toxin	Method	Matrix	Approval
VIDAS^{R} LMO2	BioMerieux	*L. monocytogenes*	ELFA	Selected food groups	AOAC 2004.02
VIDAS^{R} ICS + SLM	BioMerieux	*Salmonella* spp.	ELFA	All Foods	AOAC 2001.09 AFNOR
VIDAS^{R} Staphylococcal enterotoxin (SET2)	BioMerieux	Staphylococcal enterotoxins	ELFA	Food	AOAC 070404
VIDAS^{R} LIS	BioMerieux	*Listeria* spp.	ELFA	Food	AOAC 2004.06
VIDAS^{R} ECO	BioMerieux	*E. coli* O157:H7	ELFA	Food	AOAC 010502
Salmonella spp. SELECTA	Bioline	*Salmonella* spp.	ELISA	Food and environmental samples	AFNOR
SALMONELLA LATEX TEST	OXOID (OXOID Ltd, UK)	*Salmonella* spp.	LA	Food	
E. coli O157:H7 LATEX TEST	OXOID	*E. coli* O157:H7	LA	Food	
Salmonella spp. Screen/SE Verify^{TM}	VICAM (VICAM, USA)	*Salmonella* spp.	IMS + LA	Food	
ListerTest^{R}	VICAM	*Listeria* spp.	IMS + LA	Food	

[a]ELISA: enzyme-linked immunosorbent assay; EIA: enzyme immunosorbent assay; ELFA: enzyme-linked fluorescent assay; IMP: immunoprecipitation; IMF: immunofluorescence; IMS: immunomagnetic separation; LA: latex agglutination test.
[b]Data retrieved from manufacturers' official Internet websites.

(March and Ratnam, 1989), *Listeria* spp., *Salmonella* spp., *Shigella* spp., *S. aureus* and *Vibrio cholera*. Many of the LAT assays are performed manually, and the results are given by visual observation. However, these manual assays suffer from poor sensitivity and reproducibility. During the last years, the human eye has been replaced by spectrophotometers and nephelometers that measure the absorbed or scattered light. Also, particle counters have been developed to detect very small clumps. Angular anisotropy and quasi-elastic light scattering have also been applied (Gella *et al.*, 1991).

Reverse Passive Latex Agglutination (RPLA) is a modification of the LAT (Feng, 1997). Antibodies that are attached to latex microspheres react with soluble antigens in test tubes or in wells of microtitre plates. If binding occurs, a diffuse pattern is observed at the bottom of the tube or well. If not, a ring is observed. RPLA is more suitable for testing for toxin production (Park and Szabo, 1986). Commercially, there are tests for detection of enterotoxins and other toxins produced by *S. aureus*, *B. cereus*, *Cl. perfringens* and *E. coli.*

Generally, in a direct immunoassay, the antibodies are immobilized onto a solid support, and then the test sample is added. If binding between a specific antigen and antibody occurs, the complex can be subsequently detected. The best-known immunoassay is ELISA, which has been used as the basis for the first rapid methods developed. ELISA has been widely used (Candish, 1991; Kerdahi and Istafanos, 1997). The method of detection relies on the use of another antibody conjugated to an enzyme or a fluorescent dye specific for the antigen. The enzyme activity is quantified using a substrate that produces a coloured product (Yolken and Leister, 1982). For measuring the emission of fluorescence when a fluorescent-labelled antibody is used, a spectrofluorometer or an epilfuorescence microscope is required.

There are two basic types of ELISA: direct and indirect (Gerhardt *et al.*, 1994). The direct ELISA uses the method of directly labelling the antibody itself. Microwell plates are coated with the target antigen, and the binding of labelled antibody is quantified by a colorimetric, chemiluminescent or fluorescent end-point. Since the secondary antibody step is omitted, the direct ELISA is relatively quick and avoids potential problems of cross-reactivity of the secondary antibody with components in the antigen sample. However, the direct ELISA requires the labelling of every antibody to be used, which can be a time-consuming and expensive approach. In addition, certain antibodies may be unsuitable for direct labelling. Direct methods also lack the additional signal amplification that can be achieved with the use of a secondary antibody.

The indirect ELISA utilizes an unlabelled primary antibody in conjunction with a labelled secondary antibody. Since the labelled secondary antibody is directed against all antibodies of a given species (e.g. anti-mouse), it can be used with a wide variety of primary antibodies (e.g. all mouse monoclonal antibodies). There are commercially available anti-species labelled antibodies from many suppliers. The use of secondary antibody also provides an additional step for signal amplification, increasing the overall sensitivity of the assay.

ELISA is simple, fast and sensitive, but it still requires an enrichment step because the test needs about 10^4–10^7 cells. It does not require hazardous reagents, and it can screen many samples simultaneously in a microtitre plate using only minor sample volumes. Furthermore, the ELISA methodology can be used with 'difficult' sample matrices, which makes these tests particularly suitable for food testing. Because manual operation of ELISA is inconvenient, automated ELISA procedures have been developed (Curiale *et al.*, 1997; Kerdahi and Istafanos, 1997). BioMerieux (BioMerieux, Durham, NC) developed the VIDAS and miniVIDAS devices that are based on the Enzyme Linked Fluoroscent Assay principle, an automated multiparametric immunoassay. Each test is composed of a strip that contains all reagents required for the reaction and a solid-phase receptacle, which is a plastic pipette-like device whose inside surface is coated with the appropriate antibody or antigen, and which can be also used as a sampling device. There are VIDAS tests for *Listeria* spp., *L. monocytogenes*, *Salmonella* spp., *E. coli* O157:H7 and staphylococcal enterotoxins, and all have been validated by AOAC.

Another antibody-based method, which is based on the technology originally developed for home pregnancy tests is immunoprecipitation. This technique detects multivalent soluble antigens that react with homologous divalent antibodies. The antigens and antibodies should be in appropriate quantities in order for a visible network to occur as a precipitate. There is an equivalence zone (range of concentration of antigens and antibodies) where precipitation can occur. If one of the two components (antibody or antigen) is in excess, the precipitate dissolves. The antibodies are coated with a precipitable material such as coloured latex particles or colloidal gold. The sample is wicked to the area where the antibodies are placed. If the appropriate antigen is present, an antibody–antigen complex is formed, and it goes through the matrix to a binding zone where a second antibody is found. This technique can be performed either in a tube (i.e. solution), where a precipitation ring can be observed, or in an agar matrix (i.e. immunodiffusion). Immunoprecipitate detection devices have been described for several food pathogens (Feldsine *et al.*, 1997a, 1997b).

Immunomagnetic separation (IMS) is an elegant method in which antibodies attached to magnetic particles or beads are used to capture the antigen. A magnet is applied to one side of the tube, and the magnetic beads are pulled towards that side and become concentrated. Non-target cells and food debris are washed away, and a suspension of target cells remains. Captured antigens can be plated onto a selective agar (Skjerve *et al.*, 1990) or further tested using other assays such as ELISA tests, flow cytometry (Jung *et al.*, 2003), PCR (Rijpens *et al.*, 1999) or biosensors. The philosophy of this assay is analogous to that of using selective media but much milder. Many authors have used IMS successfully to replace enrichment steps (Mansfield and Forsythe, 1993; Dziadkowiec *et al.*, 1995). IMS can save at least one day compared with the total protocol of pre-enrichment and enrichment steps. With IMS, a 10- to 100-fold increase in concentration of the cells can be obtained, and the procedure can be performed manually or automated. Vermunt *et al.* (1992) used IMS for the isolation of *Salmonella* spp., and Hsih and Tsen (2001) applied IMS as a selective step followed by PCR for the simultaneous detection of *L. monocytogenes* and *Salmonella* spp. in dairy samples. Dynal Biotech ASA (Dynal Invitrogen Corporation, Smestad, Oslo) has developed anti-*Salmonella*, anti-*Listeria*, anti-*Legionella* Dynabeads® for the selective enrichment of these microorganisms directly from pre-enrichment samples using IMS. The method can be performed manually or can be automated using a BeadRetriever™.

Immunofluorescence is a technique that has the same principle as ELISA, with the only difference that the antibodies are attached to chemical compounds other than enzymes. The binding of an antibody to the appropriate antigen can be made visible by specific dyes called fluorochromes, which become fluorescent or emit visible light when they are exposed to UV, violet or blue light. Fluorochromes such as rhodamine B (red) or fluoresceinisothiocyanate (FITC) (green) can be attached to an antibody or to an antigen. This binding does not alter the specificity of the antibody but enables detection of the antibody–antigen complex by use of the fluorescent microscope. This test is sensitive, simple, fast and highly specific (Gerhardt *et al.*, 1994) and can be used to detect a great variety of target microorganisms (Tortorello and Gendel, 1993).

Flow cytometry (FCM) is a very powerful technique based on the same principle as immunofluorescence. Flow cytometers can detect cell sizes in the range normal for microorganisms. The instrument detects fluorescent cells that are moving in a fluid stream past an optical sensor. The flow cytometer measures the light scattered or the fluorescence emitted by the cells as they pass through the beam. The light energy is converted into an electrical signal by photomultiplier tubes. FCM is sensitive, does not require culturing or enrichment procedures and can be both qualitative and quantitative (Attfield *et al.*, 1999). It can be used as a successful tool to determine product quality (Dumain *et al.*, 1990). The bacteria can be identified by their cytometric properties or with the help of fluorochromes, which can be used independently or bound to specific antibodies or oligonucleotide probes (Fouchet *et al.*, 1993; Vesey *et al.*, 1994; Attfield *et al.*, 1999). Protein and lipid globules of milk may interfere with the staining and detection of bacteria, so sample processing is required (Gunasekera *et al.*, 2000, 2002). FCM combined with immunofluorescent labelling has been used to detect *L. monocytogenes* and other pathogenic bacteria in dairy products (Donnely and Baigent, 1986; Clarke and Pinder, 1998; Smith *et al.*, 2001).

6.9.2 Nucleic-acid-based methods

The application of molecular techniques to food microbiology has vastly increased in recent years (Olsen *et al.*, 1995; Olsen 2000). These techniques represent a new generation of rapid methods based on primary information in the nucleic acid sequence of an organism (Grant and Kroll, 1993). The whole DNA molecule contains a large amount of information for the identification of bacteria. The target of the nucleic acid-based methods is the sequence of nucleotides in the DNA of the microorganism, especially short nucleotide sequences that are unique to the microorganism. All the other methods (i.e. immunological or biochemical/enzymatic) depend on the phenotypic expression of the genotypic characteristics of the microorganism. The nucleic acid-based methods depend directly on the genotypic characteristics of microorganisms, which are far more stable (Farber, 1996). The principles of the tests are based on the hybridization of a characterized nucleic acid probe to a specific nucleic acid sequence in a test sample followed by the detection of the paired hybrid (Tothill and Magan, 2003). The two most commonly used methods are DNA hybridization and PCR. The specificity in the formation of hybrids between two complementary nucleic acid sequences is the basis for these assays. Some of the commercially available kits are presented in

Table 6.8 Some commercially available DNA-based test kits used for the microbiological analysis of dairy products.

Test kit	Manufacturer[a]	Microbe or toxin	Method	Matrix	Approval
Assurance GDS	BioControl (Biocontrol Systems SARL, France)	*E. coli* O157:H7	DNA amplification	Food	AOAC 2005.04
Probelia^R	BioControl	*Clostridium botulinum*	DNA amplification	Food	
Probelia^R	BioControl	*Salmonella* spp.	PCR	Food	
Probelia^R	BioControl	*Listeria* spp.	PCR	Food	
Probelia^R	BioControl	*E. coli* O157:H7	PCR	Food	
GENE-TRAK	Neogen (Neogen Europe, Ltd, UK)	*Campylobacter* spp.	Probes	Food	
GENE-TRAK	Neogen	*E. coli*	Probes	Food	
GENE-TRAK	Neogen	*Listeria* spp.	Probes	Food	AOAC 981201
GENE-TRAK	Neogen	*L. monocytogenes*	Probes	Food	AFNORDNA-14/2–02/95
GENE-TRAK	Neogen	*Salmonella* spp.	Probes	Food	AOAC 961101
GENE-TRAK	Neogen	*Y. enterocolitica*	Probes	Food	
BAX^R	DuPont Qualicon (DuPont Qualicon, Inc., USA)	*Salmonella* spp.	PCR	Food	AOAC 100201
BAX^R	DuPont Qualicon	*E. coli*	PCR	Meat, environmental swabs	
BAX^R	DuPont Qualicon	*L. monocytogenes*	PCR	Food and dairy	AOAC 2003.12

[a]Data retrieved from manufacturers' official Internet websites.

Table 6.8, and detailed lists are available on the AOAC's official site (AOAC, 2005b) and in BAM (US FDA, 2005).

6.9.2.1 DNA hybridization

In DNA hybridization, following an enrichment step, a known DNA probe is hybridized to a complementary nucleotide sequence of DNA or RNA, and so the presence or absence of a specific DNA sequence that is unique for a microorganism can be detected. A DNA probe is a 20–2000 single-strand nucleotide sequence prepared synthetically or biologically, and many synthetic probes consist of 20 nucleotides or less. Probes must be unique to a particular microbe or group of microorganisms and must be labelled so that the hybridization can be easily detected. Labelling is achieved using a radioisotope, an enzyme or a fluorescent compound that can be measured in small amounts. There are two basic formats: solid-phase and liquid-phase. In the solid-phase format, the target sequence is attached to a solid support (e.g. membrane filter), and in the second format, it is suspended in liquid.

The first generation of probes included probes that utilized radioactive compounds. The second generation of probes used enzymatic colour reactions to detect the presence of the microorganisms and RNA as the target molecule. Since, in a single cell, there is only one complete copy of DNA but 1000 or more copies of ribosomal RNA, this naturally amplified target offers greater sensitivity (Temmerman *et al.*, 2004). rRNA is a very important biomolecule that can be used for the discrimination between different genera, species or even subspecies. RNA–DNA hybrids are more stable than DNA–DNA hybrids, and there are also antibodies that recognize these hybrids. PCR combined with DNA hybridization in a microtitre plate is a convenient and highly sensitive and specific approach for microbial detection (Cocolin *et al.*, 1997). Examples of commercial applications of DNA hybridization technology are the Gene-Trak and GeneQuence™ assays (Neogen Europe Ltd). The GeneQuence™ kit uses an advanced DNA probe technology that contains two specific DNA elements, a capture and detector-DNA probe specific to the rRNA of the target organism and a coated solid phase. Both the capture and the detector probe must bind in order to obtain a positive result. The presence of two specific elements increases the specificity. Both kits can detect *L. monocytogenes, Listeria* spp., *E. coli* O157:H7 and *Salmonella* spp.

Probes are more specific and defined entities than antibodies. Probe sequences can be checked by sequence analysis and can be reproduced very easily. Also, nucleic acid probes are more stable and resistant during the course of an experiment. They are also very sensitive, being able to detect less than 1 μg of nucleic acid per sample.

6.9.2.2 Polymerase chain reaction (PCR)

PCR is a widely used and accepted method for the detection of many bacterial groups, including pathogens. The technique is based on selective amplification of the target DNA. The procedure has three distinct steps. During the first step (denaturation step), the target DNA is heated to about 95°C, and denaturation occurs (the double-stranded DNA becomes single-stranded DNA). Then, during the second step (annealing step), the temperature is lowered to about 55°C so that primers (oligonucleotides in the range of 15–30 nucleotides) can anneal to specific regions of the single-stranded DNA. The last step (extension step) includes synthesis of a new DNA strand and formation of new double-stranded DNA. This step is catalyzed by a special heat-stable polymerase, the TAQ enzyme from *Thermus aquaticus*, which adds complementary nucleotides to single-stranded DNA. The optimal temperature for extension is about 70°C. The three steps constitute one cycle. After one cycle, one copy of DNA becomes two couples, and following the first cycle, multiple additional cycles are performed with the same order of steps. Detection of the PCR products can be done by agarose gel electrophoresis, southern blot (Wan *et al.*, 2000) or ELISA-based systems (Daly *et al.*, 2002). PCR can detect very small amounts of genetic material that cannot be directly detected by other methods.

In practice, a food-enrichment step is needed in PCR protocols (Piknova *et al.*, 2002), as it has been shown that analysis without enrichment yields unreliable results (Aznar and Alarcon, 2003). Alternatively, a pre-concentration step is required in order to achieve lower detection limits. Although a variety of methods such as centrifugation, filtration and IMS have been used for bacterial concentration in food samples, the appropriate concentration method should be chosen for the specific food matrix or microorganism (Stevens and Jaykus, 2004).

PCR is a very sensitive, rapid and specific method, suitable for detecting microbial agents that are not detectable by culture techniques. In contrast to DNA hybridization, where comparatively large amounts of target DNA or RNA are required, PCR requires only minute amounts of target DNA. It is more specific than other methods because species-specific DNA regions exist, and specific traits of pathogenicity can be targeted. Also, it has a great potential for automation. It is applicable to microorganisms that are non-culturable or when the isolation of the microorganism is very difficult. PCR can also be used to detect sheared or partially degraded DNA that other DNA-based methods do not. Crucial parameters are the quality of the template, the target region, the primer sequences and the efficiency of amplification. A primer can be designed differently for different purposes. Primers can be specific (i.e. targeted to specific features of a microorganism) (Almeida and Almeida, 2000) or multiple, with a broader spectrum for multiple species. Careful design is required in order to obtain valuable sets of primers. The specificity of the assay depends mostly on the choice of suitable primers. PCR has been successfully applied for the detection of *Campylobacter* in raw milk and other dairy products with a sensitivity of 1–10 cells ml^{-1} or 1–10 cells g^{-1} (Wegmüller *et al.*, 1993).

Possible problems may arise from inhibitory substances in complex biological samples, which can decrease the efficiency of DNA amplification. Inhibitors influence the assay's performance and sensitivity, even in small amounts. For this reason, attentive sample processing prior to PCR is required (Rådström *et al.*, 2003). These inhibitors (enzymes or other compounds) must be removed or diluted from the reaction mix. Another problem is the contamination of PCR products, which can result in erroneous data. Aseptic technique should therefore be practised during PCR set-up and execution. Another problem is that PCR can also detect dead cells, since dead cell DNA can also be amplified (Wegmüller *et al.*, 1993; Gasanov *et al.*, 2005). Therefore, PCR should be combined with other methods that involve biological and enzymatic amplification (only living cells can be detected) or combined with tests targeting rRNA or mRNA (Jensen *et al.*, 1993). Tests targeting mRNA have gained favour, since mRNA indicates that cells are alive (mRNA is an unstable molecule, and following cell death it rapidly degrades). However, mRNA amplification must be performed under conditions that prevent its degradation. The high cost of equipment and the need for highly trained personnel constitute considerable disadvantages (Keer and Birch, 2003).

Reverse transcription PCR (RT-PCR) is a two-step reaction, whereby RNA is transcribed into complementary DNA (cDNA) and then PCR-amplified. Conventional RT-PCR includes only one specific set of primers, while multiplex RT-PCR can detect several types of target RNA sequences in a sample (Burtscher and Wuertz, 2003; Lübeck and Hoorfar, 2003). Another type of PCR that uses multiple primer sets in successive reactions and targets the same sequence is nested PCR. Both sensitivity and specificity are increased with this method (Ha *et al.*, 2002). PCR methods can be combined also with Nucleic Acid Sequence-Based Amplification (NASBA) for a more rapid and efficient detection of bacteria in complex food matrices (Cook, 2003; Stevens and Jaykus, 2004). NASBA is an isothermal amplification technique that is based on the action of three enzymes and works at a low temperature (generally 41°C). Reverse transcriptase is used in combination with a primer to produce a cDNA–RNA hybrid.

Then, an RNAse enzyme is used to remove the RNA from the hybrid, allowing the reverse transcriptase to synthesize a double-stranded cDNA. This cDNA is then used as a template for the generation of RNA transcripts by a T7 polymerase. The products are single-stranded RNA molecules, which can be detected either by agarose gel electrophoresis or by using specifically labelled oligonucleotide probes for hybridization assays combined with colometric detection systems (Keer and Birch, 2003; Gasanov *et al.*, 2005). Rodriquez-Lazaro *et al.* (2004) suggested the detection of NASBA products of *Mycobacterium avium* subsp. *paratuberculosis* in real time using molecular beacons (single-stranded nucleic acid sequences). The result is the real-time measurable fluorescence emission that is directly proportional to the concentration of the target sequence. With this method, Rodriquez-Lazaro *et al.* successfully detected 10^4 *Mycobacterium avium* subspp. *paratuberculosis* cells in 20 ml of artificially contaminated semi-skimmed milk. Real-time PCR is more rapid, sensitive and reproducible, while the risk of carryover contamination is minimized. It is the detection process that distinguishes real-time PCR from conventional PCR. The labelling of primers, oligonucleotide probes or amplicons with fluorescent molecules has expanded the role of conventional PCR from a research tool to an adaptable technology for diagnostic microbiology. These labels produce a change in signal following direct interaction with the amplicon. The signal is proportional to the amount of the amplicon during each cycle (Mackay, 2004). This type of analysis requires specialized equipment and materials that substantially increase the cost of testing. The increased speed of real-time PCR in comparison with conventional PCR is due to the lower number of cycles required and the more rapid detection of amplicons. Quantitative measurements are important in food analysis, and methods have been developed to make the tests quantitative. Idaho Technology (Salt Lake City, UT) proposed an automated sample preparation and real-time detection method for *L. monocytogenes* in milk in one working day.

Genetic methods also exist that cannot directly detect microorganisms in a food matrix, but can identify and characterize organisms to the genus, species and subspecies levels by using a pure colony of the organism. Such methods are the Riboprinter and Pulsed Field Gel Electrophoresis (PFGE). PFGE is a discriminatory and reproducible method that is considered 'the gold standard' of molecular typing methods. The bacteria are embedded in agarose plugs and then subjected to lysis and digestion with a restriction enzyme that is referred to infrequent sites. Agarose electrophoresis of the digested bacterial plugs follows in an apparatus in which the electric field is changing orientation at predetermined intervals. Thus, clear separation of very large DNA fragments can be achieved (Schwartz and Cantor, 1984; Carle *et al.*, 1986; Chu *et al.*, 1986). This method discriminates between very closely related strains of pathogenic bacteria because the whole genome of the bacteria is analysed. A disadvantage is the time required for the whole procedure (2–3 days), although faster protocols have been applied successfully (Gautom, 1997). Tenover *et al.* (1995) proposed criteria for the interpretation of the PFGE products. PFGE has been used for the investigation of food-borne bacterial outbreaks that have been caused by organisms such as *L. monocytogenes* (Miettinen *et al.*, 1999; Waak *et al.*, 2002), *Salmonella* (Louie *et al.*, 1996), antibiotic-resistant enterococci and streptococci (Barbier *et al.*, 1996; Patterson and Kelly, 1998). Olive and Bean (1999) and Lukinmaa *et al.* (2004) have published reviews on the molecular methods used for the detection and typing of bacteria. Finally, another powerful tool with high discriminatory power is the 16S or 23S RNA sequencing (Rådström *et al.*, 2003). Online databases of sequenced DNA such as the Ribosomal Database Project (MSU, 2005) permit rapid comparison of 16S RNA sequences.

6.9.3 Membranes

The first use of membranes to detect microorganisms was for testing water samples or other non-interfering substances. Pre-filters are used to eliminate the food ingredients prior to filtration as well as diluents and enzyme digestion techniques to facilitate the filtration of the pre-filtered sample.

6.9.3.1 Hydrophobic grid membrane filter

This technique was first described by Sharpe and Michaud (1974). It consists of a square filter with printed hydrophobic grids to form 1600 squares per filter. It does not require many dilutions, and it results in good separation of the microorganisms with minimum interference (Sharpe *et al.*, 1983). The reproducibility is very high, and the filter can be transported from one medium to another very easily without disturbing the colonies. A preliminary incubation step can is needed for injured cells. Two or more different reactions can be detected on the same filter. The filters must be cleaned very carefully and sterilized every time before use. It is necessary to establish a handling protocol for every food product, but once established, protocols remain constant.

Hydrophobic Grid Membrane Filters (HGMF) have been used for enumeration and detection of many bacteria in dairy products (Peterkin and Sharpe, 1980) such as *E. coli* (Entis, 1984), *L. monocytogenes* and *Salmonella* spp. (Entis, 1990). A combination with other techniques such as those using enzyme-labelled antibody (Todd *et al.*, 1988, 1993), and colony hybridization (Peterkin *et al.*, 1992) has been done successfully. Examples of HGMF are the ISO-GRID™ for the presumptive enumeration of *Listeria* spp. and *L. monocytogenes* in 24 h using LM-137 agar (Entis and Lerner, 2000), the 24-h *Salmonella* spp. Screening Test with a pre-enrichment step in SCCRAM broth (6–7 h) and an incubation step in EF-18 agar (18–24 h), and the Direct Coliform/*E. coli* enumeration method using Lactose Monensin Glucuronate (LMG) agar and buffered 4-methylumbelliferyl-β-D-glucuronide (MUG) agar for confirmation. Entis and Lerner (1998) found that counts of *E. coli* obtained from cottage cheese were significantly higher using the ISO-GRID (hydrophobic grid membrane filter) method with SD-39 agar compared with the LMG and MUG agar method (reference method).

6.9.4 Impedance

Impedance/conductance methods are among the most successful rapid techniques used both in QA microbiology and for research purposes (Harrigan, 1998). These methods are based on the detection of the decrease in the electrical impedance of the growth medium as a result of microbial metabolism and multiplication. In direct impedance technology, the change in the conductivity of a liquid culture medium serves as a measuring parameter, whereas indirect impedimetry measures the change in the electrical conductivity of a reaction solution, which occurs through the absorption of gases from the inoculated bacterial culture (Wawerla *et al.*, 1999). Impedance is the resistance to flow of an alternating current through a conducting material (e.g. growth medium). It is a vector parameter consisting of two components: conductance (G) and capacitance (C). With the growth of bacteria, nutrient macromolecules of the media break down into smaller units as a result of microbial metabolism, and the ionic concentration of the medium increases (Fung, 1994). When the ionic concentration reaches a magnitude similar to the initial ionic concentration of the medium, the conductivity increases. The time point at which the change can be visualized is called the detection time (DT). DT depends on the temperature, the medium, the generation time of the microorganisms and the presence of inhibitors in the sample or medium (Entis *et al.*, 2001). Specific growth media are usually needed in direct impedimetry. Many commercial systems are available, such as the Bactometer (BioMerieux), the Rapid Automated Bacterial Impedance Technique (Don Whitley Scientific, Microbiology International, Frederick, MD) and the Malthus V (Malthus Diagnostics, North Ridgeville, OH). All the above are computer-controlled impedance systems that can perform a rapid enumeration for a wide range of microorganisms.

The indirect impedimetry is a more recent method, based on the production of CO_2 due to bacterial growth. The CO_2 is absorbed into an alkaline solution, and the reduction in conductivity of the solution is measured. The production of CO_2 can be detected much earlier than changes of conductivity due to the breakdown of the nutrients of the media, thus it is considered a more rapid method. No specific media are required and the method has a lower detection limit (Owens *et al.*, 1989; Bolton, 1990). BioMerieux has developed the BacT/Alert 3D, which is a patented colometric sensor and detection method that detects microorganism growth by tracking CO_2 production.

Automated methods based on impedance have been used to estimate bacterial populations since 1980 (Firstenberg-Eden and Tricarico, 1983; Firstenberg-Eden *et al.*, 1984; Hancock *et al.*, 1993). However, such methods are not capable of enumerating low numbers of bacteria in samples. The use of electrical impedance technology combined with fluorescent labelling has been used for enumeration of *S. aureus* in milk (Smith *et al.*, 2001).

6.9.5 Biochemical enzymatic methods and diagnostic kits

This group of assays consists of a large variety of biochemical and enzymatic tests which include specialized and/or chromogenic media, miniaturized microbiological and diagnostic kits, quantitative enzymatic methods or simple modified conventional methods that result in savings of time, labour and materials.

Substrates containing specific chromogens can be used to rapidly screen and identify microorganisms. Such differential colour reactions have been exploited in the development of growth media for a variety of microorganisms such as *Salmonella* spp. (Rambach, 1990; Monfort *et al.*, 1994), *Listeria* spp. (Istafanos *et al.*, 2002; El Marrakchi *et al.*, 2005), *E. coli* (Foschino *et al.*, 2003) and enterococci (Merlino *et al.*,

1998). They are simple, cost-effective, easy to interpret, highly sensitive and specific. Examples of chromogenic substrates are the Fluorocult® Lauryl Sulfate broth for the determination of *E. coli* with parallel determination of coliform bacteria in milk and dairy products, the Fluorocult® LMX broth (Lauryl Sulfate-MUG-XGal) for the simultaneous detection of total coliforms and *E. coli* in food and water, and the Bactident® for the determination of *E. coli* in 30 min, all provided by Merck (Merck and Co. Inc., Rahway, NJ). Colifast® Milk estimates the number of coliforms in water and milk samples within a few hours based on the activity of D-galactosidase hydrolysing the 4-methylumbelliferyl-D-galactose that is present in the selective growth medium (Colifast® Systems ASA). Foschino *et al.* (2003) reported that the sensitivity of this medium was not sufficient for the detection of coliforms in pasteurized milk. A list of commercially available media is available online (US FDA, 2005) and a variety of microbiological and diagnostic kits are commercially available (Russel *et al.*, 1997; AOAC, 2005b).

6.9.6 ATP bioluminence

ATP bioluminence is a very rapid and sensitive method. Bioluminence is the emission of light from viable cells and is a widespread phenomenon among living organisms. The ATP-specific luciferin/luciferase reaction in the presence of oxygen and magnesium produces a bioluminence with a maximal spectral emission at 560 nm. The amount of light released is proportional to the amount of ATP, and therefore biological material, in the sample (Stanley, 1989). One problem is that the amount of light does not always correlate with the exact microbial number, and when testing milk samples, the sample needs to be treated to remove non-bacterial ATP present in somatic cells before the analysis (te Giffel *et al.*, 2001).

One successful application of the ATP-reaction is its implementation in monitoring of sanitation in the dairy production environment. Also, measuring microbial ATP is a safe way of detecting sterility failure in dairy food samples, and results are available a minimum of 48 h earlier than with traditional culture methods (Samkutty *et al.*, 2001). Millipore (Billerica, MA) has developed the Milliflex Rapid Microbiology System and AutoSpray Station, an automated system for rapid detection of microorganisms based on ATP bioluminescence that can detect viable microorganisms down to 1 cfu ml^{-1} and can deliver results in approximately one-quarter of the time of traditional methods.

6.10 Sensory evaluation of dairy products

Sensory evaluation (SE) is an exciting, constantly evolving and rapidly expanding field of food science that constitutes a valuable tool for the food industry. Sensory analysis has been defined as the scientific discipline used to measure, analyse and interpret reactions to those characteristics of foods and other materials which are perceived by the senses of sight, smell, taste, touch and hearing (Stone and Sidel, 1993). SE combines scientific principles from the areas of chemistry, physics, physiology, psychology and statistics, and has been applied for quite some time for the evaluation of dairy products (O'Mahony, 1979; Bodyfelt *et al.*, 1988). In the dairy industry, SE can be used for quality assurance, product development and marketing purposes. The SE of dairy products is governed by the same principles and relies on the same analytical methods as the SE of any other food commodity, but owing to unique sensory properties of dairy products (e.g. 'iciness' of ice cream), descriptors exist that are unique to dairy products.

Based on its goals, SE in the dairy (food) industry can be classified as Analytical Sensory Testing or Sensory Evaluation I ('SE I'), Measurement of Consumer Perception, or Sensory Evaluation II ('SE II') and Consumer Acceptance and Preference Testing. In SE I, humans are used as sensitive 'analytical instruments', analogous to laboratory instruments, to measure the sensory characteristics of foods. In SE II, the techniques, methodologies and analyses employed aim at evaluating how the consumer perceives the food. Often, this boils down to whether ordinary untrained consumers can notice differences that were noted by trained experts in SE I (O'Mahony, 1995). Therefore, in SE II, ordinary consumers of the product in question are tested (studied), and testing is usually (or preferably) conducted under ordinary consumption conditions. In contrast, in SE I, highly trained, sensitive and reliable panelists (experts) evaluate the product, working under controlled laboratory conditions, often in specially designed testing booths. These booths have specified spatial requirements and are found inside testing rooms that allow for controlled temperature, ventilation and illumination. The two SE approaches can be used in conjunction. If, for example, during product reformulation, the dairy manufacturing company's trained panelists (SE I) determine that substitution of a given ingredient in a product with a cheaper alternative results in slight changes in the product's taste, then it is of interest to the company to find out if the product's target population (the ordinary consumers of the product) would notice the difference or not. To address this question, the company would have to test the reformulated product against the target consumer population (SE II). Depending on the SE testing goal (SE I vs. SE II), different experimental designs and methods of analysis are required. Once it has been documented via an SE II approach that consumers can distinguish similar products based on existing differences in the products' sensory attributes, a third approach/use of SE, namely Consumer (Preference and Acceptance) Testing, is applied (Fig. 6.1). Consumer Testing, which resembles SE II in that ordinary untrained consumers are being tested, is the bottom-line testing for the dairy industry. Here, the aim is to evaluate how much consumers like, prefer or accept a given product with an ultimate aim at predicting product sales. Consumer Testing, however, differs from SE II in that consumers are being asked to state their preference and liking between or among similar products and not asked to distinguish possible perceptual differences (i.e. perceive or discriminate) (O'Mahony, 1995). SE II must precede Consumer Testing, as, in the likelihood that differences between products are too small for consumers to perceive, Consumer Preference and Acceptance Testing can be avoided.

The production of high-quality dairy products depends upon the strict control of all possible factors that determine their sensory properties. The sensory properties of dairy products, classified as flavour, texture and appearance, are crucial to the dairy industry because they relate directly to product quality, and together with wholesomeness, nutritional value and price, they determine consumer acceptance. SE in the dairy

Figure 6.1 Graduate students in sensory evaluation at the University of California at Davis conduct consumer testing of yoghurt. Photograph, courtesy of Professor Michael O'Mahony, Department of Food Science and Technology, UC Davis.

industry has been traditionally employed to seek and detect defects or undesirable characteristics in dairy products using point scales and score cards. 'Expert' tasters grade products starting with a maximum score and subsequently deducting points for product-defects. However, newer and modern applications of SE for the dairy industry include the detection of differences in sensory characteristics between two or more similar products, the description and quantification of differences in sensory characteristics, and the evaluation of product acceptability and preference by the consumer (Delahunty, 2002). As an example, one common use of SE in the dairy industry is the evaluation of the results of reformulations in dairy products. Product reformulation (e.g. addition of a new ingredient or substitution of an ingredient with a cheaper alternative) is frequently practised by food manufacturers as a direct consequence of the need to compete in the market. In order for the industry to evaluate a reformulated product, it is necessary to determine, characterize and quantify the product's sensory attributes that are affected as a result of the change in product composition. Modern SE schemes combined with multivariate statistical methods enable the detection, objective characterization and quantification of such alterations in product attributes. Sidel and Stone (1993) have discussed the evolution of SE in the food industry.

Different testing methodologies should be used in sensory analysis depending on the industry's specified goals and the food product under consideration. Discrimination (difference) testing, such as the Paired Comparison (ISO, 1983), the Duo-Trio (ISO, 2004a) and the Triangle test (ISO, 2004b), is used to determine whether there is any perceptible difference between products in a given sensory characteristic. Scaling techniques usually employ a nine-point hedonic or intensity scale on which tasters or consumers are asked to denote their liking or the perceived intensity for a given attribute of a product, respectively. Scaling is useful in studying the relationship between perceived intensity and physical intensity for one or more characteristics of food attributes.

Nowadays, descriptive sensory analysis (DSA) is a highly promising SE approach for the food industry. The term encompasses a variety of techniques that aim to characterize foods based on all their sensory attributes as opposed to a single attribute at a time. Instead of using a single or only a handful of experts, descriptive evaluation is conducted by a panel of evaluators who work together and ultimately reach a consensus. DSA is advantageous compared with the defect-oriented terminology systems that are traditionally used by the dairy industry (Claassen and Lawless, 1992; Lawless and Claassen, 1993). DSA has been used to characterize changes in sensory properties of dairy products that result after the addition of supplements (Campbell *et al.*, 2003) or modification in the concentration of existing constituents (Phillips *et al.*, 1995), or after changes in technological parameters (Hannon *et al.*, 2005). There are several different methods of DSA: the Quantitative Descriptive Analysis (QDA) (Stone *et al.* 1974), the Flavour Profile Method (Cairncross and Sjostrom, 1950), the Texture Profile Method (Brandt *et al.*, 1963), the Spectrum Method™ and the Free-Choice Profiling method. QDA has been used to characterize conventionally pasteurized milk (Quinones *et al.*, 1998), ultrapasteurized milk (Chapman *et al.*, 2001), ice cream (Roland *et al.*, 1999), cheese (Ordonez *et al.*, 1998) and yoghurt (Jaworska *et al.*, 2005). The choice of a DSA method and therefore the type of panel that is used (use and level of training of 'experts', or use of typical product consumers) and the mechanics of the testing approach adopted differ according to industry goals and the product to be tested. Murray *et al.* (2001) have discussed the implementation of a descriptive sensory programme in light of recent developments in the field.

Piggott *et al.* (1998) have reviewed and commented on the different available sensory methodologies together with the most commonly used statistical designs. Among other statistical methods of analysis for sensory output data, PCA is a multivariate statistical approach that is probably used in SE more than in any other scientific domain. PCA allows for key attributes of (dairy) products to be objectively determined and described, by reducing a large set of correlated variables (product descriptors) to a smaller set of orthogonal (independent) variables (or principal components) that are linear combinations of the original variables and collectively account for (explain) most of the variability of the original data set (Westad *et al.*, 2003).

Among the different dairy products, cheeses present the highest degree of variability in terms of their sensory attributes. This is a direct consequence of the variety in the starting materials (milk origin, type and microflora) and in the technologies involved in cheese manufacture (type of starter cultures, salt content and degree of ripening). Cheeses, and in particular Cheddar cheese, have been evaluated and characterized in a number of SE studies (Nielsen and Zannoni, 1998; Murray and Delahunty, 2000; Rehman *et al.*, 2000; Barcenas *et al.*, 2001a, 2001b, 2003, 2004; Foegeding *et al.*, 2003; Pinho *et al.*, 2004; Coker *et al.*, 2005; Retiveau *et al.*, 2005). SE has also proved to be a valuable tool for PDO cheeses, as it can be used to define the necessary attributes to characterize traditional cheeses and help protect such products from adulterations and imitations (Elortondo *et al.*, 1999).

For a thorough and comprehensive coverage of the concepts, techniques, applications and statistical analyses used in SE, the interested reader can refer to classic SE textbooks (O'Mahony, 1986; Lawless and Heymann, 1999; Meilgaard *et al.*, 1999; Stone and Sidel, 2004). Standardized protocols pertaining to various aspects of conducting SE have been developed from standardization organizations and are available at a cost. Examples of such protocols are the Standards concerning the general strategy for sensory analysis (ISO, 2005), the design of test rooms (ISO, 1998b), the selection, training and monitoring of assessors (ISO, 1991, 1992a, 1993, 1994a) and the conduct of sensory tests (IDF, 1997b; ISO, 1982, 1983, 1985, 1987, 1988, 1992b, 1994b, 1994c, 1999b, 1999c, 2002, 2003b, 2003c, 2003d, 2004a, 2004b, 2004c).

It is the authors' humble opinion that dairy manufacturers will benefit by hiring sensory science graduates to set up organized SE programmes in their companies.

Acknowledgements

We would like to acknowledge the kind contribution of Professor Michael O'Mahony in the Department of Food Science and Technology, UC Davis, who provided the photograph in this section.

References

Abd El-Salam, M.H., Al-Khamy, A.F. and El-Etriby, H. (1986) Evaluation of the Milkoscan 104 A/B for determination of milk fat, protein and lactose in milk of some mammals. *Food Chemistry* **19**, 213–224.

Adams, M.R. and Moss, M.O. (eds) (1995) *Food Microbiology*. The Royal Society of Chemistry, Cambridge.

Addeo, F., Nicolai, M.A., Chianise, L., Moio, L., Musso, S.S., Bocca, A. and Del Giovine, L. (1995) A control method to detect bovine milk in ewe and water buffalo cheese using immunoblotting. *Milchwissenschaft* **50**, 83–85.

Almeida, P.F. and Almeida, R.C.C. (2000) A PCR protocol using *inl* gene as a target for specific detection of *Listeria monocytogenes*. *Food Control* **11**, 97–101.

Althaus, R., Molina, M.P., Rodriguez, M. and Fernandez, N. (2001) Detection limits of β-lactam antibiotics in ewe milk by Penzym enzymatic test. *Journal of Food Protection* **64**, 1844–1847.

Althaus, R., Torres, A., Peris, C., Beltran, M.C., Fernandez, N. and Molina, M.P. (2003b) Accuracy of BRT and Delvotest microbial inhibition tests as affected by composition of ewe's milk. *Journal of Food Protection* **66**, 473–478.

Althaus, R.L., Torres, A., Montero, A., Balasch, S. and Molina, M.P. (2003a) Detection limits of antimicrobials in ewe milk by Delvotest photometric measurements. *Journal of Dairy Science* **86**, 457–463.

Andrew, S.M. (2000) Effect of fat and protein content of milk from individual cows on the specificity rates of antibiotic residue screening tests. *Journal of Dairy Science* **83**, 2992–2997.

Andrew, S.M., Frobish, R.A., Paape, M.J. and Maturin, L.J. (1997) Evaluation of selected antibiotic residue screening tests for milk from individual cows and examination of factors that affect the probability of false-positive outcomes. *Journal of Dairy Science* **80**, 3050–3057.

Andrews, W.H. and Hammack, T.S. (1998) Food sampling and preparation of sample homogenate. In *Bacteriological Analytical Manual*, 8th edn. Food and Drug Administration, Center for Food Safety and Applied Nutrition, College Park, MD.

Angeletti, R., Gioacchini, A.M., Seraglia, R., Piro, R. and Traldi, P. (1998) The potential of matrix-assisted laser desorption/ionization mass spectrometry in the quality control of water buffalo mozzarella cheese. *Journal of Mass Spectroscopy* **33**, 525–531.

Angelidis, A.S., Farver, T.B. and Cullor, J.S. (1999) Evaluation of the Delvo-X-Press assay for detecting antibiotic residues in milk samples from individual cows. *Journal of Food Protection* **62**, 1183–1190.

Angelidis, A.S., Smith, L.T. and Smith, G.M. (2002) Elevated carnitine accumulation by *Listeria monocytogenes* impaired in glycine betaine transport is insufficient to restore wild-type cryotolerance in milk whey. *International Journal of Food Microbiology* **75**, 1–9.

Angelidis, A.S., Chronis, E.N., Papageorgiou, D.K., Kazakis, I.I., Arsenoglou, K.C. and Stathopoulos, G.A. (2006) Non-lactic acid, contaminating microbial flora in ready-to-eat foods: A potential food-quality index. *Food Microbiology* **23**, 95–100.

Anguita, G., Martin, R., Garcia, T., Morales, P., Haza, A.I., Gonzalez, I., Sanz, B. and Hernandez, P.E. (1996) Immunostick ELISA for detection of cow's milk in ewe's milk and cheese using a monoclonal antibody against bovine β-casein. *Journal of Food Protection* **59**, 436–437.

Anguita, G., Martin, R., Garcia, T., Morales, P., Haza, A.I., Gonzalez, I., Sanz, B. and Hernandez, P.E. (1997) A competitive enzyme-linked immunosorbent assay for detection of bovine milk in ovine and caprine milk and cheese using a monoclonal antibody against bovine β-casein. *Journal of Food Protection* **60**, 64–66.

Anonymous (1995) Milk lost to drug residues totals less than 0.05% of nation's production. *Dairy Quality Assurance Quest* **1**, 7.

AOAC (2005a) Association of Official Analytical Chemists. http://www.aoac.org/testkits/kits-antibiotics. HTM (accessed 15 July 2005).

AOAC (2005b) Association of Official Analytical Chemists. http://www.aoac.org/testkits/microbiologykits/htm (accessed 15 July 2005).

Ardö, Y. and Polychroniadou, A. (1999) Laboratory manual for chemical analysis of cheese. *COST 95 Improvement of the Quality of the Production of Raw Milk Cheeses*. European Communities, Luxembourg.

Arizcum, C., Barcina, Y. and Torre, P. (1997) Identification and characterization of proteolytic activity of *Enterococcus* spp. isolated from milk and Roncal and Idiazábal cheese. *International Journal of Food Microbiology*, **38**, 17–24.

Aschaffenburg, R. and Dance, J.E. (1968) Detection of cow's milk in goat's milk by gel electrophoresis. *Journal of Dairy Research* **35**, 383–384.

Asperger, H. (1993) Quality assurance of media in the microbiological laboratory. In *Analytical Quality Assurance and Good Laboratory Practice in Dairy Laboratories*, Proceedings of an International Seminar, Sonthofen, Germany, 18–20 May 1992, IDF Special Issue No. 9302, pp. 85–92. International Dairy Federation, Brussels.

Attfield, P., Gunasekera, T., Boyd, A., Deere, D. and Veal, D. (1999) Applications of flow cytometry to microbiology of food and beverage industries. *Australasian Biotechnology* **9**, 159–166.

AVMA (2005) American Veterinary Medical Association. http://www.avma.org/scienact/amduca/amduca2.asp (accessed 2 August 2005).

Aznar, R. and Alarcon, B. (2003) PCR detection of *Listeria monocytogenes*: a study of multiple factors affecting sensitivity. *Journal of Applied Microbiology* **95**, 958–966.

Bania, J., Ugorski, M., Polanowski, A. and Adamczyk (2001) Application of polymerase chain reaction for detection of goats' milk adulteration by milk of cow. *Journal of Dairy Research* **68**, 333–336.

Barbier, N., Saulnier, P., Chachaty, E., Dumontier, S. and Andremont, A. (1996) Random amplified polymorphic DNA typing versus pulsed-field gel electrophoresis for epidemiological typing of vancomycin-resistant enterococci. *Journal of Clinical Microbiology* **34** (5), 1096–1099.

Barcenas, P., de San Roman, R.P., Elortondo, F.J.P. and Albisu, M. (2001b) Consumer preference structures for traditional Spanish cheeses and their relationship with sensory properties. *Food Quality and Preference* **12**, 269–279.

Barcenas, P., Elortondo, F.J.P. and Albisu, M. (2003) Sensory changes during ripening of raw ewes' milk cheese manufactured with and without the addition of a starter culture. *Journal of Food Science* **68**, 2572–2578.

Barcenas, P., Elortondo, F.J.P. and Albisu, M. (2004) Projective mapping in sensory analysis of ewes milk cheeses: a study on consumers and trained panel performance. *Food Research International* **37**, 723–729.

Barcenas, P., Elortondo, F.J.P., Salmeron, J. and Albisu, M. (2001a) Sensory profile of ewe's milk cheeses. *Food Science and Technology International* **7**, 347–353.

Battelli, G. and Pellegrino, L. (1994) Detection of non-dairy fat in cheese by gas chromatography of triglycerides. *Italian Journal of Food Science* **4**, 407–419.

Baumgartner, C., Landgraf, A. and Buermeyer, J. (2003) pH determination: new applications of Mid-infra-red spectrometry for the analysis of milk and milk products. *Bulletin of the IDF* **383**, 23–28.

Baxter, G.A., Ferguson, J.P., O'Connor, M.C. and Elliott, C.T. (2001) Detection of streptomycin residues in whole milk using an optical immunobiosensor. *Journal of Agricultural and Food Chemistry* **49**, 3204–3207.

Baylis, C.L. (ed) (2003) *Manual of Microbiological Methods for the Food and Drinks Industry*, Guideline No. 43, 4th edn, Campden and Chorleywood Food Research Association Group, Chipping Campden, UK.

Becker, M., Zittlau, E. and Petz, M. (2004) Residue analysis of 15 penicillins and cephalosporins in bovine muscle, kidney and milk by liquid chromatography-tandem mass spectrometry. *Analytica Chimica Acta* **520**, 19–32.

Bergere, J.L. and Lenoir, J. (2000) Cheese manufacturing accidents and cheese defects. In Eck, A. and Gillis, J.-C. (eds), Davies, G. and Murphy, P.M. (trans.), *Cheesemaking – From Science to Quality Assurance*, 2nd edn, pp. 477–508. Intercept, Andover, UK.

Berruga, M.I., Yamaki, M., Althaus, R.L., Molina, M.P. and Molina A. (2003) Performances of antibiotic screening tests in determining the persistence of penicillin residues in ewe's milk. *Journal of Food Protection* **66**, 2097–2102.

Biering, G., Karlsson, S., Clark, N.C., Jonsdottir, K.E., Ludvigsson, P., and Steingrimsson, O. (1989) Three cases of neonatal meningitis caused by *Enterobacter sakazakii* in powdered milk. *Journal of Clinical Microbiology* **27**, 2054–2056.

Biggs, D.A. (1978) Instrumental infrared estimation of fat, protein and lactose in milk: Collaborative study. *Journal of the Association of Official Analytical Chemists* **61**, 1015–1020.

Bintsis, T. and Papademas, P. (2002) Microbiological quality of white-brined cheeses: a review. *International Journal of Dairy Technology* **55**, 113–120.

Bodyfelt, F.W., Tobias, J. and Trout, G.M. (1988) *The Sensory Evaluation of Dairy Products*. Van Nostrand Reinhold International, New York.

Bogialli, S., Capitolino, V., Curini, R., Di Corcia, A., Nazzari, M. and Sergi, M. (2004) Simple and rapid liquid chromatography-tandem mass spectrometry confirmatory assay for determining amoxicillin and ampicillin in bovine tissues and milk. *Journal of Agricultural and Food Chemistry* **52**, 3286–3291.

Bogialli, S., Curini, R., Di Corcia, A., Lagana, A., Mele, M. and Nazzari, M. (2005) Simple confirmatory assay for analyzing residues of aminoglycoside antibiotics in bovine milk: hot water extraction followed by liquid chromatography-tandem mass spectrometry. *Journal of Chromatography A* **1067**, 93–100.

Boison, J.O. and MacNeil, J.D. (1995) New test kit technology. In *Chemical Analysis for Antibiotics Used in Agriculture* (ed. Oka, H.), pp. 77–119. AOAC International, Arlington, VA.

Bolton, F.J. (1990) An investigation of indirect conductimetry for detection of some food-borne bacteria. *Journal of Applied Bacteriology* **69**, 655–661.

Bonwick, G. and Smith, C.J. (2004) Immunoassays: their history, development and current place in food science and technology. *International Journal of Food Science and Technology* **39**, 817–827.

Bordin, G., Raposo, F.C., de la Calle, B. and Rodriguez, A.R. (2001) Identification and quantification of major bovine milk proteins by liquid chromatography. *Journal of Chromatography A* **928**, 63–76.

Botsoglou, N.A. and Fletouris, D.J. (2001) *Drug Resides in Foods. Pharmacology, Food Safety, and Analysis*. Marcel Dekker, New York.

Bottero, M.T., Civera, T., Anastasio, A., Turi. R.M. and Rosati, S. (2002) Identification of cow's milk in buffalo cheese by duplex polymerase chain reaction. *Journal of Food Protection* **65**, 362–366.

Bottero, M.T., Civera, T., Nucera, D., Rosati, S., Sacchi, P. and Turi, R.M. (2003) A multiplex polymerase chain reaction for the identification of cows', goats' and sheep's milk in dairy products. *International Dairy Journal* **13**, 277–282.

Bradley, R.L. Jr, Arnold, E. Jr, Barbano, D.M., Semerad, R.G., Smith, D.E. and Vines, B.K. (1993) Chemical and physical methods. In *Standard Methods for the Examination of Dairy Products*, 16th edn, Chapter 15, pp. 433–531. American Public Health Association, Washington, DC.

Branciari, R., Nijman, I.J., Plas, M.E., Di Antonio, E. and Lenstra, A. (2000) Species origin of milk in Italian mozzarella and Greek feta cheese. *Journal of Food Protection* **63**, 408–411.

Brandt, M.A., Skinner, E.Z. and Coleman, J.A. (1963) Texture profile method. *Journal of Food Science* **28**, 404–409.

Brazis, A.R. (1991) Methods for estimating colony forming units. *Bulletin of the IDF* **256**, 4–8.

Bruno, F., Curini, R., Di Corcia, A., Nazzari, M. and Samperi, R. (2001) Solid-phase extraction followed by liquid chromatography-mass spectrometry for trace determination of β-lactam antibiotics in bovine milk. *Journal of Agricultural and Food Chemistry* **49**, 3463–3470.

Burtscher, C. and Wuertz, S. (2003) Evaluation of the use of PCR and reverse transcriptase PCR for the detection of pathogenic bacteria in biosolids from anaerobic digestors and aerobic composters. *Applied and Environmental Microbiology* **69**, 4618–4627.

Cacciatore, G., Petz, M., Rachid, S., Hakenbeck, R. and Bergwerff, A.A. (2004) Development of an optical biosensor assay for detection of β-lactam antibiotics in milk using the penicillin-binding protein 2x*. *Analytica Chimica Acta* **520**, 105–115.

Cairncross, S.E. and Sjostrom, L.B. (1950) Flavour profiles: a new approach to flavour problems. *Food Technology* **4**, 308–311.

Campbell, W., Drake, M.A. and Larick, D.K. (2003) The impact of fortification with conjugated linoleic acid (CLA) on the quality of fluid milk. *Journal of Dairy Science* **86**, 43–51.

Candish, A.A.G. (1991). Immunological methods in food microbiology. *Food Microbiology* **8**, 1–14.

Carle, G.F., Frank, M. and Olson, M.V. (1986). Electrophoretic separation of large DNA molecules by periodic inversion of the electric field. *Science* **232**, 65–68.

Carlsson, A., Bjorck, L. and Persson, K. (1989) Lactoferrin and lysozyme in milk during acute mastitis and their inhibitory effect in Delvotest P. *Journal of Dairy Science* **72**, 3166–3175.

Carson, M.C. and Heller, D.N. (1998) Confirmation of spectinomycin in milk using ion-pair solid-phase extraction and liquid chromatography-electrospray ion trap mass spectrometry. *Journal of Chromatography B* **718**, 95–102.

Cartoni, G., Coccioli, F., Jasionowska, R. and Masci, M. (1998) Determination of cow milk in ewe milk and cheese by capillary electrophoresis of the whey protein fractions. *Italian Journal of Food Science* **10**, 317–327.

Cartoni, G., Coccioli, F., Jasionowska, R. and Masci, M. (1999) Determination of cows' milk in goats' milk and cheese by capillary electrophoresis of the whey protein fractions. *Journal of Chromatography A* **846**, 135–141.

Catinella, S., Traldi, P., Pinelli, C. and Dallaturca, E. (1996) Matrix-assisted laser desorption/ionization mass spectrometry: a valid analytical tool in the dairy industry. *Rapid Communications in Mass Spectrometry* **10**, 1123–1127.

Cattaneo, T.M.P., Nigro, F. and Greppi, G.F. (1996) Analysis of cow, goat and ewe milk mixtures by capillary zone electrophoresis (CZE): preliminary approach. *Milchwissenschaft* **51**, 616–619.

Centeno, J.A., Menendez, S. and Rodriquez-Otero, J.L. (1996) Main microbial flora present as natural starters in Gebreiro raw cow's milk cheese (Northwest Spain). *International Journal of Food Microbiology*, **33**, 307–313.

Centeno, J.A., Menendez, S., Hermida, M.A. and Rodriquez-Otero, J.L. (1999) Effects of the addition of Enterococcus faecalis in Gebreiro cheese manufacture. *International Journal of Food Microbiology*, **48**, 97–101.

Cerquaglia, O. and Avellini, P. (2004) A rapid γ-casein isoelectrofocusing method for detecting and quantifying bovine milk used in cheese making: application to sheep cheese. *Italian Journal of Food Science* **16**, 447–455.

Chapman, K.W., Lawless, H.T. and Boor, K.J. (2001) Quantitative descriptive analysis and principal component analysis for sensory characterization of ultrapasteurized milk. *Journal of Dairy Science* **84**, 12–20.

Chen, R.-K., Chang, L.-W., Chung, Y.-Y., Lee, M.-H. and Ling, Y.-C. (2004) Quantification of cow milk adulteration in goat milk using high-performance liquid chromatography with electrospray ionization mass spectrometry. *Rapid Communications in Mass Spectrometry* **18**, 1167–1171.

Chu,G., Vollrath, D. and Davis, R.W. (1986) Separation of large DNA molecules by contour-clamped homogeneous electric fields. *Science* **234**, 1582–1585.

Claassen, M. and Lawless, H.T. (1992) Comparison of descriptive terminology systems for sensory evaluation of fluid milk. *Journal of Food Science* **57**, 596–600, 621.

Clarke, R.G. and Pinder, A.C. (1998) Improved detection of bacteria by flow cytometry using a combination of antibody and viability markers. *Journal of Applied Microbiology* **84**, 577–584.

Cocolin, L., Manzano, M., Cantoni, C. and Comi, G. (1997) A PCR-microplate capture hybridization method to detect *Listeria monocytogenes* in blood. *Molecular and Cell Probes* **11**, 453–455.

Cogan, T.M. (1972) Susceptibility of cheese and yogurt starter bacteria to antibiotics. *Applied Microbiology* **23**, 960–965.

Coker, C.J., Crawford, R.A., Johnston, K.A., Singh, H. and Creamer, L.K. (2005) Towards the classification of cheese variety and maturity on the basis of statistical analysis of proteolysis data – a review. *International Dairy Journal* **15**, 631–643.

Coni, E., Di Pasquale, M., Coppolelli, P. and Bocca, A. (1994) Detection of animal fats in butter by differential scanning calorimetry: a pilot study. *Journal of the American Oil Chemists' Society* **71**, 807–810.

Contreras, A., Paape, M.J., Di Carlo, A.L., Miller, R.H. and Rainard, P. (1997) Evaluation of selected antibiotic residue screening tests for milk from individual goats. *Journal of Dairy Science* **80**, 1113–1118.

Cook, N. (2003) The use of NASBA for the detection of microbial pathogens in food and environmental samples. *Journal of Microbiological Methods* **53**, 165–174.

Cozzolino, R., Passalacqua, S., Salemi, S and Garozzo, D. (2002) Identification of adulteration in water buffalo mozzarella and in ewe cheese by using whey proteins as biomarkers and matrix-assisted laser desorption/ionization mass spectrometry. *Journal of Mass Spectroscopy* **37**, 985–991.

Cozzolino, R., Passalacqua, S., Salemi, S Malvagna, P., Spina, E. and Garozzo, D. (2001) Identification of adulteration in milk by matrix-assisted laser desorption/ionization time-of-flight mass spectrometry. *Journal of Mass Spectroscopy*, **36**, 1031–1037.

Cullor, J.S. and Chen, J. (1991) Evaluating two new OTC milk screening ELISAs: do they measure up? *Veterinary Medicine* **86**, 845–850.

Cullor, J.S. (1992) Tests for identifying antibiotic residues in milk: how well do they work? *Veterinary Medicine* **87**, 1235–1241.

Cullor, J.S. (1993) Antibiotic residue tests for mammary gland secretions. *Veterinary Clinics of North America: Food Animal Practice* **9**, 609–620.
Cullor, J.S. (1994) Testing the tests intended to detect antibiotic residues in milk. *Veterinary Medicine* **89**, 462–472.
Cullor, J.S., Van Eenennaam, A., Dellinger, J., Perani, L., Smith, W. and Jensen, L. (1992). Antibiotic residue assays: can they be used to test milk from individual cows? *Veterinary Medicine* **87**, 477–494.
Cullor, J.S., Van Eenennaam, A., Gardner, I., Perani, L., Dellinger, J., Smith, W., Thompson, T., Payne, M.A., Jensen, L. and Guterbock, W.M. (1994) Performance of various tests used to screen antibiotic residues in milk samples from individual animals. *Journal of AOAC International* **77**, 862–870.
Curiale, M.S., Gangar, V. and Gravens, C. (1997) VIDAS enzyme-linked fluorescent immunoassay for detection of *Salmonella* in foods: collaborative study. *Journal of AOAC International* **80**, 491–504.
Cutting, J.H., Kiessling, W.M., Bond, F.L., McCarron, J.E., Kreuzer, K.S., Hurlbut, J.A. and Sofos, J.N. (1995) Agarose gel electrophoretic detection of six β-lactam antibiotic residues in milk. *Journal of AOAC International* **78**, 663–667.
D'Aoust, J.Y., Sewell, A.M. and Greco, P. (1991) Commercial latex agglutination kits for the detection of foodborne *Salmonella*. *Journal of Food Protection* **54**, 725–730.
Daeseleire, E., De Ruyck, H. and van Renterghem, R. (2000) Confirmatory assay for the simultaneous detection of penicillins and cephalosporins in milk using liquid chromatography/tandem mass spectrometry. *Rapid Communications in Mass Spectrometry* **14**, 1404–1409.
Daly, P., Collier, T. and Doyle, S. (2002) PCR-ELISA detection of *Escherichia coli* in milk. *Letters in Applied Microbiology* **34**, 222–226.
de Jong, N., Visser, S. and Olieman, C. (1993) Determination of milk proteins by capillary electrophoresis. *Journal of Chromatography A* **652**, 207–213.
de Vuyst, L., Foulquie Moreno, M.R. and Revets, H. (2002) Screening for enterocins and detection of hemolycin and vancomycin resistance in enterococci of different origins. *International Journal of Food Microbiology*, **2635**, 1–20.
de Zayas-Blanco, F., Garcia-Falcon, M.S. and Simal-Gandara, J. (2004) Determination of sulfamethazine in milk by solid phase extraction and liquid chromatographic separation with ultraviolet detection. *Food Control* **15**, 375–378.
Delahunty, C.M. (2002) Sensory evaluation. In *Encyclopedia of Dairy Sciences*, Roginski, H., Fuguay, J.W. and Fox, P.F. (eds), pp. 106–110. Academic Press, The Netherlands, Elsevier Academic Press.
Dewdney, J.M., Maes, L., Raynaud, J.P., Blanc, F., Scheid, J.P, Jackson, T., Lens, S. and Verschueren, C. (1991) Risk assessment of antibiotic residues of β-lactams and macrolides in food products with regard to their immuno-allergic potential. *Food and Chemical Toxicology* **29**, 477–483.
Di Corcia, A. and Nazzari, M. (2002) Liquid chromatographic-mass spectrometric methods for analysing antibiotic and antibacterial agents in animal food products. *Journal of Chromatography A* **974**, 53–89.
Donnely, C.W. and Baigent, G.J. (1986) Method for flow cytometric detection of *Listeria monocytogenes* in milk. *Applied and Environmental Microbiology* **52**, 689–695.
Dudrikova, E., Jozef, S. and Jozef, N. (1999) Liquid chromatographic determination of tylosin in mastitic cow's milk following therapy. *Journal of AOAC International* **82**, 1303–1307.
Dumain, P.P., Desnouveaux, R., Bloch, L., Fuhrmann, B., De Colombel, E., Pessis, M.C. and Valery, S. (1990) Use of flow cytometry for yeast and mould detection in process control of fermented milk products: the ChemFlow system – A factor study. *Biotechnology Forum Europe* **7**, 224–229.
Dziadkowiec, D., Mansfield, L.P. and Forsyth, S.J. (1995) The detection of *Salmonella* in skimmed milk powder enrichments using conventional methods and immunomagnetic separation. *Letters in Applied Microbiology* **20**, 361–364.

EC (1990) Regulation 2377/90 laying down a Community procedure for the establishment of maximum residue limits of veterinary medicinal products in foodstuffs of animal origin. *Official Journal of the European Community* **L224**, 1–8.
EC (1996) Regulation 1081/96. Reference method for the detection of cows' milk and caseinate in cheeses from ewes' milk, goats' milk and buffaloes' milk or mixtures of ewes', goats', and buffaloes' milk. *Official Journal of the European Commission* **L142**, 289–295.
EC (2000) Directive 2000/54/EC of the European Parliament and of the Council of 18 September 2000, on the protection of workers from risks related to exposure to biological agents at work. *Official Journal of the European Union* **L262**, 21–45.
EC (2001) Commission Regulation (EC) No. 213/2001 of 9 January 2001 laying down detailed rules for the application of Council Regulation (EC) No. 1255/1999 as regards methods for the analysis and quality evaluation of milk and milk products and amending Regulations (EC) No. 277/1999 and (EC) No. 2799/1999. *Official Journal of the European Union* **L37**, 1–99.
EC (2002) Commission Decision 2002/657 implementing Council Directive 96/23 concerning the performance of analytical methods and the interpretation of results. *Official Journal of the European Community* **L221**, 8–36.
EC (2004) Directive 2004/10/EC of the European Parliament and of the Council of 11 February 2004, on the harmonization of laws, regulations and administrative provisions relating to the application of the principles of good laboratory practice and the verification of their applications for tests on chemical substances. *Official Journal of the European Union* **L50**, 44–59.
El Marrakchi, A., Boum'handi, N. and Hamama, A. (2005) Performance of a new chromogenic plating medium for the isolation of *Listeria monocytogenes* from marine environments. *Letters in Applied Microbiology* **40**, 87–91.
Elortondo, P.F.J., Barcenas, P., Casas, C., Salmeron, J. and Albisu, M. (1999) Development of standardized sensory methodologies: some applications to Protected Designation of Origin cheeses. *Sciences des Aliments* **19**, 543–558.
Entis, P. and Lerner, I. (1998) Enumeration of beta-glucuronidase-positive *Escherichia coli* in foods by using the ISO-GRID method with SD-39 agar. *Journal of Food Protection* **61**(7), 913–916.
Entis, P. and Lerner, I. (2000) Twenty-four-hour direct presumptive enumeration of *Listeria monocytogenes* in food and environmental samples using the ISO-GRID method with LM-137 agar. *Journal of Food Protection* **63**(3), 354–363.
Entis, P. (1984) Enumeration of total coliforms, fecal coliforms and *Escherichia coli* in foods by hydrophobic grid membrane filter: collaborative study. *Journal of AOAC* **67**, 812–823.
Entis, P. (1990) Improved hydrophobic grid membrane filter method, using EF-18 agar, for detection of *Salmonella* in foods: collaborative study. *Journal of AOAC* **73**, 734–742.
Entis, P., Fung, D.Y.C., Griffiths, M.W., Mclntyre, L., Russell, S., Sharpe, A.N. and Tortorello, M.L. (2001) Rapid methods for detection, identification, and enumeration. In: *Compendium of Methods for the Microbiological Examination of Foods*, 4th edn, Downes, F.P. and Ito, K. (eds), pp. 89–126. American Public Health Association, Washington, DC.
Espeja, E., Garcia, M.C. and Marina, M.L. (2001) Fast detection of added soybean proteins in cow's, goat's, and ewe's milk by perfusion reversed-phase high-performance liquid chromatography. *Journal of Separation Science*, **24**, 856–864.
EURACHEM/CITAC (2002) *Guide to Quality in Analytical Chemistry – An Aid to Accreditation.* EURACHEM/CITAC. http://www.citac.cc/publications (accessed 7 November 2005).
EURACHEM/CITAC (2003) *Traceability in Chemical Measurement.* EURACHEM/CITAC. Available on-line, http://www.citac.cc/publications (accessed 7 November 2005).
Evers, J.M. (1999) Determination of fat in butter – A review. *Bulletin of the IDF* **340**, 3–15.
Fanton, C., Delogu, G., Maccioni, E., Podda, G., Seraglia, R. and Traldi, P. (1998) Matrix-assisted laser desorption/ionization mass spectrometry in the dairy industry 2. The protein fingerprint of ewe cheese and its application to detection of adulteration by bovine milk. *Rapid Communications in Mass Spectrometry* **12**, 1569–1573.

Farber, J.M. (1996) An introduction to the Hows and Whys of molecular typing. *Journal of Food Protection* **59**, 1091–1101.

Feldsine, P.T., Forgey, R.L., Falbo-Nelson, M.T. and Brunelle, S.L. (1997a) *Escherichia coli* O157:H7 visual immunoprecipitate assay: a comparative validation study. *Journal of AOAC International* **80**, 43–48.

Feldsine, P.T., Lienau, A.H., Forgey, R.L. and Calhoon, R.D. (1997b) Visual immunoprecipitate assay (VIP) for *Listeria monocytogenes* and related *Listeria* species detection in selected foods: collaborative study. *Journal of AOAC International* **80**, 791–805.

Feligini, M., Bonizzi, I., Curik, V.C., Parma, P., Greppi, G.F. and Enne, G. (2005) Detection of adulteration in Italian mozzarella cheese using mitochondrial DNA templates as biomarkers. *Food Technology and Biotechnology* **43**, 91–95.

Feng, P. (1997) Impact of molecular biology on the detection of foodborne pathogens. *Molecular Biotechnology* **7**, 267–278.

Ferguson, J.P., Baxter, G.A., McEvoy, J.D.G., Stead, S., Rawlings, E. and Sharman, M. (2002) Detection of streptomycin and dihydrostreptomycin residues in milk, honey and meat samples using an optical biosensor. *Analyst* **127**, 951–956.

Ferreira, I.M.P.L.V.O. and Cacote, H. (2003) Detection and quantification of bovine, ovine and caprine milk percentages in protected denomination of origin cheeses by reversed-phase high-performance liquid chromatography of beta-lactoglobulins. *Journal of Chromatography A* **1015**, 111–118.

Firstenberg-Eden, R. and Tricarico, M.K. (1983) Impedimetric determination of total, mesophilic and psychotrophic counts in raw milk. *Journal of Food Science* **48**, 1750–1754.

Firstenberg-Eden, R., Van Sise, M.L., Zindulis, J. and Kahn, P. (1984) Impedimetric estimation of coliforms in dairy products. *Journal of Food Science* **49**, 1449–1452.

Fitts, J.E. and Laird, D. (2004) Direct microscopic methods for bacteria or somatic cells. In *Standard Methods for the Examination of Dairy Products*, Wehr, H.M. and Frank, J.F. (eds), 17th edn, pp. 269–280. American Public Health Association, Washington, DC.

Flowers, R.S., Andrews, W., Donnelly, C.W. and Koenig, E. (1993) Pathogens in milk and milk products. In *Standard Methods for the Examination of Dairy Products*, 16th edn, pp. 103–212, American Public Health Association, Washington, DC.

Foegeding, E.A., Brown, J., Drake, M. and Daubert, C.R. (2003) Sensory and mechanical aspects of cheese texture. *International Dairy Journal* **13**, 585–591.

Foschino, R., Colombo, S., Crepaldi, V. and Baldi, L. (2003) Comparison between Colifast® Milk and the standard method for the detection of coliforms in pasteurized milk. *Lait* **83**, 161–166.

Foster, E.M. (1990) Emerging pathogens. In *Proceedings of the XXIII International Dairy Congress*, 8–12 October 1990, Montreal, Canada.

Fouchet, P., Jayat, C., Hechard, Y., Ratinaud, M.-H. and Frelat, G. (1993) Recent advances of flow cytometry in fundamental and applied microbiology. *Biology of Cell* **78**, 95–109.

Foulquie Moreno, M.R., Callewaert, R., Devreese, B., Van Beeumen, J. and De Vuyst, L. (2003) Isolation and biochemical characterization of enterocins produced by enterococci from different sources. *Journal of Applied Microbiology*, **94**, 214–229.

Fox, J.R., Duthie, A.H. and Wulff, S. (1988) Precision and sensitivity of a test for vegetable fat adulteration of milk fat. *Journal of Dairy Science* **71**, 574–581.

Frank, J.F., Christen, G.L. and Bullerman, L.B. (1993) Test for groups of microorganisms. In *Standard Methods for the Examination of Dairy Products*, 16th edn, Chapter 8, pp. 271–286, American Public Health Association, Washington, DC.

Fung, D.Y.C. (1994) Rapid methods and automation in food microbiology: a review. *Food Reviews International*, **10**, 357–375.

Garde, S., Gaya, P., Medina, M. and Nunez, M. (1997) Acceleration of flavour formation in cheese by a bacteriocin-producing adjunct lactic culture. *Biotechnological Letters*, **19**, 1011–1014.

Gardner, I.A., Cullor, J.S., Galey, F.D., Sischo, W., Salman, M., Slenning, B., Erb, H.N. and Tyler, J.W. (1996) Alternatives for validation of diagnostic assays used to detect antibiotic residues in milk. *Journal of the American Veterinary Medical Association*, **209**, 46–52.

Garfield, F.M., Hirsch, J.H. and Klesta, E.J. Jr (2000) *Quality Assurance Principles for Analytical Laboratories*, 3rd edn. AOAC International, Arlington, VA.

Gasanov, U., Hughes, D. and Hansbro, P.M. (2005) Methods for the isolation and identification of *Listeria* spp. and *Listeria monocytogenes*: a review. *FEMS Microbiology Reviews*, **29**, 851–875.

Gaudin, V. and Pavy, M.-L. (1999) Determination of sulfamethazine in milk by biosensor immunoassay. *Journal of AOAC International* **82**, 1316–1320.

Gaudin, V. and Maris, P. (2001) Development of a biosensor-based immunoassay for screening of chloramphenicol residues in milk. *Food and Agricultural Immunology* **13**, 77–86.

Gaudin, V., Maris, P., Fuselier, R., Ribouchon, J.-L., Cadieu, N. and Rault, A. (2004) Validation of a microbiological method: the STAR protocol, a five-plate test, for the screening of antibiotic residues in milk. *Food Additives and Contaminants* **21**, 422–433.

Gaudin, V., Cadieu, N. and Sanders, P. (2005) Results of a European proficiency test for the detection of streptomycin/dihydrostreptomycin, gentamicin and neomycin in milk by ELISA and biosensor methods. *Analytica Chimica Acta* **529**, 273–283.

Gautom, R.K. (1997) Rapid pulsed-field gel electrophoresis protocol for typing of *Escherichia coli* O157:H7 and other Gram-negative organisms in one day. *Journal of Clinical Microbiology* **35**, 2977–2980.

Gella, F.J., Serra, J. and Gener, J. (1991) Latex agglutination in immunodiagnosis. *Pure and Applied Chemistry* **63**, 1131–1134.

Gentili, A., Perret, D. and Marchese, S. (2005) Liquid chromatography-tandem mass spectrometry for performing confirmatory analysis of veterinary drugs in animal-food products. *Trends in Analytical Chemistry* **24**, 704–733.

Gerhardt, P., Murray, R.G.E., Wood, W.A. and Krieg, N.R. (1994) *Methods for General and Molecular Bacteriology*, pp. 123–126. American Society for Microbiology, Washington, DC.

Gibbons-Burgener, S.N., Kaneene, J.B., Lloyd, J.W., Leykam, J.F. and Erskine, R.J. (2001) Reliability of three bulk-tank antimicrobial residue detection assays used to test individual milk samples from cows with milk clinical mastitis. *American Journal of Veterinary Research* **62**, 1716–1720.

Giraffa, G. (1995) Enterococcal bacteriocins: their potential as anti-Listeria factors in dairy technology. *Food Microbiology*, **12**, 291–299.

Grace, V., Houghtby, G.A., Rudnick, H., Whaley, K. and Lindamood, J. (1993) Sampling dairy and related products. In *Standard Methods for the Examination of Dairy Products*, 16th edn, pp. 59–83. American Public Health Association, Washington, D.C.

Gracias, K.S. and McKillip, J.L. (2004) A review of conventional detection and enumeration methods for pathogenic bacteria in food. *Canadian Journal of Microbiology* **50**, 883–890.

Grant, K.A. and Kroll, R.G. (1993) Molecular biology techniques for the rapid detection and characterization of foodborne bacteria. *Food Science and Technology Today* **7**, 80–88.

Gunasekera, T.S., Attfield, P.V. and Veal, D.A. (2000) A flow cytometry for rapid detection and enumeration of total bacteria in milk. *Applied and Environmental Microbiology* **66**, 1228–1232.

Gunasekera, T.S., Veal, D.A. and Attfield, P.V. (2002) Potential for broad applications of flow cytometry and fluorescence techniques in microbiological and somatic cell analysis of milk. *International Journal of Food Microbiology* **85**, 269–279.

Gustavsson, E., Bjurling, P. and Sternesjo, A. (2002). Biosensor analysis of penicillin G in milk based on the inhibition of carboxypeptidase activity. *Analytica Chimica Acta* **468**, 153–159.

Gustavsson, E. and Sternesjo, A. (2004) Biosensor analysis of β-lactams in milk: comparison with microbiological, immunological, and receptor-based screening methods. *Journal of AOAC International* **87**, 614–620.

Gustavsson, E., Degelaen, J., Bjurling, P. and Sternesjo, A. (2004) Determination of β-lactams in milk using a surface plasmin resonance-based biosensor. *Journal of Agricultural and Food Chemistry* **52**, 2791–2796.

Guy, P.A., Royer, D., Mottier, P., Gremaud, E., Perisset, A. and Stadler, R.H. (2004) Quantitative determination of chloramphenicol in milk powders by isotope liquid chromatography coupled to tandem mass spectrometry. *Journal of Chromatography A* **1054**, 365–371.

Ha, K.S., Park, S.J., Seo, S.J., Park, J.H. and Chung, D.H. (2002) Incidence and polymerase chain reaction assay of *Listeria monocytogenes* from raw milk in Gyeongnam Province of Korea. *Journal of Food Protection* **65**, 111–115.

Halbert, L.W., Erskine, R.J., Bartlett, P.C. and Johnson II, G.L. (1996) Incidence of false-positive results for assays used to detect antibiotics in milk. *Journal of Food Protection* **59**, 886–888.

Hancock, I., Bointon, B.M. and McAthey, P. (1993) Rapid detection of *Listeria* species by selective impedimetric assay. *Letters in Applied Microbiology* **16**, 311–314.

Hanna, M.O., Stewart, J.C., Carpenter, Z.L. and Vanderzant, C. (1977) Heat resistance of *Yersinia enterocolitica* in skim milk. *Journal of Food Science* **42**, 1134–1136.

Hannon, J.A., Wilkinson, M.G., Delahunty, C.M., Wallace, J.M., Morrissey, P.A. and Beresford, T.P. (2005) Application of descriptive sensory analysis and key chemical indices to assess the impact of elevated ripening temperatures on the acceleration of Cheddar cheese ripening. *International Dairy Journal* **15**, 263–273.

Harding, F. (1995) Adulteration of milk. In *Milk Quality*, Harding, F. (ed.), pp. 60–74. Chapman & Hall, London.

Harrigan, W.F. (1998) *Laboratory Methods in Food Microbiology*, 3rd edn. Academic Press, London.

Harik-Khan, R. and Moats, W.A. (1995) Identification and measurement of β-lactam antibiotic residues in milk: integration of screening kits with liquid chromatography. *Journal of AOAC International* **78**, 978–986.

Hawkins, R.E., Lissner, C.R. and Sanford, J.P. 1991. *Enterobacter sakazakii* bacteremia in an adult. *South Medical Journal* **84**, 793–795.

Haza, A.I., Morales, P. Martin, R., Garcia, T., Anguita, G., Gonzales, I., Sanz, B. and Hernandez, P.E. (1996) Development of monoclonal antibodies against caprine αS2-casein and their potential for detecting the substitution of ovine milk by caprine milk by an indirect ELISA. *Journal of Agricultural and Food Chemistry* **44**, 1756–1761.

Haza, A.I., Morales, P. Martin, R., Garcia, T., Anguita, G., Gonzales, I., Sanz, B. and Hernandez, P.E. (1997) Use of a monoclonal antibody and two enzyme-linked immunosorbent assay formats for detection and quantification of the substitution of caprine milk for ovine milk. *Journal of Food Protection* **60**, 973–977.

Haza, A.I., Morales, P. Martin, R., Garcia, T., Anguita, G., Sanz, B. and Hernandez, P.E. (1999) Detection and quantification of goat's cheese in ewe's cheese using monoclonal antibody and two ELISA formats. *Journal of the Science of Food and Agriculture* **79**, 1043–1047.

Heller, D.N. and Ngoh, M.A. (1998) Electrospray ionization and tandem ion trap mass spectrometry for the confirmation of seven β-lactam antibiotics in bovine milk. *Rapid Communications in Mass Spectrometry* **12**, 2031–2040.

Heller, D.N., Clark, S.B. and Righter, H.F. (2000) Confirmation of gentamicin and neomycin in milk by weak cation-exchange extraction and electrospray ionization/ion trap tandem mass spectrometry. *Journal of Mass Spectrometry* **35**, 39–49.

Hernandez, M., Borrull, F. and Calull, M. (2003) Analysis of antibiotics in biological samples by capillary electrophoresis. *Trends in Analytical Chemistry* **22**, 416–427.

Herrero-Martinez, J.M., Simo-Alfonso, E.F., Ramis-Ramos, G., Gelfi, C. and Righetti, P.G. (2000) Determination of cow's milk in non-bovine and mixed cheeses by capillary electrophoresis of whey proteins in acidic isoelectric buffers. *Journal of Chromatography A* **878**, 261–271.

Hill, B.M. (1991a) Microscopic (or direct) methods for estimation of bacteria. *Bulletin of the IDF* **256**, 17–20.

Hill, B.M. (1991b) Direct epifluorescent filter technique (DEFT). *Bulletin of the IDF* **256**, 20–24.

Hillerton, J.E., Halley, B.I., Neaves, P. and Rose, M.D. (1999) Detection of antimicrobial substances in individual cow milk and quarter milk samples using Delvotest microbial inhibitor tests. *Journal of Dairy Science* **82**, 704–711.

Holstege, D.M., Puschner, B., Whitehead, G. and Galey, F.D. (2002) Screening and mass spectral confirmation of β-lactam antibiotic residues in milk using LC-MS/MS. *Journal of Agricultural and Food Chemistry* **50**, 406–411.

Hooi, R., Barbano, D.M., Bradley, R.L., Budde, D., Bulthaus, M., Chettiar, M., Lynch, J. and Reddy, R. (2004) Chemical and physical methods. In *Standard Methods for the Examination of Dairy Products*, Wehr, H.M. and Frank, J.F. (eds), 17th edn, pp. 363–536. American Public Health Association, Washington, DC.

Horwitz, W. (ed.) (2005) *Official Methods of Analysis of AOAC International*, 18th edn. Association of Official Analytical Chemists, Gaithersburg, MD.

Hsih, H.Y. and Tsen, H.Y. (2001) Combination of immunomagnetic separation and polymerase chain reaction for the simultaneous detection of *Listeria monocytogenes* and *Salmonella* spp. in food samples. *Journal of Food Protection* **64**, 1744–1750.

Huis in't Veld, J. and Hofstra, H. (1991) Biotechnology and the quality assurance of foods. *Food Biotechnology* **5**, 313–322.

Hunter, G.J.E. (1949) The effect of penicillin in milk on the manufacture of Cheddar cheese. *Journal of Dairy Research* **16**, 235–241.

Hurley, I.P., Coleman, R.C., Ireland, H.E. and Williams, J.H.H. (2004a) Measurement of bovine IgG by indirect competitive ELISA as a means of detecting milk adulteration. *Journal of Dairy Science* **87**, 543–549.

Hurley, I.P., Ireland, H.E., Coleman, R.C. and Williams, J.H.H. (2004b) Application of immunological methods for the detection of species adulteration in dairy products. *International Journal of Food Science and Technology* **39**, 873–878.

Huth, S.P., Warholic, P.S., Devou, J.M., Chaney, L.K. and Clark, G.H. (2002) ParalluxTM β-lactam: a capillary-based fluorescent immunoassay for the determination of penicillin-G, ampicillin, amoxicillin, cloxacillin, cephapirin, and ceftiofur in bovine milk. *Journal of AOAC International* **85**, 355–364.

ICMSF (1978) *Microorganisms in Foods 1. Their Significance and Methods of Enumeration*, 2nd edn. International Commission on Microbial Specifications for Foods, University of Toronto Press, Toronto.

ICMSF (1986) *Microorganisms in Foods 2. Sampling for Microbiological Analysis: Principles and Specific Applications*, 2nd edn. International Commission on Microbiological Specifications for Foods. University of Toronto Press, Toronto.

ICMSF (2002) *Microorganisms in Foods 7. Microbiological Testing in Food Safety Management*. International Commission on Microbiological Specifications for Foods. University of Toronto Press, Toronto.

IDF (1970) FIL-IDF Standard 54:1970, *Detection of Vegetable Fat in Milk Fat by Gas–Liquid Chromatography of Sterols*. International Dairy Federation, Brussels.

IDF (1987a) IDF Standard 009C:1987, *Dried Milk, Dried Whey, Dried Buttermilk and Dried Butter Serum – Determination of Fat Content (Röse Gottlieb Method)*. International Dairy Federation, Brussels.

IDF (1987b) IDF Standard 013C:1987, *Evaporated Milk and Sweetened Condensed Milk – Determination of Fat Content (Röse Gottlieb Method)*. International Dairy Federation, Brussels.

IDF (1987c) IDF Standard 016C:1987, *Cream – Determination of Fat Content (Röse Gottlieb Method)*. International Dairy Federation, Brussels.

IDF (1987d) IDF Standard 022B:1987, *Skimmed Milk, Whey and Buttermilk – Determination of Fat Content (Röse Gottlieb Method)*. International Dairy Federation, Brussels.

IDF (1988) IDF Standard 88A:1988, *Cheese and Processed Cheese Products – Determination of Chloride Content – Potentiometric Method.* International Dairy Federation, Brussels.

IDF (1991a) *IDF Standard 100B:1991, Milk and milk products – Enumeration of microorganisms (Colony count technique at 30°C).* International Dairy Federation, Brussels.

IDF (1991b) Detection and confirmation of inhibitors in milk and milk products. *Bulletin* of *the IDF* **258.** International Dairy Federation, Brussels.

IDF (1992) IDF Standard 136A:1992, *Milk and Milk Products – Sampling – Inspection by Variables.* International Dairy Federation, Brussels.

IDF (1993) Inhibitory substances in milk – current analytical practice. *Bulletin* of *the IDF* **283.** International Dairy Federation, Brussels.

IDF (1995a) IDF Standard 50C:1995, *Milk and Milk Products – Guidance on Sampling.* International Dairy Federation, Brussels.

IDF (1995b) IDF Standard 148A:1995, *Milk – Enumeration of Somatic Cells.* International Dairy Federation, Brussels.

IDF (1997a) IDF Standard 145A:1997, *Milk and Milk Products – Enumeration of Coagulase-Positive Staphylococci (Colony Count Technique).* International Dairy Federation, Brussels.

IDF (1997b) IDF Standard 099C:1997, *Sensory Evaluation of Dairy Products by Scoring.* International Dairy Federation, Brussels.

IDF (1998a) IDF Standard 181:1998, *Dried Milk Products – Enumeration of* Bacillus cereus *(Most Probable Number Technique).* International Dairy Federation, Brussels.

IDF (1998b) IDF Standard 73B:1998, *Enumeration of Coliforms.* International Dairy Federation, Brussels.

IDF (1998c) IDF Standard 83A:1998, *Milk and Milk Products – Standard Methods for the Detection of Thermonuclease Produced by Coagulase-Positive Staphylococci in Milk and Milk-Based Products.* International Dairy Federation, Brussels.

IDF (1999) IDF Standard 183:1999, *Guidance for the Standardized Evaluation of Microbial Inhibitor Tests.* International Dairy Federation, Brussels.

IDF (2000) IDF Standard 12C:2000, *Butter – Determination of Salt Content – Mohr Method.* International Dairy Federation, Brussels.

IDF (2001a) IDF Standard 122:2001, *Milk and Milk Products –General Guidance for the Preparation of Test Samples, Initial Suspensions and Decimal Dilutions for Microbiological Examination.* International Dairy Federation, Brussels.

IDF (2001b) IDF Standard 93:2001, *Milk and Milk Products – Detection of* Salmonella *spp.* International Dairy Federation, Brussels.

IDF (2002a) IDF Standard 108:2002, *Milk – Determination of Freezing Point.* International Dairy Federation, Brussels.

IDF (2002b) IDF Standard 184:2002, *Milkfat – Determination of the Fatty Acid Composition by Gas–Liquid Chromatography.* International Dairy Federation, Brussels.

IDF (2002c) IDF Standard 153:2002, *Butter, Fermented Milks and Fresh Cheese – Enumeration of Contaminating Microorganisms (Colony Count Technique at 30°C).* International Dairy Federation, Brussels.

IDF (2002d) IDF Standard 60:2002, *Milk and Milk-Based Products – Detection of Coagulase-Positive Staphylococci – Most Probable Number Technique.* International Dairy Federation, Brussels.

IDF (2004a) IDF Standard 113:2004, *Milk and Milk Products – Sampling – Inspection by Attributes.* International Dairy Federation, Brussels.

IDF (2004b) IDF Standard 179:2004, *Butter – Determination of Salt Content – Potentiometric Method.* International Dairy Federation, Brussels.

IDF (2004c) IDF Standard 94:2004, *Milk and Milk Products – Enumeration of Colony Forming Units of Yeasts and/or Moulds.* International Dairy Federation, Brussels.

ISO (1982) ISO Standard 5497:1982, *Sensory Analysis. Methodology – Guidelines for the Preparation of Samples for Which Direct Sensory Analysis Is Not Feasible.* International Organization for Standardization, Geneva.

ISO (1983) ISO Standard 5495:1983, *Sensory Analysis. Methodology – Paired Comparison Test.* International Organization for Standardization, Geneva.
ISO (1985) ISO Standard 6564:1985, *Sensory Analysis. Methodology – Flavour Profile Methods.* International Organization for Standardization, Geneva.
ISO (1987) ISO Standard 8588:1987, *Sensory Analysis. Methodology – 'A' – 'Not A' Test.* International Organization for Standardization, Geneva.
ISO (1988) ISO Standard 8587:1988, *Sensory Analysis. Methodology – Ranking.* International Organization for Standardization, Geneva.
ISO (1991) ISO Standard 3972:1991, *Sensory Analysis. Methodology – Method of Investigating Sensitivity of Taste.* International Organization for Standardization, Geneva.
ISO (1992a) ISO Standard 5496:1992, *Sensory Analysis. Methodology – Initiation and Training of Assessors in the Detection and Recognition of Odours.* International Organization for Standardization, Geneva.
ISO (1992b) ISO Standard 5492:1992, *Sensory Analysis. Vocabulary*. International Organization for Standardization, Geneva.
ISO (1993) ISO Standard 8586–1:1993, *Sensory Analysis. General Guidance for the Selection, Training and Monitoring of Assessors. Part 1: Selected Assessors.* International Organization for Standardization, Geneva.
ISO (1994a) ISO Standard 8586–2:1994, *Sensory Analysis. General Guidance for the Selection, Training and Monitoring of Assessors. Part 2: Experts.* International Organization for Standardization, Geneva.
ISO (1994b) ISO Standard 11035:1994, *Sensory Analysis. Identification and Selection of Descriptors for Establishing a Sensory Profile by a Multidimensional Approach.* International Organization for Standardization, Geneva.
ISO (1994c) ISO Standard 11036:1994, *Sensory Analysis. Methodology – Texture Profile.* International Organization for Standardization, Geneva.
ISO (1998a) ISO Standard 11290:1998, *Microbiology of Food and Animal Feeding Stuffs – Horizontal Method for the Detection and Enumeration of Listeria monocytogenes*. International Organization of Standardization, Geneva.
ISO (1998b) ISO Standard 8589:1998, *Sensory Analysis. General Guidance for the Design of Test Rooms.* International Organization for Standardization, Geneva.
ISO (1999a) ISO Standard 17025:1999, *General Requirements for the Competence of Testing and Calibration Laboratories*. International Organization of Standardization, Geneva.
ISO (1999b) ISO Standard 11037:1999, *Sensory Analysis. General Guidance and Test Method for Assessment of the Colour of Foods*. International Organization for Standardization, Geneva.
ISO (1999c) ISO Standard 11056:1999, *Sensory Analysis. Methodology – Magnitude Estimation Method*. International Organization for Standardization, Geneva.
ISO (2002) ISO Standard 13301:2002, *Sensory Analysis. Methodology – General Guidance for Measuring Odour, Flavour and Taste Detection Thresholds by a Three-Alternative Forced-Choice (3–AFC) procedure.* International Organization for Standardization, Geneva.
ISO (2003a) ISO Standard 18330:2003, *Milk and Milk Products – Guidelines for the Standardized Description of Immunoassays or Receptor Assays for the Detection of Antimicrobial Residues.* International Organization for Standardization, Geneva.
ISO (2003b) ISO Standard 13302:2003, *Sensory Analysis. Methods for Assessing Modifications to the Flavour of Foodstuffs Due to Packaging.* International Organization for Standardization, Geneva.
ISO (2003c) ISO Standard 13299:2003, *Sensory Analysis. Methodology – General Guidance for Establishing a Sensory Profile.* International Organization for Standardization, Geneva.
ISO (2003d), ISO Standard 4121:2003, *Sensory Analysis. Guidelines for the Use of Quantitative Response Scales.* International Organization for Standardization, Geneva.
ISO (2004a) ISO Standard 10399:2004, *Sensory Analysis. Methodology – Duo Trio Test.* International Organization for Standardization, Geneva.
ISO (2004b) ISO Standard 4120:2004, *Sensory Analysis. Methodology – Triangle Test.* International Organization for Standardization, Geneva.

ISO (2004c) ISO Standard 16820:2004, *Sensory Analysis. Methodology – Sequential Analysis.* International Organization for Standardization, Geneva.
ISO (2005) ISO Standard 6658:2005, *Sensory Analysis. Methodology – General Guidance.* International Organization for Standardization, Geneva.
Istafanos, P., James, L. and Hunt, J. (2002) Comparison of visual immunoassay and chromogenic culture medium for the presence of *Listeria* spp. in foods. *Journal of AOAC International* **85**, 1201–1203.
Iverson, J.L. and Sheppard, A.J. (1989) Detection of adulteration in cow, goat, and sheep cheeses utilizing gas–liquid chromatographic fatty acid data. *Journal of Dairy Science* **72**, 1707–1712.
Jaworska, D., Waszkiewicz-Robak, B., Kolanowski, W. and Swiderski, F. (2005) Relative importance of texture properties in the sensory quality and acceptance of natural yoghurts. *International Journal of Dairy Technology* **58**, 39–46.
Jensen, M.A., Webster, J.A. and Straus, N. (1993) Rapid identification of bacteria on the basis of polymerase chain reaction-amplified ribosomal DNA spacer polymorphisms. *Applied and Environmental Microbiology* **59**, 945–952.
Jung, Y.S., Frank, J.F. and Brackett, R.E. (2003) Evaluation of antibodies for immunomagnetic separation combined with flow cytometry detection of *Listeria monocytogenes*. *Journal of Food Protection* **66**, 1283–1287.
Kaminarides, S.E. and Koukiassa, P. (2002) Detection of bovine milk in ovine yoghurt by electrophoresis of para-κ-casein. *Food Chemistry* **78**, 53–55.
Kaminarides, S.E., Kandarakis, J.G. and Moschopoulou, E. (1995) Detection of bovine milk in ovine Halloumi cheese by electrophoresis of α_{s1}-casein. *The Australian Journal of Dairy Technology* **50**, 58–61.
Kamm, W., Dionisi, F., Hischenhuber, C., Schmarr, H.-G. and Engel, K.-H. (2002) Rapid detection of vegetable oils in milk fat by on-line LC-GC analysis of β-sitosterol as marker. *European Journal of Lipid Science and Technology* **104**, 756–761.
Kang, J.-H. and Kondo, F. (2001) Occurrence of false-positive results of inhibitor on milk samples using the Delvotest SP assay. *Journal of Food Protection* **64**, 1211–1215.
Kang, J.H., Jin, J.H. and Kondo, F. (2005) False-positive outcome and drug residue in milk samples over withdrawal times. *Journal of Dairy Science* **88**, 908–913.
Karwoski, M. (1996) Automated direct and indirect methods in food microbiology: a literature review. *Food Reviews International* **12**, 155–174.
Katz, S.E. and Siewierski, M. (1995) *Bacillus stearothermophilus* disc assay: a review. *Journal of AOAC International* **78**, 1408–1415.
Keer, J.T. and Birch, L. (2003) Molecular methods for the assessment of bacterial viability. *Journal of Microbiological Methods* **53**, 175–183.
Kennedy, D.G., McCracken, R.J., Cannavan, A. and Hewitt, S.A. (1998) Use of liquid chromatography-mass spectrometry in the analysis of residues of antibiotics in meat and milk. *Journal of Chromatography A* **812**, 77–98.
Kenner, B.A., Clark, H.F. & Kabler, P.W. (1961) Fecal Streptococci. I. Cultivation and enumeration of streptococci in surface waters. *Applied Microbiology* **9**, 15–20.
Kerdahi, K.F. and Istafanos, P.F. (1997) Comparative study of colometric and fully automated enzyme-linked immunoassay system for rapid screening of *Listeria* spp. in foods. *Journal of AOAC International* **80**, 1139–1142.
Kirk, R.S. and Sawyer, R. (1991) *Pearson's Composition and Analysis of Foods*, 9th edn. Longman Scientific and Technical, Harlow, UK.
Krusa, M., Torre, M. and Marina, M.L. (2000) A reversed-phase high-performance liquid chromatographic method for the determination of soya bean proteins in bovine milks. *Analytical Chemistry* **72**, 1814–1818.
Lachowsky, W.M., McNab, W.B., Griffiths, M. and Odumeru, J. (1997) A comparison of the Bactoscan 8000S to the three cultural methods for enumeration of bacteria in raw milk. *Food Research International* **30**, 273–280.

Lara-Villoslada, F., Olivares, M. and Xaus, J. (2005) The balance between caseins and whey proteins in cow's milk determines its allergenicity. *Journal of Dairy Science* **88**, 1654–1660.

Lawless, H.T. and Claassen, M.R. (1993) Validity of descriptive and defect-oriented terminology systems for sensory analysis of fluid milk. *Journal of Food Science* **58**, 108–112, 119.

Lawless, H.T. and Heymann, H. (1999) *Sensory Evaluation of Food: Principles and Practices*. Aspen, Gaithersburg, MD.

Lefier, D. (1996) Analytical methods for the determination of the urea content in milk. *Bulletin of the IDF* **315**, 35–38.

Levieux, D. and Venien, A. (1994) Rapid, sensitive two-site ELISA for detection of cows' milk in goats' or ewes' milk using monoclonal antibodies. *Journal of Dairy Research* **61**, 91–99.

Lindeberg, J. (1996) Capillary electrophoresis in food analysis. *Food Chemistry* **55**, 73–94.

Litopoulou-Tzanetaki, E. (1990) Changes in the number of lactic acid bacteria during ripening of Kefalotyri cheese. *Journal of Food Science* **55**, 111–113.

Litopoulou-Tzanetaki, E., Tzanetakis, N. and Vafopoulou-Mastrojiannaki, A. (1993) Effect of type of lactic starter on microbiological, chemical and sensory characteristics of Feta cheese. *Food Microbiology* **10**, 31–41.

Lockley, A.K. and Bardsley, R.G. (2000) DNA-based methods for food authentication. *Trends in Food Science and Technology* **11**, 67–77.

Loomans E.E.M.G., van Wiltenburg, J., Koets, M. and van Amerongen, A. (2003) Neamin as an immunogen for the development of a generic ELISA detecting gentamicin, kanamycin, and neomycin in milk. *Journal of Agricultural and Food Chemistry* **51**, 587–593.

Lopez-Calleja, I., Gonzalez, I., Fajardo, V., Rodriguez, M.A., Hernandez, P.E., Garcia, T. and Martin, R. (2004) Rapid detection of cows' milk in sheeps' and goats' milk by a species-specific polymerase chain reaction technique. *Journal of Dairy Science* **87**, 2839–2845.

Lopez-Calleja, I., Gonzalez, I., Fajardo, V., Martin, I., Hernandez, P.E., Garcia, T. and Martin, R. (2005a) Application of polymerase chain reaction to detect adulteration of sheep's milk with goats' milk. *Journal of Dairy Science* **88**, 3115–3120.

Lopez-Calleja, I., Gonzalez Alonso, I., Fajardo, V., Rodriguez, M.A., Hernandez, P.E., Garcia, T. and Martin, R. (2005b) PCR detection of cows' milk in water buffalo milk and mozzarella cheese. *International Dairy Journal* **15**, 1122–1129.

López-Diaz, T.M., Santos, J.A., González, C.J., Moreno, B. and García, M.L. (1995) Bacteriological quality of traditional Spanish blue cheese. *Milchwissenschaft* **50**, 503–505.

Louie, M., Jayaratne, P., Luchsinger, I., Devenish, J., Yao, J., Schlech, W. and Simor, A. (1996) Comparison of ribotyping, arbitary primed PCR, and pulsed field gel electrophoresis for molecular typing of *Listeria monocytogenes*. *Journal of Clinical Microbiology* **34**, 15–19.

Lübeck, P.S. and Hoorfar, J. (2003) PCR technology and applications to zoonotic food-borne bacterial pathogens. In *Methods in Molecular Biology: PCR Detection of Microbial Pathogens: Methods and Protocols,* Sachse, K. and Frey, J. (eds), pp. 65–84. Humana Press, Totowa, NJ.

Lück, H. (1991) Dye reduction tests. *Bulletin of the IDF* **256**, 31–34.

Lukinmaa, S., Nakari, U.-M., Eklund, M. and Siitonen, A. (2004) Application of molecular genetic methods in diagnostics and epidemiology of food-borne pathogens. *APMIS* **112**, 908–929.

Mabrook, M.F. and Petty, M.C. (2003) A novel technique for the detection of added water to full fat milk using single frequency admittance measurements. *Sensors and Actuators B* **96**, 215–218.

Macaulay, D.M. and Packard, V.S. (1981) Evaluation of methods used to detect antibiotic residues in milk. *Journal of Food Protection* **44**, 696–698.

Mackay, I.M. (2004) Real-time PCR in the microbiology laboratory. A review. *Clinical Microbiology and Infection* **10**, 190–212.

Mafra, I., Ferreira, I.M.P.L.V.O., Faria, M.A. and Oliveira, B.P.P. (2004) A novel approach to the quantitation of bovine milk in ovine cheeses using a duplex polymerase chain reaction method. *Journal of Agricultural and Food Chemistry* **52**, 4943–4947.

Malin, E.L., Basch, J.J., Shieh, J.J., Sullivan, B.C. and Holsinger, V.H. (1994) Detection of adulteration of buttermilk powder by gel electrophoresis. *Journal of Dairy Science* **77**, 2199–2206.

Mansfield, L.P. and Forsythe, S.J. (1993) Immunomagnetic separation as an alternative to enrichment broths for *Salmonella* detection. *Letters in Applied Microbiology* **16**, 122–125.

Manso, M.A., Cattaneo, T.M., Barzaghi, S., Olieman, C. and Lopez-Fandino, R. (2002) Determination of vegetal proteins in milk powder by sodium dodecyl sulfate-capillary gel electrophoresis: interlaboratory study. *Journal of AOAC International* **85**, 1090–1095.

Marazuela, M.D. and Moreno-Bondi, M.C. (2004) Multiresidue determination of fluoroquinolones in milk by column liquid chromatography with fluorescence and ultraviolet absorbance detection. *Journal of Chromatography A* **1034**, 25–32.

March, S.B. and Ratnam, S. (1989) Latex agglutination test for detection of *Escherichia coli* serotype O157. *Journal of Clinical Microbiology* **27**, 1675–1677.

Martlbauer, E., Usleber, E., Schneider, E. and Dietrich, R. (1994) Immunochemical detection of antibiotics and sulfonamides. *Analyst* **119**, 2543–2548.

Maudet, C. and Taberlet, P. (2001) Detection of cows' milk in goats' cheeses inferred from mitochondrial DNA polymorphism. *Journal of Dairy Research* **68**, 229–235.

Mayer, H.K., Heidler, D. and Rockenbauer, C. (1997) Determination of the percentages of cows', ewes' and goats' milk in cheese by isoelectric focusing and cation-exchange HPLC of γ- and para-κ-caseins. *International Dairy Journal* **7**, 619–628.

Mayer, H.K. (2005) Milk species identification in cheese varieties using electrophoretic, chromatographic and PCR techniques. *International Dairy Journal* **15**, 595–604.

Meilgaard, M., Civille, G.V. and Carr, T. (1999) *Sensory Evaluation Techniques*, 3rd edn. CRC Press, New York.

Merlino, J., Funnell, G.R., Parkinson, D.L., Siarakas, S. and Veal, D. (1998) Enzymatic chromogenic identification and differentiation of enterococci. *Australian Journal of Medical Science* **19**, 76–81.

Miettinen, M.K., Siitonen, A., Heiskanen, P., Haajanen, H., Björkroth, K.J. and Korkeala, H.J. (1999) Molecular epidemiology of an outbreak of febrile gastroenteritidis caused by *Listeria monocytogenes* in cold-smoked rainbow trout. *Journal of Clinical Microbiology* **37**, 2358–2360.

Mitchell, J.M., Griffiths, M.W., McEwen, S.A., McNab, W.B. and Yee, A.J. (1998) Antimicrobial drug residues in milk and meat: causes, concerns, prevalence, regulations, tests, and test performance. *Journal of Food Protection* **61**, 742–756.

Moats, W.A. and Romanowski, R.D. (1998) Multiresidue determination of β-lactam antibiotics in milk and tissues with the aid of high-performance liquid chromatographic fractionation for clean up. *Journal of Chromatography A* **812**, 237–247.

Moats, W.A. (1999) Confirmatory test results on milk from commercial sources that tested positive by β-lactam antibiotic screening tests. *Journal of AOAC International* **82**, 1071–1076.

Moatsou, G. and Anifantakis, E. (2003) Recent developments in antibody-based analytical methods for the differentiation of milk from different species. *International Journal of Dairy Technology* **56**, 133–138.

Moatsou, G., Hatzinaki, A., Psathas, G. and Anifantakis, E. (2004) Detection of caprine casein in ovine Haloumi cheese. *International Dairy Journal* **14**, 219–226.

Molina, E., Fernandez-Fournier, A., De Frutos, M. and Ramos, M. (1996) Western blotting of native and denatured bovine β-lactoglobulin to detect addition of bovine milk in cheese. *Journal of Dairy Science* **79**, 191–197.

Molina, M.P., Althaus, R.L., Balasch, S., Torres, A., Peris, C. and Fernandez, N. (2003a) Evaluation of screening test for detection of antimicrobial residues in ewe milk. *Journal of Dairy Science* **86**, 1947–1952.

Molina, M.P., Althaus, R.L., Molina, A. and Fernandez, N. (2003b) Antimicrobial agent detection in ewes' milk by the microbial inhibitor test brilliant black reduction test-BRT AiM®. *International Dairy Journal* **13**, 821–826.

Molkentin, J. and Precht, D. (1998) Comparison of gas-chromatographic methods for analysis of butyric acid in milk fat and fats containing milk fat. *Zeitschrift für Lebensmitteluntersuchung und -Forschung A* **206**, 213–216.

Molkentin, J. and Precht, D. (2000) Validation of a gas-chromatographic method for the determination of milk fat contents in mixed fats by butyric acid analysis. *European Journal of Lipid Science and Technology* **102**, 194–201.

Monfort, P., Le Gal, D., Le Saux, J.C., Piclet, G., Raquenes, P., Boulben, S. and Plusquellec, A. (1994) Improved rapid method for isolation and enumeration of salmonella from bivalves using Rambach agar. *Journal of Microbiological Methods* **19**, 67–79.

Mostert, J.F. and Jooste, P.J. (2002) Quality control in the dairy industry. In *Dairy Microbiology Handbook*, R.K. Robinson (Ed.), 3rd edn, pp. 655–736. John Wiley, New York.

MSU (2005) Ribosomal Database Base, Michigan State University. http://rdp.cme.msu.edu (accessed 22 August 2005).

Murray, J.M. and Delahunty, C.M. (2000) Selection of standards to reference terms in a cheddar-type cheese flavor language. *Journal of Sensory Studies* **15**, 179–199.

Murray, J.M., Delahunty, C.M. and Baxter, I.A. (2001) Descriptive sensory analysis: past, present and future. *Food Research International* **34**, 461–471.

Negroni, L., Bernard, H., Clement, G., Chatel, J.M., Brune, P., Frobert, Y., Wal, L.M. and Grassi, J. (1998) Two-site enzyme immunometric assays for determination of native and denatured β-lactoglobulin. *Journal of Immunological Methods* **220**, 25–37.

Nielsen, R.G. and Zannoni, M. (1998) Progress in developing an international protocol for sensory profiling of hard cheese. *International Journal of Dairy Technology* **51**, 57–64.

Nierman, P. (2004) Methods to detect abnormal milk. In *Standard Methods for the Examination of Dairy Products,* Wehr, H.M. and Frank, J.F. (eds), 17th edn, pp. 281–292. American Public Health Association, Washington, DC.

Niessen, W.M.A. (1998) Analysis of antibiotics by liquid chromatography-mass spectrometry. *Journal of Chromatography A* **812**, 53–75.

Nouws, J., van Egmond, H., Smulders, I., Loeffen, G., Schouten, J. and Stegeman, H. (1999) A microbiological assay system for assessment of raw milk exceeding EU maximum residue levels. *International Dairy Journal* **9**, 85–90.

Nouws, J.F.M., Loeffen, G., Schouten, J., Van Egmond, H., Keukens, H. and Stegeman, H. (1998) Testing of raw milk for tetracycline residues. *Journal of Dairy Science* **81**, 2341–2345.

O'Donnell, R., Holland, J.W., Deeth, H.C. and Alewood, P. (2004) Milk proteomics. *International Dairy Journal* **14**, 1013–1023.

O'Mahony, M. (1979) Psychophysical aspects of sensory analysis of dairy products: a critique. *Journal of Dairy Science* **62**, 1954–1962.

O'Mahony, M. (1986) *Sensory Evaluation of Food: Statistical Methods and Procedures*. Marcel Dekker, New York.

O'Mahony, M. (1995) Sensory measurement in food science: fitting methods to goals. *Food Technology* **72**, 72, 74, 76–78, 80–82.

OECD (1998) OECD Principles of Good Laboratory Practice, In: *OECD Series on Principles of Good Laboratory Practice and Compliance Monitoring*, Number 1. Organization for Economic Co-operation and Development, Paris.

Oka, H., Harada, K., Ito, Y. and Ito, Y. (1998) Separation of antibiotics by counter-current chromatography. *Journal of Chromatography A* **812**, 35–52.

Olive, D.M. and Bean, P. (1999) Principles and applications of methods for DNA-based typing of microbial organisms. *Journal of Clinical Microbiology* **37**, 1661–1669.

Olsen, J.E. (2000) DNA-based methods for detection of food-borne bacterial pathogens. *Food Research International* **33**, 257–266.

Olsen, J.E., Aabo, S. and Hill, W. (1995) Probes and polymerase chain reaction for detection of food-borne bacterial pathogens. *International Journal of Food Microbiology* **28**, 1–78.

Ordonez, A.I., Ibanez, F.C., Torre, P. and Barcina, Y. (1998) Application of multivariate analysis to sensory characterization of ewes' milk cheese. *Journal of Sensory Studies* **13**, 45–55.

Owens, J.D., Thomas, D.S., Thompson, P.S. and Timmerman, J.W. (1989) Indirect conductimetry: a novel approach to the conductimetric enumeration of microbial populations. *Letters in Applied Microbiology* **9**, 245–249.

Papademas, P. and Bintsis, T. (2002) Microbiology of ice cream and related products, In *Dairy Microbiology Handbook*, R.K. Robinson (ed.), 3rd edn, pp. 213–260. John Wiley, New York.

Park, E.E. and Szabo, R. (1986) Evaluation of the reverse passive latex agglutination (RPLA) test kit for detection of staphylococcal enterotoxins A, B, C, and D in food. *Canadian Journal of Microbiology* **32**, 723–727.

Patterson, J.E. and Kelly, C.C. (1998) Pulsed-field gel electrophoresis as an epidemiologic tool for enterococci and streptococci. *Methods in Cell Science* **20**, 233–239.

Perez, N., Gutierrez, R., Noa, M., Diaz, G., Luna, H., Escobar, I. and Munive, Z. (2002) Liquid chromatographic determination of multiple sulfonamides, nitrofurans, and chloramphenicol residues in pasteurized milk. *Journal of AOAC International* **85**, 20–24.

Peterkin, P.I. and Sharpe, A.N. (1980) Membrane filtration of dairy products for microbiological analysis. *Applied and Environmental Microbiology* **39**, 1138–1143.

Peterkin, P.I., Idziak, E.S. and Sharpe, A.N. (1992) Use of hydrophobic grid-membrane filter DNA probe method to detect *Listeria monocytogenes* in artificially-contaminated foods. *Food Microbiology* **9**, 155–160.

Pettipher, G.L., Mansell, R., McKinnon, C.H. and Cousins, C.M. (1980) Rapid membrane filtration. Epifluorescent microscope technique for direct enumeration of bacteria in raw milk. *Applied and Environmental Microbiology* **39**, 423–429.

Phillips, L.G., McGiff, M.L., Barbano, D.M. and Lawless, H.T. (1995) The influence of fat on the sensory properties, viscosity, and color of lowfat milk. *Journal of Dairy Science* **78**, 1258–1266.

Piggott, J.R., Simpson, S.J. and Williams, S.A.R. (1998) Sensory analysis. *International Journal of Food Science and Technology* **33**, 7–18.

Piknova, L., Štefanoviča, A., Drahovská, H., Sásik, M. and Kuchta, T. (2002) Detection of *Salmonella* in food, equivalent to ISO 6579, by a three-days polymerase chain reaction-based method. *Food Control* **13**, 191–194.

Pinho, O., Mendes, E., Alves, M.M. and Ferreira, I.M.P.L.V.O. (2004) Chemical, physical, and sensorial characteristics of 'Terrincho' ewe cheese: changes during ripening and intravarietal comparison. *Journal of Dairy Science* **87**, 249–257.

Plath, A., Krause, I. and Einspanier, R. (1997) Species identification in dairy products by three different DNA-based techniques. *Zeitschrift für Lebensmitteluntersuchung und -Forschung A* **205**, 437–441.

Prager, M.J. (1989) Differential characteristics of fatty acids in cheese from milk of various animal species by capillary gas chromatography. *Journal of AOAC International* **72**, 418–421.

Prentice, G.A. and Landridge, E.W. (1992) Laboratory control in milk product manufacture. In *The Technology of Dairy Products,* R. Early (ed.), pp. 247–271. Blackie and Sons, London.

Quinones, H.J., Barbano, D.M. and Philips, L.G. (1998) Influence of protein standardization by ultrafiltration on the viscocity, color and sensory properties of 2 and 3.3% milks. *Journal of Dairy Science* **81**, 884–894.

Rådström, P., Knutsson, R., Wolffs, P., Dahlenborg, M. and Löfström C. (2003) Pre-PCR processing of food samples. In *Methods in Molecular Biology: PCR Detection of Microbial Pathogens: Methods and Protocols,* Sachse, K. and Frey, J. (eds), pp. 31–50. Humana Press, Totowa, NJ.

Rambach, A. (1990) New plate medium for facilitated differentiation of *Salmonella* spp. from *Proteus* spp. and other enteric bacteria. *Applied and Environmental Microbiology* **56**, 301–303.

Ramirez, A., Gutierrez, R., Diaz, G., Gonzalez, C., Perez, N., Vega, S. and Noa, M. (2003) High-performance thin-layer chromatography-bioautography for multiple antibiotic residues in cow's milk. *Journal of Chromatography B* **784**, 315–322.

Ramos, M. and Juarez, M. (1986) Chromatographic, electrophoretic and immunological methods for detecting mixtures of milk from different species. *Bulletin of the IDF* **202**, 175–187.

Rea, S., Chikuni, K., Branciari, R., Sancamayya, R.S., Ranucci, D. and Avellini, P. (2001) Use of duplex polymerase chain reaction (duplex-PCR) technique to identify bovine and water buffalo milk used in making mozzarella cheese. *Journal of Dairy Research* **68**, 689–698.

Recio, I., Amigo, L. and Lopez-Fandino, R. (1997a) Assessment of the quality of dairy products by capillary electrophoresis of milk proteins. *Journal of Chromatography B* **697**, 231–242.

Recio, I., Perez-Rodriguez, M.L., Ramos, M. and Amigo, L. (1997b) Capillary electrophoretic analysis of genetic variants of milk proteins from different species. *Journal of Chromatography A* **768**, 47–56.

Recio, I., Garcia-Risco, M.R., Lopez-Fandino, R., Olano, A. and Ramos, M. (2000) Detection of rennet whey solids in UHT milk by capillary electrophoresis. *International Dairy Journal* **10**, 333–338.

Recio, I., Garcia-Risco, M.R., Amigo, L., Molina, E., Ramos, M. and Martin-Alvares, P.J. (2004) Detection of milk mixtures in Haloumi cheese. *Journal of Dairy Science* **87**, 1595–1600.

Rehman, S-U., Banks, J.M., Brechany, E.Y., Muir, D.D., McSweeney, P.L.H. and Fox, P.F. (2000) Influence of ripening temperature on the volatiles profile and flavour of cheddar cheese made from raw or pasteurized milk. *International Dairy Journal*, **10**, 55–65.

Retiveau, A., Chambers, D.H. and Esteve, E. (2005) Developing a lexicon for the flavor description of French cheeses. *Food Quality and Preference* **16**, 517–527.

Richter, W., Krause, I., Graf, C., Sperrer, I., Schwarzer, C. and Klostermeyer, H. (1997) An indirect competitive ELISA for the detection of cow's milk and caseinate in goats' and ewes' milk and cheese using polyclonal antibodies against bovine γ-caseins. *Zeitschrift für Lebensmitteluntersuchung und -Forschung A* **204**, 21–26.

Riediker, S. and Stadler, R.H. (2001) Simultaneous determination of five β-lactam antibiotics in bovine milk using liquid chromatography coupled with electrospray ionization tandem mass spectrometry. *Analytical Chemistry* **73**, 1614–1621.

Riediker, S., Diserens, J-M. and Stadler, R.H. (2001) Analysis of β-lactam antibiotics in incurred raw milk by rapid tests methods and liquid chromatography coupled with electrospray ionization tandem mass spectrometry. *Journal of Agricultural and Food Chemistry* **49**, 4171–4176.

Rijpens N., Herman, L., Vereecken, F., Jannes, G., De Smedt, J. and De Zutter, L. (1999) Rapid detection of stressed *Salmonella* spp. in dairy and egg products using immunomagnetic separation and PCR. *International Journal of Food Microbiology* **46**, 37–44.

Rodriquez-Lazaro, D., Lloyd, J., Herrewegh, A., Ikonomopoulos, J., D'Agostino, M., Pla, M., and Cook, N. (2004) A molecular beacon-based real-time NASBA assay for detection of *Mycobacterium avium* subsp. *paratuberculosis* in water and milk. *FEMS Microbiology Letters* **237**, 119–126.

Roland, A.M., Phillips, L.G. and Boor, K. (1999) Effects of fat content on the sensory properties, melting, color, and hardness of ice cream. *Journal of Dairy Science* **82**, 32–38.

Ronald, M.P., Bitri, L. and Besancon, P. (1993) Polyclonal antibodies with predetermined specificity against bovine α_{s1}-casein: application to the detection of bovine milk in ovine milk and cheese. *Journal of Dairy Research*, **60**, 413–420.

Ruegg, P.L. and Tabone, T.J. (2000) The relationship between antibiotic residue violations and somatic cell counts in Wisconsin dairy herds. *Journal of Dairy Science* **83**, 2805–2809.

Russel, S.M., Cox, N.A., Bailey, J.S. and Fung, D.Y.C. (1997) Miniaturized biochemical procedures for identification of bacteria. *Journal of Rapid Methods and Automation of Microbiology* **5**, 169–178.

Samkutty, P.J., Gough, R.H., Adkinson, R.W., McGrew, P. (2001) Rapid assessment of the bacteriological quality of raw milk using ATP. *Journal of Food Protection* **64**, 208–212.

Sanchez, L., Perez, M.D., Puyol, P. and Calvo, M. (2002) Determination of vegetal proteins in milk powder by enzyme-linked immunosorbent assay: interlaboratory study. *Journal of AOAC International* **85**, 1390–1397.

Sarantinopoulos, P., Andrighetto, Ch., Georgalaki, M.D., Rea, M.C., Lombardi, A., Cogan, T.M., Kalantzopoulos, G. and Tsakalidou, E. (2001) Biochemical properties of enterococci relevant to their technological performance. *International Dairy Journal*, **11**, 621–647.
Sarantinopoulos, P., Kalantzopoulos, G. and Tsakalidou, E. (2002). Effect of *Enterococcus faecium* on microbiological, physicochemical and sensory characteristics of Greek Feta cheese. *International Journal of Food Microbiology* **76**, 93–105.
Sato, T., Kawano, S. and Iwamoto, M. (1990) Detection of foreign fat adulteration of milk fat by near infrared spectroscopic method. *Journal of Dairy Science* **73**, 3408–3413.
Saville, W.J.A., Wittum, T.E. and Smith, K.L. (2000) Association between measures of milk quality and risk of violative antimicrobial residues in grade-A raw milk. *Journal of the American Veterinary Medical Association* **217**, 541–545.
Schalm, O.W. and Noorlander, D.O. (1957) Experiments and observations leading to development of the California mastitis test. *Journal of the American Veterinary Association* **130**, 199–207.
Schenck, F.J. and Callery, P.S. (1998) Chromatographic methods of analysis of antibiotics in milk. *Journal of Chromatography A* **812**, 99–109.
Schmidt, M.A.E., Radovic, B.S., Lipp, M., Harzer, G. and Anklam, E. (1999) Characterization of milk samples with various protein contents by pyrolysis-mass spectrometry (Py-MS). *Food Chemistry* **65**, 123–128.
Schwartz, D.C. and Cantor, C.R. (1984) Separation of yeast artificial chromosome-sized DNAs by pulsed field gradient gel electophoresis. *Cell* **37**, 6–75.
Senyk, G.F., Davidson, J.H., Brown, J.M., Hallstead, E.R. and Sherbon, J.W. (1990). Comparison of rapid tests used to detect antibiotic residues in milk. *Journal of Food Protection* **53**, 158–164.
Shaikh, B. and Moats, W.A. (1993) Liquid chromatographic analysis of antibacterial drug residues in food products of animal origin. *Journal of Chromatography A* **643**, 369–378.
Sharpe, A.N. and Michaud, G.L. (1974) Hydrophobic grid-membranes filters: new approach to microbiological enumeration. *Applied Microbiology* **28**, 223–225.
Sharpe, A.N., Diotte, M.P., Dudas, I., Malcolm, S. and Peterkin, P.I. (1983) Colony counting on hydrophobic grid-membrane filters. *Canadian Journal of Microbiology* **29**, 797–802.
Sheppard, A.J., Shen, C.S. and Rudolf, T.S. (1985) Detection of vegetable oil adulteration in ice cream. *Journal of Dairy Science* **68**, 1103–1108.
Siciliano, R., Rega, B., Amoresano, A. and Pucci, P. (2000) Modern mass spectrometric methodologies in monitoring milk quality. *Analytical Chemistry* **72**, 408–415.
Sidel, J.L. and Stone, H. (1993) The role of sensory evaluation in the food industry. *Food Quality and Preference* **4**, 65–73.
Simonne, A.H., Simonne, E.H., Eitenmiller, R.R., Mills, H.A. and Cresman III, C.P. (1997) Could the Dumas method replace the Kjeldahl digestion for nitrogen and crude protein determinations in foods. *Journal of the Science of Food and Agriculture* **73**, 39–45.
Sischo, W.M. and Burns, C.M. (1993) Field trial of four cowside antibiotic residue screening tests. *Journal of the American Veterinary Medical Association* **202**, 1249–1254.
Sischo, W.M. (1996) Quality milk and tests for antibiotic residues. *Journal of Dairy Science* **79**, 1065–1073.
Sivakesava, S. and Irudayaraj, J. (2002) Rapid determination of tetracycline in milk by FT-MIR and FT-NIR spectroscopy. *Journal of Dairy Science* **85**, 487–493.
Skjerve, E., Rorvik, E.L.M. and Olsvik, O. (1990) Detection of *Listeria monocytogenes* in foods by immunomagnetic separation. *Applied Environmental Microbiology* **56**, 3478–3481.
Smith, E.M., Mason, D.J., Medley, G.F., Dow, C.S. and Green, L.E. (2001) Preliminary results of a method for the rapid detection of *Staphylococcus aureus* in milk. In *Proceedings of the British mastitis Conference*, p. 100. Institute for Animal Health/Milk Development Council, Garstang, UK.
Smittle, R.B. and Okrend, A. (2001) Laboratory Quality Assurance. In *Compendium of Methods for the Examination of Foods*, 4th edn, Downes, F.P. and Ito, K. (eds), pp. 1–11. American Public Health Association, Washington, DC.

Söderhjelm, P. and Lindqvist, B. (1980) The ammonia content of milk as an indicator of its biological deterioration or ageing. *Milchwissenschaft* **35**, 541–543.

Sørensen, H., Sørensen, S., Bjergegaard, C. and Michaelsen, S. (1998) High performance capillary electrophoresis. In *Chromatography and Capillary Electrophoresis in Food Analysis*, Belton, P.S. (ed.), pp. 208–277. Royal Society of Chemistry, London.

Sørensen, L.K., Elbaek, T.H. and Hansen, H. (2003) Determination of chloramphenicol in bovine milk by liquid chromatography/tandem mass spectrometry. *Journal of AOAC International* **86**, 703–706.

Sørensen, L.K. (2004) Use of routine analytical methods for controlling compliance of milk and milk products with compositional requirements. *Bulletin of the IDF* **390**, 42–49.

Stanley, P.E. (1989) A concise beginner's guide to rapid microbiology using adenosine triphosphate (ATP) and bioluminescence. In *ATP Luminescence: Rapid Methods in Microbiology*, Stanley, P.E., McCarthy, B.J. and Smither, R. (eds), pp. 1–7. Blackwell Scientific, Oxford.

Stead, D.A. (2000) Current methodologies for the analysis of aminoglycosides. *Journal of Chromatography B* **747**, 69–93.

Sternesjo, A., Mellgren, C. and Bjorck, L. (1995) Determination of sulfamethazine residues in milk by a surface resonance-based biosensor assay. *Analytical Biochemistry* **226**, 175–181.

Stevens, K.A. and Jaykus, L.-A. (2004) Direct detection of bacterial pathogens in representative dairy products using a combined bacterial concentration–PCR approach. *Journal of Applied Microbiology* **97**, 1115–1122.

Stoba-Wiley, C.M. and Readnour, R.S. (2000) Determination of tilmicosin residues in cow and sheep milk by liquid chromatography. *Journal of AOAC International* **83**, 555–562.

Stone, H., Sidel, J., Oliver, S., Woolsey, A. and Singleton, R.C. (1974) Sensory evaluation by quantitative descriptive analysis. *Food Technology* **28**, 24, 26, 28–29, 32, 34.

Stone, H. and Sidel, J.L. (2004) *Sensory Evaluation Practices*, 3rd edn. Food Science and Technology, International Series, Elsevier Academic Press, San Diego, CA.

Stone, H. and Sidel, J.L. (1993) *Sensory Evaluation Practices*, 2nd edn. Academic Press, New York.

Strange, E.D., Malin, E.L., van Hekken, D.L. and Basch, J.J. (1992) Chromatographic and electrophoretic methods used for analysis of milk proteins. *Journal of Chromatography* **624**, 81–102.

Suhren, G. and Heeschen, W. (1991) Determination of pyruvate and other metabolites. *Bulletin of the IDF* **256**, 37–41.

Suhren, G., Reichmuth, J. and Heeschen, W. (1991) Bactoscan technique. *Bulletin of the IDF* **256**, 24–30.

Suhren, G. and Knappstein, K. (2004) Detection of residues of antibiotics in milk of treated cows by screening methods. *Milchwissenschaft* **59**, 656–660.

Sundlof, S.F. (1994) Human health risks associated with drug residues in animal-derived foods. *Journal of Agromedicine* **12**, 5–20.

Sundlof, S.F., Kaneene, J.B. and Miller, R.A. (1995) National survey on veterinarian drug use in lactating dairy cows. *Journal of the American Veterinary Medical Association* **207**, 347–352.

Swaminathan, B., Harmon, M.C. and Mehlamn, I.J. (1982) *Yersinia enterocolitica*: a review. *Journal of Applied Bacteriology*, **52**, 151–183.

Takeba, K., Fujinuma, K., Miyazaki, T. and Nakazawa, H. (1998) Simultaneous determination of β-lactam antibiotics in milk by ion-pair liquid chromatography. *Journal of Chromatography A* **812**, 205–211.

te Giffel, M.C., Meeuwisse, J. and de Jong, P. (2001) Control of milk processing based on rapid detection of microorganisms. *Food Control* **12**, 305–309.

Temmerman, R., Huys, G. and Swings, J. (2004) Identification of lactic acid bacteria: culture-dependent and culture-independent methods. *Trends in Food Science* **15**, 348–359.

Tenover, F.C., Arbeit, R.D., Goering, R.V., Mickelsen, P.A., Murray, B.E., Persing, D.H. and Swaminathan, B. (1995) Interpreting chromosomal DNA restriction patterns produced

by Pulsed Field Gel Electrophoresis: criteria for bacterial subtyping. *Journal of Clinical Microbiology* **33**, 2233–2239.
Timms, R. (1980) Detection and quantification of non-milk fat in mixtures of milk and non-milk fats. *Journal of Dairy Research* **47**, 295–303.
Todd, E.C.D., Szabo, R.A., Peterkin, P., Sharpe, A.N., Parrington, L., Bundle, D., Gidney, M.A.J. and Perry, M.B. (1988) Rapid hydrophobic grid membrane filter-enzyme-labeled antibody procedure for identification and enumeration of *Escherichia coli* O157 in foods. *Applied and Environmental Microbiology* **54**, 2536–2540.
Todd E.C.D., MacKenzie, J.M. and Peterkin, P.I. (1993) Development of an enzyme-linked antibody hydrophobic grid membrane filter method for detection of *Salmonella* in foods. *Food Microbiology* **10**, 87–99.
Torre, M., Cohen, M.E., Corzo, N., Rodriguez, M.A. and Diez-Masa, J.C. (1996) Perfusion liquid chromatography of whey proteins. *Journal of Chromatography A* **729**, 99–111.
Torri Tarelli, G., Carminati, D. & Giraffa, G. (1994) Production of bacteriocins active against *Listeria monocytogenes* and *Listeria innocua* from dairy enterococci. *Food Microbiology* **11**, 243–252.
Tortorello, M., and Gendel, S.M. (1993) Fluorescent antibodies applied to direct epifluorescent filter techniques for microscopic enumeration of *Escherichia coli* O157:H7 in milk and juice. *Journal of Food Protection* **56**, 672–677.
Tothill, I.E. and Magan, N. (2003) Rapid detection methods for microbial contamination. In *Rapid and On-Line Instrumentation for Food Quality Assurance,* Tothill, I.E. (ed.), pp. 136–155. Woodhead, Cambridge.
Tremblay, L., Laporte, M.F., Leonil, J. Dupont, D. and Paquin, P. (2003) Quantitation of proteins in milk and milk products. In *Advanced Dairy Chemistry, Volume 1: Proteins,* 3rd edn, pp. 49–138. Fox, P.F. and McSweeney, P.L.H. (eds). Kluwer Academic/Plenum, London.
Tsakalidou, E., Manolopoulou, E., Tsilibari, V., Georgalaki, M. and Kalantzopoulos, G. (1993) Esterolytic activities of *Enterococcus durans* and *Enterococcus faecium* strains isolated from Greek cheese. *Netherlands Milk Dairy Journal* **47**, 145–150.
Tyler, J.W., Cullor, J.S., Erskine, R.J., Smith, W.L., Dellinger, J.D. and McClure, K. (1992) Milk antimicrobial drug residue assay results in cattle with experimental, endotoxin-induced mastitis. *Journal of the American Veterinary Medical Association* **201**, 1378–1384.
Ulberth, F. (2003) Milk and dairy products. In *Food Authenticity and Traceability*, Lees, M. (ed.), pp. 357–377. CRC Press, London.
Urmenyi, A.M.C. and Franklin, A.W. (1961) Neonatal death from pigmented coliform infection. *Lancet* **1**, 313–315.
US FDA (2003) Grade 'A' Pasteurized Milk Ordinance, Appendix N. Food and Drug Administration, Rockville, MD, U.S.A.
US FDA (2004) Center for Food Safety and Applied Nutrition. 2003 annual report of the National Milk Drug Residue Database. http://www.cfsan.fda.gov/~ear/milkrp03.html (accessed 3 November 2005).
US FDA (2005) US. Food and Drug Administration, Bacteriological Analytical Manual (BAM). http://www.cfsan.fda.gov/~ebam/bam-toc.htm (accessed 15 July 2005).
van Acker, J., De Smet, F., Muyldermans, G., Bougatef, A., Naessens, A. and Lauwers, S. (2001) Outbreak of necrotizing enterocolitis associated with *Enterobacter sakazakii* in powdered milk formula. *Journal of Clinical Microbiology* **39**, 293–297.
van Bruijnsvoort, M., Ottink, S.J.M., Jonker, K.M. and de Boer, E. (2004) Determination of streptomycin and dihydrostreptomycin in milk and honey by liquid chromatography with tandem mass spectrometry. *Journal of Chromatography A* **1058**, 137–142.
van den Bijgaart, H. (2003) Urea. New applications of mid-infra-red spectrometry for the analysis of milk and milk products. *Bulletin of the IDF* **383**, pp. 2–15.

van Eenennaam, A.L., Cullor, J.S., Perani, L., Gardner, I.A., Smith, W.L., Dellinger, J. and Guterbock, W.M. (1993) Evaluation of milk antibiotic residue screening tests in cattle with naturally occurring clinical mastitis. *Journal of Dairy Science* **76**, 3041–3053.

van Hekken, D.L. and Thompson, M.P. (1992) Application of PhastSystem to the resolution of bovine milk proteins on urea polyacrylamide gel electrophoresis. *Journal of Dairy Science* **75,** 1204–1210.

van Rhijn, J.A., Lasaroms, J.J.P., Berendsen, B.J.A. and Brinkman, U.A.T. (2002) Liquid chromatographic-tandem mass spectrometric determination of selected sulphonamides in milk. *Journal of Chromatography A* **960**, 131–133.

van Riel, J. and Olieman, C. (1995) Determination of caseinomacropeptide with capillary electrophoresis and its application to the detection and estimation of rennet why solids in milk and buttermilk powder. *Electrophoresis* **16**, 529–533.

Vasavada, P.C. (1993) Rapid methods and automation in dairy microbiology. *Journal of Dairy Science* **76**, 3101–3113.

Veledo, M.T., de Frutos, M. and Diez-Masa, J.C. (2005) Analysis of trace amounts of bovine *β*-lactoglobulin in infant formulas by capillary electrophoresis with on-capillary derivatization and laser-induced fluorescence detection. *Journal of Separation Science* **28**, 941–947.

Veloso, A.C.A., Teixeira, N. and Ferreira, I.M. (2002) Separation and quantification of the major casein fractions by reverse-phase high-performance liquid chromatography and urea-polyacrylamide gel electrophoresis detection of milk adulterations. *Journal of Chromatography A* **967**, 209–218.

Veloso, A.C.A., Teixeira, N., Peres A.M., Mendonca, A. and Ferreira, I.M.P.L.V.O. (2004) Evaluation of cheese authenticity and proteolysis by HPLC and urea-polyacrylamide gel electrophoresis. *Food Chemistry* **87**, 289–295.

Vermunt, A.E.M., Franken, A.A.J. and Beumer, R.R. (1992) Isolation of *Salmonella* by immunomagnetic separation. *Journal of Applied Bacteriology* **72**, 112–118.

Vesey, G., Narai, J., Ashbolt, N., Williams, K. and Veal, D. (1994) Detection of specific microrganisms in environmental samples using flow cytometry. *Methods in Cell Biology* **42**, 489–522.

Volitaki, A.J. and Kaminarides, S.E. (2001) Detection of bovine milk in ovine Halloumi cheese by HPLC analysis of cheese caseins hydrolysed by plasmin. *Milchwissenschaft* **56**, 207–210.

Waak, E., Tham, W. and Danielsson-Tham, M.-L. (2002) Prevalence and fingerprinting of *Listeria monocytogenes* strains isolated from raw whole milk in farm bulk tanks and in dairy plant receiving tanks. *Applied and Environmental Microbiology* **68**, 3366–3370.

Wagstaffe, P.J. (1993) Metrological requirements for analytical quality assurance and the need for reference materials. In *Analytical Quality Assurance and Good Laboratory Practice in Dairy Laboratories*, Proceedings of an International Seminar, 18–20 May 1992, Sonthofen, Germany, IDF Special Issue No. 9302, pp. 41–52. International Dairy Federation, Brussels.

Wan, J., King, K., Craven, H., McAuley, C., Tan, S.E. and Coverty, M.J. (2000) Probelia™ PCR system for rapid detection of *Salmonella* in milk powder and ricotta cheese. *Letters in Applied Microbiology* **30**, 267–271.

Wawerla, M., Stolle, A., Schalch, B. and Eisgruber, H. (1999) Impedance microbiology: applications in food hygiene. *Journal of Food Protection* **62**, 1488–1496.

Webber, G., Lauwaars, M. and van Schaik, M. (2000) Inventory of IDF/ISO/AOAC international adopted methods of analysis and sampling for milk and milk products. *Bulletin of the IDF* **350**, 3–41.

Weber, C.C., Link, N., Fux, C., Zisch, A.H., Weber, W. and Fussenegger, M. (2005) Broad-spectrum protein biosensors for class-specific detection of antibiotics. *Biotechnology and Bioengineering* **89**, 9–17.

Wegmüller, B., Lüthy, J. and Candrian, U. (1993) Direct polymerase chain reaction detection of *Campylobacter jejuni* and *Campylobacter coli* in raw milk and dairy products. *Applied and Environmental Microbiology* **59**, 2161–2165.

Westad, F., Hersleth, M., Lea, P. and Martens, H. (2003) Variable selection in PCA in sensory descriptive and consumer data. *Food Quality and Preference* **14**, 463–472.

WHO (2003) *Laboratory Biosafety Manual* (2nd edn). World Health Organization, Geneva.

Yolken, R.H. and Leister, F.J. (1982) Comparison of fluorescent and colorigenic substrates for enzyme immunoassay. *Journal of Clinical Microbiology* **15**, 757–760.

Zeece, M. (1992) Capillary electrophoresis: a new analytical tool for food science. *Trends in Food Science and Technology* **3**, 6–10.

Zeng, S.S., Escobar, E.N. and Brown-Crowder, I. (1996) Evaluation of screening tests for detection of antibiotic residues in goat milk. *Small Ruminant Research* **21**, 155–156.

Zeng, S.S., Hart, S., Escobar, E.N. and Tesfai, K. (1998) Validation of antibiotic residue tests for dairy goats. *Journal of Food Protection* **61**, 344–349.

Zwald, A.G., Ruegg, P.L., Kaneene, J.B., Warnick, L.D., Wells, S.J., Fossler, C. and Halbert, L.W. (2004) Management practices and reported antimicrobial usage on conventional and organic dairy farms. *Journal of Dairy Science* **87**, 191–201.

Chapter 7

Dealing with Environmental Issues

Trevor J. Britz, Corné Lamprecht and Gunnar O. Sigge

7.1 Introduction

Environmental issues have been on the international agenda for several decades, but the success of an environmental quality scheme depends on a set of well-designed national and international laws and regulations. However, when a market is 'economically driven', it is very difficult to have such legislation accepted and even more difficult to have it implemented. The dairy industry is still generally considered to be the largest generator of food processing wastes in many countries. A typical European dairy generates approximately 500 m^3 of wastewater per day (Demirel *et al.* 2005). During the manufacture of dairy products, processing plants generate significant quantities of wastewater with relatively high organic loads on a daily basis. In addition, this industry uses large water volumes for factory cleaning and hygienic requirements. Besides environmental damage from wastewater discharges, the presence of compounds like milk solids as part of the wastewater stream represents a significant economical loss of valuable product for the processing plant.

Although the dairy industry is not commonly associated with severe environmental problems, it must continually consider its environmental impact – particularly as dairy pollutants are mainly of organic origin. In general, wastes from the dairy industry contain high concentrations of proteins, carbohydrates, lipids, nitrogen, phosphate, suspended solids; high biological oxygen demand (BOD) and chemical oxygen demand (COD); and high fats, oils and grease (FOG) contents. These all require 'specialty' treatment so as to prevent or minimize environmental problems, and this thus increases the complexity of the treatment process. Furthermore, dairy waste streams are also characterized by large variations in pH and wide fluctuations in flow rates as a result of the discontinuity in the production cycles of the different products. These all contribute to wastes of different quality and quantity which, if not treated, will lead to increased disposal costs and severe pollution problems. To enable the dairy industry to contribute to environmental management, an efficient and cost-effective wastewater management strategy is critical.

7.2 Dairy wastewaters: sources and composition

7.2.1 General composition of dairy wastewaters

Dairy factory processing wastewaters are usually a combination generated during cleaning operations, product or by-product spills or leaks resulting from equipment malfunction and operational errors, residual milk or dairy products present in tanks and piping before cleaning, condensates from evaporation processes as well as pressings and brines from cheese factories (Britz *et al.* 2004; Wendorff 2001).

In terms of biodegradability, dairy wastewaters are generally considered to be complex substrates since the main contributors to their organic loads are easily degradable carbohydrates (mainly lactose) as well as less biodegradable proteins and lipids (Demirel *et al.* 2005). Although volatile fatty acids are absent in whole milk, the presence of especially acetate and propionate in dairy wastewaters has been observed and is probably due to microbial fermentation either during cheese production or in wastewater lines (Danalewich *et al.* 1998). In addition, dairy industry wastewaters typically exhibit extreme pH variations, which are mainly the result of the general use of acid and alkaline cleaners and sanitizers during Cleaning-In-Place (CIP) operations (Danalewich *et al.* 1998). Alkaline cleaners in particular can also result in high sodium concentrations in dairy industry wastewaters (Demirel *et al.* 2005). Wastewater flow rates of different factories are also highly variable, depending on the type of product (many factories produce a variety of dairy products), the time of day (more wastewater will be produced during cleaning operations) as well as the season (the amount of milk received for processing will typically be higher in summer than in winter) (Danalewich *et al.* 1998; Demirel *et al.* 2005)

In general, the proportions of carbohydrates, proteins and FOG in dairy wastewaters can be very different depending on the dairy products produced by a factory as well as the methods of operation (Vidal *et al.* 2000). Some of the reported COD values for typical dairy products are listed in Table 7.1. Wastage of any of these products could increase the strength of the dairy factory wastewater. The organic load of the wastewater to be treated can be greatly reduced by implementing waste minimization strategies during production. Before waste-prevention plans can be implemented, however, the specific wastewater sources during production need to be identified.

7.2.2 Milk reception and storage areas

This type of wastage could be the result of overflows, inadequate hose connections, pipeline leaks as well as inadequate drainage of milk transport trucks and storage

Table 7.1 Chemical-oxygen-demand (COD) values for dairy products and domestic sewage.

Dairy product	COD (mg l^{-1})	Reference
Milk	147,000	Wendorff *et al.* (1997)
	150,000	Danalewich *et al.* (1998)
	190,000	Odlum (1990)
	210,000	Steffen *et al.* (1989
Skim milk	100,000	Steffen *et al.* (1989)
	120,000	Odlum (1990)
	121,400	Wendorff *et al.* (1997)
Evaporated or condensed milk	364,790	Danalewich *et al.* (1998)
	378,000	Hale *et al.* (2003)
Milk powder	950,000	Hale *et al.* (2003)
Whey powder	929,000	Hale *et al.* (2003)
Whey	80,000	Odlum (1990)
	75,000	Steffen *et al.* (1989)
Cream (30%)	860,000	Odlum (1990)
	860,000	Steffen *et al.* (1989)
	750,000	Hale *et al.* (2003)
Domestic sewage	500	Odlum (1990)
	500	Steffen *et al.* (1989)

tanks before CIP starts. Wastewater generated in this area would contain mostly whole milk components such as milk fat, milk protein and lactose (Hale *et al.* 2003). Pollution loads could however be minimized by the proper drainage of tanker trucks and storage tanks, properly connected and well-stitched hoses, self-draining pipes that have been installed at a slight angle, level controls in storage tanks to prevent overflows and the use of well-adapted CIP processes for the cleaning of tankers and storage tanks where the final rinse water could be reused for the initial rinse of the next cleaning process (Hale *et al.* 2003; Tetra Pak 2003).

7.2.3 Heat processing of milk

Wastewaters arising from pasteurization and sterilization processes of liquid milk and cream are normally the results of cleaning operations of heat exchangers where most of the organic loading of cleaning water occurs at the product and water interfaces. This happens at the beginning and end of the process where, to prevent the accidental adulteration of milk with water, the heat exchanger is purged with milk at the beginning of the process, and the milk is purged with water for cleaning at the end of the process. Wastewater generated in this area would contain mostly whole milk components such as milk fat, milk protein and lactose. Pollution loads could be minimized by proper management of heat-treatment processes: restricting the number of process interruptions and changeovers, and ensuring that that storage tanks for the treated product are adequately sized to accommodate filling process interruptions. The interface milk/water fraction could also be recycled into other products or used as animal feed (Hale *et al.* 2003).

7.2.4 Production of evaporated milk products

The production of both sterilized unsweetened evaporated milk and sweetened condensed milk involves the concentration of pasteurized milk in evaporators and the heat treatment of the product immediately before or after packaging. The only difference in the production of sweetened condensed milk is the addition of sugar to the evaporated milk just after evaporation (Steffen *et al.* 1989; Hale *et al.* 2003). The organic loads of wastewaters generated during these processes are mostly influenced by the amount of milk that is flushed from the processing equipment at the start and end of production, as well as the amount of organic and inorganic compound deposits that formed as a result of the heating processes during milk evaporation and are removed by cleaning. Sugar and product spillages during packaging could also contribute to the COD load of cleaning waters (Hale *et al.* 2003). The amount of deposits formed during the heating processes could be minimized by running evaporators at the lowest possible temperature (Tetra Pak 2003).

7.2.5 Production of powdered dairy products

The production of powdered dairy products such as milk and whey powders is a two-step process that involves the concentration of liquid pasteurized milk or whey by evaporation, after which the products are spray-dried. Milk powder can be produced directly after pasteurization, but before whey can be evaporated and dried, casein fines and milk fat need to be recovered, after which the whey should be thermalized before concentrating and drying. As a result of these processing differences, wastewaters arising from the above-mentioned drying practices could differ significantly. Milk powder wastewaters would contain mostly milk protein, lactose and milk fat, while whey powder wastewaters would contain whey proteins and lactose, as well as casein fines and milk fat losses generated during whey separation. The bulk of the wastewaters are normally generated during start-up and shut-down of evaporation and drying processes as well as from cleaning operations. In general, the organic load of wastewaters could be minimized by emptying evaporators properly before CIP starts as well as between each CIP step. Dry products collected from product spillages and the cleaning of spray towers should also be removed and treated as solid waste (Hale *et al.* 2003; Tetra Pak 2003). Condensate generated during evaporation processes could be used in cooling processes or for feeding the boiler.

7.2.6 *Cheese manufacture*

The bulk of the wastewater from this section is generated when equipment is flushed at the end of a processing run, as well as during CIP operations. Wastewaters would typically contain milk, milk fat, brine, whey and cheese fines. Spillages of starter milk, whey and pieces of curd during cheese-making operations could also contribute significantly to the wastewater organic load. Wastage in this section could be reduced by ensuring that equipment such as open cheese vats is correctly sized and not filled to the rim with milk. Optimal curd cutting would also greatly reduce the milk fat and cheese fines content of whey. Whey should be collected sparingly and used in commercial applications. Curd spillages should be collected before factory floors are washed and disposed of as solid waste (Hale *et al.* 2003; Tetra Pak 2003).

7.2.7 *Butter manufacture*

As with cheese manufacturing, most of the wastewater in this section is generated when residual product is flushed from equipment and floors, and when equipment is hosed down at the end of production, as well as during CIP operations. Butter tends to stick to surfaces, and therefore wastewater generated in this section would contain mostly cream and butter components, buttermilk and high concentrations of milk fat. The amount of cream and butter that could enter cleaning wastewaters could be reduced by manually scraping accessible production surfaces before cleaning starts (Hale *et al.* 2003; Tetra Pak 2003). The timing of equipment rinses could also be optimized to reduce wastewater volumes.

7.2.8 *Yoghurt manufacture*

Once again most of the wastewater is generated during cleaning processes of production equipment such as fermentation vats, heat exchangers and ingredients' storage vessels. The wastewaters would typically contain dilute yoghurt, milk fat, heat deposits from heat exchangers, pieces of fruit and diluted fruit preserve, diluted sugars (including lactose) and stabilizers as well as flavour compounds (Tamime and Robinson 1999; Hale *et al.* 2003). The high viscosity of yoghurt contributes significantly to the organic load of the cleaning waters, as it tends to cling to production surfaces. Spillages of milk powder (on average, 9.5 times more milk solids than milk) and fruit concentrates (COD loads of ≥500 g mL^{-1}) can also increase the organic load (Hale *et al.* 2003). Water consumption could be reduced by proper CIP management, such as recovery of the final rinse water and using burst rinsing instead of continuous rinsing when cleaning yoghurt vats (Tamime and Robinson 1999). Various strategies may also be implemented to reduce the organic load of the wastewater: spillages of dry ingredients could be collected and treated as solid waste; drainage times of yoghurt vats and pipes could be extended, and heat deposition especially during the heating of sweetened yoghurt, could be reduced by more gradual heating.

7.3 Treatment options

Dairy wastewaters have a highly variable composition, mainly due to the differing volumes and flow rates, which are dependent on the factory size and operation shifts. The varying pH and suspended solid (SS) content, primarily due to the type of cleaning regime employed, makes the choice of an effective wastewater treatment system difficult. Due to the fact that dairy wastewaters are highly biodegradable, they can usually be treated quite effectively in biological wastewater treatment systems. They can, however, also pose a potential environmental hazard if they are not treated properly (Burton 1997). Dairy industries are normally faced with three main wastewater treatment options: (1) discharge to and subsequent treatment at a nearby sewage-treatment plant; (2) removal of semi-solid and special wastes from the site by waste disposal contractors or (3) the treatment in an on-site, wastewater treatment plant (Gough and McGrew 1993; Robinson 1994). Due to the increasing costs involved in the first two options as well as the increasingly stringent allowable levels of suspended solids (SS), BOD and COD in discharged wastewaters, an increasing number of dairy industries have to consider the third option of treating their wastewaters on-site. The chosen treatment should meet the required demands and reduce the costs associated with long-term industrial wastewater discharge.

7.3.1. Direct discharge to a sewage-treatment works

Municipal sewage-treatment works are normally designed to handle a specific organic concentration and can withstand occasional peak loads. Certain components found in dairy waste streams may, however, present problems. Fat is such a component and can adhere to the walls of the main system and cause settling and clogging problems in the sedimentation tanks. Another potential problem is the possible toxicity of long-chain fatty acids to biological treatment systems (Vidal *et al.* 2000). Some form of on-site pre-treatment is therefore advisable to minimize the fat content of the dairy wastewater that can be mixed with the sanitary wastewater going to the sewage-treatment works (Rusten *et al.* 1990).

Sanitary wastewater and processing wastewaters at dairies are often not separated. This is a cause for major concern during the final disposal of generated sludge, as this may contain pathogenic microorganisms. Dairies that employ on-site treatment of their wastewaters are thus advised to separate sanitary and processing wastewaters, and dispose of the sanitary wastewater by piping directly to a sewage-treatment works. The sewage-treatment works normally levy a fee, based on combinations of the wastewater flow rate, BOD_5 mass, COD concentration, suspended solids present and/or the total phosphate discharged per day (Danalewich *et al.* 1998). The specific regulations differ from country to country, but are normally based on similar principles and parameters.

7.3.2 Pre-treatment options

Wastes from dairy-processing operations are mostly composed of organic material in solution or colloidal suspension. Although some larger suspended solids, sand and foreign material may be present, dairy-processing wastewaters are not difficult to treat in conventional wastewater-treatment processes. Due to variations in wastewater volume, organic load, temperature, pH, nutrient levels and fat content, some form of pre-treatment is required to standardise the wastewater before biological treatment (Nadais *et al.* 2001). The most common pre-treatments employed with regard to dairy-processing wastewater are screening, sedimentation, flotation (removal of FOG), chemical precipitation and wastewater stabilization.

7.3.2.1 Screening

Screening of the wastewater should be performed as quickly as possible to reduce the amount of COD leaching that takes place as a result of solid solubilization. Screens are intended to remove large particles or debris, thus reducing potential damage to pumps, as well as downstream clogging (Droste 1997). A variety of screen designs are possible, from wire screens and grit chambers with an aperture size of 9.5 mm (Wendorff 2001) to finely spaced, mechanically brushed or inclined screens of 40 mesh (about 0.39 mm) (Hemming 1981). Screens can be cleaned either manually or mechanically and the screened material disposed of at a landfill site. The grid chamber or sand trap follows the screening (Tetra Pak 2003) and is a basin where the coarse separation of sand and other heavy materials takes place by natural sedimentation. The sediment is pumped away and disposed of separately.

7.3.2.2 Equilibration tanks

Waste stabilization normally includes pH correction and equilibration of the volume, organic load, nutrients and detergents. Large pH variations exist in dairy-processing wastewater from different dairy plants. These differences may be attributed to the different cleaning strategies employed. Acidic cleaners used for the removal of mineral deposits and acid based sanitizers can result in a pH of 1.5–6.0, while alkaline detergents used for the saponification of lipids and the effective removal of proteinacous substances, would typically have a pH of 10–14 (Bakka 1992; Graz and McComb 1999). The optimum pH range for biological treatment plants is 6.5–8.5 (IDF 1984; Eckenfelder 1989). Extreme pH values can be highly detrimental to any biological treatment facility, not only for the negative effect that it will have on the microbial community, but also due to the increased corrosion of pipes that will occur at pH values below 6.5 and above 10 (Tetra Pak 2003). Some form of pH adjustment as a pre-treatment step is, thus, strongly advised before such wastewaters are discharged to the drain or further treated on-site. In most cases, flow balancing and pH adjustment are performed in the same equilibration tank. According to the International Dairy Federation (IDF 1984) a near-neutral pH is usually obtained, when waters used in different production processes are combined. If pH correction needs to be carried out in the equilibration tank, the most commonly used chemicals are H_2SO_4, HNO_3, NaOH, CO_2 or lime.

Since discharged dairy wastewaters can also vary greatly with respect to volume, strength, temperature and nutrient levels, equilibration is a prime requirement for any subsequent biological process to operate efficiently (Hemming 1981). pH adjustment and flow balancing can be achieved by keeping wastewater in equilibration tank for at least 6–12 h (Wendorff 2001). During this time, residual oxidants can react completely with solid particles, neutralizing cleaning solutions. The stabilized wastewater can then be treated with a variety of different options (Wendorff 2001). Equilibration tanks should be adequately mixed to obtain proper blending of the contents and to prevent solids from settling. This is usually achieved by the use of mechanical aerators (Odlum 1990). The size of the equilibration tank is also critical and should be accurately determined so that it can effectively handle a dairy factory's daily flow during the peak season. It is also recommended that an equilibration tank be large enough to allow a few hours' extra capacity to handle unforeseen peak loads and not discharge shock loads to public sewers or on-site biological treatment plants (IDF 1984).

7.3.2.3 FOG removal

The presence of FOG in dairy-processing wastewater can cause a variety of problems in biological wastewater treatment systems, especially in anaerobic systems. Fats and proteins present in the wastewaters have a low biodegradability coefficient. Oil and grease adsorbed on the surface of the sludge may limit transport of soluble substrates to the biomass and consequently reduce the rate of substrate conversion (Cammarota *et al.* 2001; Nadais *et al.* 2001). Continuous operation has been shown to cause problems in the form of scum and sludge layers on the liquid surface of the reactors with subsequent biomass washout. It is therefore important to reduce, if not essential to remove, the FOG completely prior to further treatment. According to the IDF (1997), factories processing whole milk, such as milk-separation plants as well as cheese and butter plants, whey-separation factories and milk-bottling plants, experience severe problems with FOG (EIPPCB 2005). The processing of skim milk seldom presents problems in this respect.

As mentioned previously, wastewater equilibration is recommended for dairy-processing plants. The FOG-treatment unit may be placed before or after the equilibration tank. It is generally accepted (IDF 1997) that wastewater equilibration should precede FOG removal, as large fat globules can accumulate in the tank as the discharged wastewater cools down, and suspended fats aggregate during the retention period. If the equilibration tank is placed after the FOG removal unit, the unit should be large enough to accommodate the maximum anticipated flow from the processing plant. General FOG removal systems include the following.

Gravity traps are effective, self-operating and easy to construct. Wastewater flows through a series of cells, and the FOG mass, which usually floats on top, is removed by retention within the cells. Baffles or plate separators are often included to increase the residence time and separation efficiency (EIPPCB 2005). Some disadvantages include frequent monitoring and cleaning to prevent FOG build-up and decreased removal efficiency at pH values above 8 (IDF 1997). Excessively hot water can cause fats to be carried through and may also melt pre-collected fats, thereby reducing the removal efficiency.

The *Dissolved Air Flotation* (DAF) process is where air is dissolved in the dairy wastewater at pressures above atmospheric pressure (300–600 kPa) (Droste 1997) in enclosed units. The wastewater, now supersaturated with air with respect to atmospheric pressure is then fed into a tank at atmospheric pressure. Bubble formation occurs as the pressure is reduced. The formed bubbles entrap flocs or adhere to flocs and buoy them to the surface, while heavy solids form a sediment (Robinson 1994; Jameson 1999; Tamime and Robinson 1999; Tetra Pak 2003; EIPPCB 2005). One of the drawbacks of the DAF is that only SS and free FOG can be removed. Dissolved air flotation is often combined with a chemical treatment, used to improve the flocculation of suspended material, thereby increasing the flotation system's efficiency. The addition of chemicals, such as polymers, aluminium sulfate, ferric chloride or polyelectrolyte, thus improves the adhesion of bubbles to flocs (Rusten *et al.* 1993). A skimmer collects the floating solids, and a portion of the wastewater is recycled through the air-dissolution tank. pH correction might also be necessary prior to the flotation treatment, since a pH of around 6.5 is required for efficient FOG removal (IDF 1997). FOG waste should be removed and disposed of according to approved methods, as it will become odorous if stored in an open tank. It is an unstable waste material that should preferably not be mixed with sludge from biological and chemical treatment processes, since it is very difficult to dewater. DAF units require regular maintenance, and the running costs are usually fairly high.

Air flotation is a more economical version of the DAF. Air bubbles are introduced directly into the flotation tank containing the untreated wastewater, by means of a cavitation aerator coupled to a revolving impeller or by blowing air through canvas or other porous material (Jameson 1999). A variety of different patented air flotation systems are available on the market and have been reviewed by the IDF (1997). These include the 'Hydrofloat', the 'Robosep', vacuum flotation, electroflotation and the 'Zeda' systems.

During the *Enzymatic Hydrolysis of FOG*, enzymatic preparations from babassu cakes containing lipase produced by a strain of *Penicillium restrictum* are used as a dairy-processing wastewater pre-treatment (Cammarota *et al.* 2001; Leal *et al.* 2002). The enzymatically pre-treated dairy wastewater was treated further by anaerobic digestion. Higher COD reduction efficiencies as well as reactor wastewaters of better quality were obtained from a laboratory-scale UASB reactor compared with the results of a UASB reactor treating the same wastewater without prior enzymatic hydrolysis treatment.

7.3.3 Aerobic biological systems

Biological wastewater treatment is achieved by many different microorganisms carrying out a complex, systematic, continuous, sequential attack on the organic compounds found in wastewater (Grismer *et al.* 2002). A wide variety of biological wastewater treatment technologies are available, but the selection of the most appropriate and cost-effective technology can be problematic, since wastewaters respond differently to each treatment process (Hayward *et al.* 2000).

Aerobic treatment systems vary in sophistication from simple lagoons to high-rate activated sludge processes (Wayman 1996). All of these aerobic processes, however, work on the same basic principles but vary in efficiency and implementation cost (Tchobanoglous and Schroeder 1991). Degradation is facilitated by bacteria that thrive in oxygen-rich environments, and pollutants are broken down to carbon dioxide, water, nitrates, sulfates and biomass. Problems normally associated with aerobic processes include foaming and poor solid–liquid separation.

7.3.3.1 Activated sludge systems

The first activated sludge treatment system had its origin in England at the turn of the twentieth century (Bitton 1999). In this system, degradation of wastewater relies on contact between the organic matter in the wastewater and high concentrations of microorganisms in the presence of dissolved oxygen (Nazaroff and Alvarez-Cohen 2001). The microorganisms, mostly aerobic heterotrophic bacteria, are suspended within the water of the reactor. These organisms use the organic material for oxidation while producing carbon dioxide and generating energy to their own advantage with the synthesis of new biomass.

A typical activated sludge unit consists of a reactor and cell separator or clarifier. Oxygen is supplied to the reactor, and this is where the reduction in BOD occurs. The supply of oxygen has a dual function: to maintain the microbial life and to keep the liquor in a mixed regime (Gilde and Aly 1976; Tchobanoglous and Burton 1991). The necessary oxygen can be supplied in the form of relatively pure oxygen or as compressed air (Droste 1997; Bitton 1999).

The water leaving the reactor contains high concentrations of microorganisms from the reactor (Nazaroff and Alvarez-Cohen 2001). In the cell separator, sedimentation is used to separate the water from the bacterial flocs. The water is then discharged, and the sludge extracted at the bottom of the clarifier is either returned to the inlet of the reactor or processed as waste sludge. The clarifier tanks of an activated sludge process function as powerful selectors for microorganisms which grow in a biomass form suitable for settling. Organisms which grow as single cells or as small clumps of a few cells do not settle well enough to be separated from the wastewater and are carried out of the process. Organisms which grow as larger clumps or as floc particles settle and are returned to the aeration tank and, thus, are able to propagate in the process (Kim *et al.* 1998).

Since the efficiency of the process is dependent on the maintenance of the microbial consortium, bacterial flocs are favoured above individual organisms, as they are larger and thus have better settling characteristics (Nazaroff and Alvarez-Cohen 2001). The flocs usually consist of members of the following genera: *Zooglea*, *Pseudomonas*, *Flavobacterium*, *Alcaligenes*, *Bacillus*, *Achromobacter*, *Corynebacterium*, *Comomonas*, *Brevibacterium*, *Acinetobacter* and filamentous microorganisms (Bitton 1999). The excess growth of filamentous organisms and the binding of water give rise to a bulking sludge with very poor settling characteristics (Tchobanoglous and Burton 1991). The cause of bulking is often related to the physical and chemical characteristics of the wastewater, the treatment plant design and limitations and plant operation. Bulking sludge is undesirable and costly to dispose of, and must therefore be prevented as far as possible.

The main disadvantages of activated sludge systems (ASSs) are the large initial capital and operational costs, and the fact that the operators must also be skilled for optimal operation. One problem, especially for the dairy industry, is the presence of excessive dissolved solid concentrations, which are often not efficiently removed (Green and Kramer 1979). Another disadvantage is the relatively large sludge mass which is produced that must be processed and disposed of creating a major operating expense (Bitton 1999; Garrido *et al.* 2001).

The ASS process, however, has advantages: it occupies little space while handling high organic loading rates, the operator can exercise control over the process, it can tolerate shock loads and no odour or fly problems are normally experienced. The activated sludge process is furthermore efficient in BOD, suspended solids (SS) and phosphorus removal, and may be very efficient in nitrification. The waste sludge is normally stable and can be put onto drying beds (Droste 1997).

7.3.3.2 Sequencing batch reactors

Sequencing batch reactors (SBRs) rely on the same principles for aeration and sedimentation as applied in the ASS process. The major difference is that only one tank is used in which both processes take place. The process works in five steps: fill, react (aeration), sedimentation/clarification, decant and idle. A portion of the settled sludge is retained in the reactor between runs so as to sustain an active biological community. It is, however, important that there is cell wastage between runs in order to maintain a nominally steady-state microbial population (Tchobanoglous and Burton 1991; Nazaroff and Alvarez-Cohen 2001).

Sequencing batch reactors can also function anaerobically (Nazaroff and Alvarez-Cohen 2001). To promote this condition, alternate electron acceptors are added in place of the aeration step. If so desired, SBRs can be operated as sequential anaerobic/aerobic modes to promote a better degradation phase under both conditions.

The SBR design has been shown to be a cost-effective treatment process for dairy-processing wastewaters with COD reductions of 91–97% being achieved (Eroglu *et al.* 1992; Samkutty *et al.* 1996). Li and Zhang (2002) achieved reduction efficiencies of 80% COD, 63% in total solids, 66% in volatile solids, 75% Kjeldahl nitrogen and 38% in total nitrogen in a SBR operated at an HRT of 24 h and a dairy wastewater COD of 10 g l^{-1}. Torrijos *et al.* (2001) operated a SBR successfully at an organic loading rate (OLR) of 0.50 kg COD m^{-3} day^{-1} and obtained treatment levels > 97% with a wastewater from a small cheese-making dairy.

7.3.3.3 Trickling filters

Trickling filters, also known as percolating filters, were first introduced in the 1890s and generally consist of four major components (Tchobanoglous and Schroeder 1985; Tchobanoglous and Burton 1991; Barnett *et al.* 1994; Bitton 1999). A circular or rectangular tank contains filter medium (stones, crushed limestone, ceramic material, treated wood, hard coal or even a plastic medium) with a bed depth of 1.0–2.5 m, which provides a large surface area to maximize microbial attachment and growth. It should also provide sufficient void space for air diffusion as well as allowing sloughed microbial biofilm to pass through. Selection of filter media is based on factors such as specific surface area, void space, unit weight, media configuration and size, and cost. The smaller the size, the higher the surface area for microbial attachment and growth, but the lower the percentage of void space. Plastic media (polyvinylchloride or polypropylene) are mainly used in high-rate trickling filters. They have a low bulk density and offer optimum surface areas (85–140 m^2 m^{-3}) and a much higher void space (up to 95%) than other filter media. Thus, filter clogging is considerably minimized when these filter media are used. Plastic is also a light material that requires less heavily reinforced concrete tanks than do stone media. Therefore, the biological tower reactors containing these materials can be as high as 6–10 m. The wastewater distributor allows a uniform hydraulic load over the filter material. It has one to four arms, and its configuration and speed depend on the filter media used. Hydraulic loads of about 5 m^3 m^{-2} day^{-1} can be obtained in low-rate filters. Wastewater is percolated or trickled over the filter and provides nutrients for the growth of microorganisms on the filter surface.

A drain-tank is normally included and is necessary for collection of treated wastewater and biological solids (microbial biomass) that have been sloughed off the biofilm material as well as for the introduction of air. A final clarifier, also called a humus tank, for separation of solids from the treated wastewater is also necessary.

A trickling filter essentially converts soluble organic matter to biomass, which is further removed via settling in the final clarifier. Typical OLRs are 0.5 kg BOD m^{-3} day^{-1} and may vary from 0.1 to 0.4 for low-rate filters to 0.5–1.0 kg BOD m^{-3} day^{-1} for high-rate filters. This parameter is important to the performance of the trickling filter and may dictate the hydraulic loading onto the filter. BOD removal by trickling filters is approximately 85% for low-rate filters and 65–75% for high-rate filters (Bitton 1999). Trickling filters are attractive to small communities because of their easy operation, low maintenance costs and reliability. They have been used to treat toxic industrial wastewaters and are able to withstand the extreme shock loads of toxic inputs. Furthermore, the sloughed biofilms can be easily removed by sedimentation (Tchobanoglous and Schroeder 1985). High organic loading may lead to filter clogging as a result of excessive growth of slime bacteria in/on the biofilms. Excessive biofilm growth can also cause odour problems in trickling filters. Clogging restricts air circulation, thus resulting in low availability of oxygen to biofilm microorganisms. However, modifications have helped improve the BOD removal of trickling filters (Tchobanoglous and Burton 1991; Barnett *et al.* 1994).

Treatment of dairy wastewaters in trickling filters can be hampered by blockages, as a result of the precipitation of ferric hydroxide and carbonates. This also leads to a decrease in microbial activity. Blockages can also be caused by the high fat and protein content of the wastewater (Maris *et al.* 1984). It is thus generally recommended that the organic loading for dairy wastewaters does not exceed 0.28–0.30 kg BOD m^{-3} and that recirculation be employed (Herzka and Booth 1981). Kessler (1981) reported a 92% BOD reduction in a dairy wastewater treated in a trickling filter. The final wastewater was, however, still too high and necessitated further treatment in an oxidation pond.

An inherent problem is that trickling filters can be blocked by precipitated ferric hydroxide and carbonates, with concomitant reduction of microbial activity. In the case of overloading with dairy wastewater, the medium becomes blocked with heavy biological and fat films. Maris *et al.* (1984) reported that biological filters are not appropriate for the treatment of high-strength wastewaters, as filter blinding by organic deposition on the filter medium is generally found.

It is generally recommended that organic loading for dairy wastewaters not exceed 0.28–0.30 kg BOD m^{-3} and that recirculation be employed (Herzka and Booth 1981). A 92% BOD removal of a dairy wastewater was reported by Kessler (1981), but since the BOD of the final wastewater was still too high, it was further treated in an oxidation pond.

7.3.3.4 Rotating biological contactors

The rotating biological contactor (RBC) is another example of fixed a film bioreactor, introduced at the beginning of the twentieth century to the United States (Tchobanoglous and Burton 1991). An RBC consists of a series of disks that are mounted on a horizontal shaft and rotate slowly in the wastewater. The corrugated disks act as a support medium for attached microbial cultures. The disks are approximately 40% submerged in the wastewater (Barnett *et al.* 1994; Droste 1997; Bitton 1999). At any time, the submerged portion of the disk removes BOD as well as dissolved oxygen. The rotation of the disks provides aeration as well as shear force that causes sloughing of the biofilm from the disk surface. Increased rotation improves oxygen transfer and enhances the contact between attached biomass and wastewater. Several RBC stages are usually employed in a staggered arrangement. The advantages offered by RBC are short residence times, low operation and maintenance costs, and production of a readily dewatered sludge that settles rapidly. Operational problems, such as shaft failure, media breakage, bearing failures and odour problems, have been encountered. Odour problems are most frequently caused by excessive organic loadings, particularly in the first stage. Reduction efficiencies for COD of 85%, treating a dairy wastewater at an OLR of 500 g COD m^{-3} h^{-1}, have been achieved (Rusten *et al.* 1992).

The RBC process offers several advantages over the activated sludge process for use in dairy wastewater treatment. The power input required is lower; the process is easy to operate and requires very little maintenance. Pumping, aeration and recycling of solids are not required, and this results in a lower operator input.

7.3.3.5 Lagoon technology

Lagoon systems or stabilization ponds have been employed for the treatment of wastewaters for many years. The early lagoons were merely holding ponds or evaporation basins created at sites chosen as convenient as possible to the processing plant (Tchobanoglous and Burton 1991). Lagoon systems are used largely as secondary and tertiary wastewater biological oxidation treatments for small municipalities and isolated rural industries. Their attractions are their low capital investment and low operational costs plus relative ease of operation (Tchobanoglous and Schroeder 1985). They are effective if properly designed and maintained. However, they require a relatively large amount of flat land in proportion to the amount of waste load being treated, and they require a prolonged retention time (usually weeks or months) for proper operation. Therefore, they are not well suited where land availability is at a premium or for large municipal and industrial waste loads (Bitton 1999) or where water is in short supply. Lagoons may function as both biological oxidation and solid sedimentation systems. They convert dissolved, suspended and settled solids to volatile gasses (CO_2, CH_4), water and biomass. Lagoons are suitable for seasonal operations, and they can also be designed to treat heavy pollution loads (Kilani 1992). Because of this, lagoons are very popular with small- to medium-sized food processors (Green and Kramer 1979; Droste 1997), including dairy-processing wastewaters (Pearson 1996).

Waste-treatment lagoons can be grouped into several different types, based on their design, function and type of metabolic regime. The organic loading rate is the primary determinant of the metabolic regime (Green and Kramer 1979).

7.3.3.5.1 Storage lagoons. Storage lagoons are utilized primarily for operations of a seasonal nature, where it is possible to store the waste from an entire season's production. Prior to discharge into the lagoon, the wastewater is screened to remove the gross solids and most suspended solids. Normally, the wastewater is stored for a period of 90–120 days and then discharged during high river flows. The discharge is regulated to avoid substantial increases in the BOD or reduction of the dissolved oxygen in the receiving stream. During storage, sedimentation of suspended solids and anaerobic decomposition of the organic matter take place (Azad 1976). The major advantages of storage lagoons include a low initial cost, minimal sludge handling problems and availability of storage for the entire season's operations with controllable discharge timing. However, severe odour problems are associated with these lagoons, and frequently the reduction in the pollution load is low (Tchobanoglous and Schroeder 1985). Furthermore, large areas of land are required (Holdsworth 1970; Sterrit and Lester 1982). In addition, the enforcement of more strict discharge requirements makes this type of lagoon unsuitable as the only method for treatment of dairy-processing wastes.

7.3.3.5.2 Oxidation ponds or facultative ponds. These lagoons are the most common and are used to treat wastewaters with intermediate loading rates (45–90 kg BOD_5 ha^{-1} day^{-1} at temperatures above 15°C) and have facultative anaerobic and aerobic zones. The depth may vary from 0.9 to 2.4 m. The greater depth allows for the development of an anaerobic bottom layer and an aerobic surface zone. The oxygen necessary for the aerobic zone is provided by natural aeration and algal photosynthesis (Azad 1976; Green and Kramer 1979). Large solids settle out to form the anaerobic sludge layer, while the soluble and colloidal organic materials are oxidized by the aerobic bacteria using the oxygen produced by the surface growing algae. Carbon dioxide produced in the organic oxidation serves as a carbon source for the algae. Anaerobic breakdown of the solids in the sludge layer results in the production of dissolved organics and gases such as CO_2, H_2S and CH_4, which are either oxidized by the aerobic bacteria or vented to the atmosphere (Tchobanoglous and Schroeder 1985). BOD removals range from 70 to 95%, with longer retention times resulting in the higher removal rates. Oxidation ponds, however, require large land areas, and their performance depends on the climatological conditions. Severe odour problems have been associated with the operation of these lagoons, especially in colder climates (Azad 1976). In addition, excessive amounts of blue-green algae are often discharged, resulting in wastewaters of unacceptable quality. The performance of facultative ponds becomes less reliable if the algal biomass drops below 300 g chlorophyll a.l^{-1}, since below this biomass concentration, there can be negative net oxygen production in the pond. At higher temperatures (25–35°C), the rate of increase in algal biomass production slows, and the solubility of oxygen in ponds also decreases significantly (Pearson 1996). Yoghurt waste was treated successfully in a laboratory-scale facultative pond, removing 70 and 80% of total and soluble COD, respectively, at retention times of 7.9 days and organic loading rates as high as 450 kg COD ha^{-1} day^{-1} (Kilani 1992). A synthetic milk wastewater was also treated in a laboratory-scale oxidation/facultative pond system at loading rates of 300 kg BOD_5 ha^{-1} day^{-1} and achieved a 90% COD removal (Al-Khateeb and Tebbutt 1992).

7.3.3.5.3 Aerobic or maturation ponds. Aerobic or maturation ponds are very shallow (30–45 cm) and receive wastewater from facultative ponds. The organic loading rate is low, and aerobic conditions exist throughout the pond depth. There are always some algae present. These ponds are designed to polish the final wastewater by settling suspended solids (SS) and stabilizing the low concentration of influent soluble organics (Droste 1997).

7.3.3.5.4 Aerated lagoons. Aerated lagoons are commonly used as an efficient means of wastewater treatment, relying on little sophisticated technology and minimal, albeit regular, maintenance. Their low capital and operating costs and ability to handle fluctuating organic and hydraulic loads have been valued for years in rural regions and in many tropical countries wherever land is available at reasonable costs (Nameche and Vasel 1998). Oxygen is supplied to the lagoon by means of a mechanical or diffused aeration unit, and a high degree of mixing is provided in order to maintain the solids in suspension (Droste 1997). Aeration is, therefore, not dependent on natural conditions (temperature, winds or sunlight), and the amount of aeration can be controlled (Green and Kramer 1979). The retention time required to attain the desired degree of BOD removal is significantly less than the facultative lagoons because a higher equilibrium level of biological solids is maintained. The aerobic lagoon is analogous to an activated sludge lagoon with extended aeration without sludge return. The wastewater from the lagoon is identical to the liquid in the aeration basin and contains the biological solids and the remaining soluble BOD. Therefore, in order to remove these solids and reduce the suspended BOD, a settling tank would be required before final discharge of the wastewater (Droste 1997). Aerated lagoons are usually operated at high organic loadings because of the low biological solids maintained in the system. Aerated lagoons have been used to treat a variety of food-processing and dairy wastes where BOD removals of up to 95% have been obtained (Sterrit and Lester 1982; Bitton 1999). However, application of excessively high loadings may result in sludge-bulking problems, and solids separation becomes a serious operational problem. Faulty or inefficient operations may also give rise to odour problems.

7.3.3.6 Natural treatment systems

In the natural environment, physical, chemical and biological processes occur when water, soil, plants, microorganisms and the atmosphere interact (Bitton 1999). Natural treatment systems are designed to take advantage of these processes to provide wastewater treatment. The processes involved include many of those used in mechanical or in-plant treatment systems – sedimentation, filtration, gas transfer, adsorption, ion exchange, chemical precipitation, oxidation and reduction, and biological conversion and degradation, plus processes unique to natural systems, like photosynthesis, photooxidation and plant uptake (Tchobanoglous and Burton 1991; Barnett *et al.* 1994; Bitton 1999). In natural systems, the processes occur at 'natural' rates and tend to occur simultaneously in a single 'eco-system reactor', as opposed to mechanical systems in which processes occur sequentially in separate reactors or tanks at accelerated rates as a result of energy input. Four main types of natural treatment systems have been identified.

7.3.3.6.1 Slow-rate irrigation. This is the most frequently used land-treatment system and involves the application of wastewater to vegetated land to provide treatment and to meet the needs of the vegetation growth. Crop water requirements, soil characteristics and climate factors interact to determine the wastewater application rate. Hydraulic loading rates are the limiting-loading rate constraint. Removal of the BOD and soluble solids (SS) may be excellent after the water has percolated through 1.5 m of soil (Droste 1997). Sprinkler or ridge-and-furrow systems are usually used for distributing the wastewater. The pre-treatment required is screening and grit removal, but primary sedimentation is desirable. As some storage is usually included as part of slow-rate systems, sedimentation is nearly always provided. The applied wastewater is reduced by either evapotranspiration or percolation vertically and horizontally through the soil profile. Any surface runoff is usually collected and reapplied to the system. In most cases, the percolate will enter the underlying groundwater, but in some cases, the percolate may be intercepted by natural surface waters or recovered by means of under drains or recovery wells. The relatively low application rates combined with the presence of vegetation and the active soil ecosystem provide slow-rate systems. Some limitations of slow-rate irrigation systems are land cost and high operating cost of transport of wastewater to the treatment site (Bitton 1999). Average hydraulic application rates are in the order of 0.5–6 m $year^{-1}$, and average BOD loadings are 370–1830 kg BOD ha^{-1} $year^{-1}$.

7.3.3.6.2 Rapid infiltration. Rapid infiltration systems are used where highly permeable sandy soils are available and where it is an objective to replenish groundwater (Droste 1997). Loading occurs over a period of several days and is followed by a drainage period of 1–2 weeks. There is usually no cover crop, and the surface is occasionally scarified to break up material plugging the surface layer (Tchobanoglous and Schroeder 1985; Bitton 1999). Rapid infiltration is often designed for final polishing after primary and secondary treatment. Nitrification or nitrification–denitrification will be achieved along with phosphorus removal. Application periods range from less than a day to 2 weeks, and the drying period is at least as long as the application period, ranging up to 20 times the application period. The treatment potential of rapid infiltration systems is somewhat less than slow-rate systems because of the lower retention capacity of permeable soils and the relatively higher hydraulic loading rates (Tchobanoglous and Burton 1991). Average hydraulic application rates are in the order of 6–125 m $year^{-1}$, and the average BOD loading is 8000–46,000 kg BOD ha^{-1} $year^{-1}$.

7.3.3.6.3 Overland flow. In the overland flow systems, pre-treated wastewater is distributed across the upper portions of carefully graded vegetated slopes and allowed to flow over the slope surfaces to runoff collection ditches at the bottom of the slopes. Percolation through the soil profile is, therefore, a minor hydraulic pathway. A portion of the applied water will also be lost to evapotranspiration (Tchobanoglous and Burton 1991). The most suitable soils are clay or loamy-clay soils with low permeability, to limit wastewater percolation through the soil profile. The bacterial community that grows on the thatch that accumulates at the soil surface carries out biological bioxidation, nitrification, and denitrification. Nutrients (N and P) and BOD, SS and pathogens are removed as wastewater flows down the slope. This type of system can achieve 95–99% removal of BOD and SS (Tchobanoglous and Schroeder 1985; Droste 1997; Bitton 1999). The two major restrictions to overland flow are difficulty in maintaining consistent quality in the renovated water and site-preparation cost (Hammer 1986). Average hydraulic application rates are in the order of 3–20 m $year^{-1}$, and the average BOD loading is 2000–7500 kg BOD ha^{-1} $year^{-1}$ (Droste 1997).

7.3.3.7 Wetlands

Wetlands are inundated land areas with water depths of less than 0.6 m that support the growth of emergent plants such as cattail, bulrush, reeds and sedges.

The vegetation provides surfaces for the attachment of bacterial films, aids in the filtration and adsorption of wastewater constituents, transfers oxygen into the water column and controls the growth of algae by restricting the penetration of sunlight. Both natural and constructed wetlands have been used for wastewater treatment (Tchobanoglous and Schroeder 1985; Tchobanoglous and Burton 1991). Wetlands are designed to remove conventional pollutants of BOD, SS and nutrients. Heavy metals can also be removed to a significant extent. The most common application of wetlands is for the polishing of secondary wastewaters (Droste 1997). Biological oxygen demand removal by wetland systems is higher during the warmer seasons when the metabolic activity of both bacteria and plants is not inhibited by cooler temperatures. Hydraulic loading rate is the primary control variable for wetland systems. Recommended loading rates are 500 m^3 ha^{-1} day^{-1} or less and should achieve effective (>60%) removal of BOD and total suspended solids (TSS). BOD_5 removal may be as high as 110 kg ha^{-1} day^{-1}. The principal nitrogen-removal mechanism in wetland systems is bacterial nitrification/denitrification and is thus also a function of climatic conditions. Two types of wetland systems have been designed. Surface flow systems are similar to natural wetlands, and a free water surface is maintained. The other alternative is a sub-surface flow system where the water flows through a permeable medium. Treatment is generally better in the subsurface flow systems, and these do not have mosquito problems.

Constructed wetlands have predominantly been used to treat dairy parlour wastewaters, and many successful applications can be found in the literature (Tanner *et al.* 1994; Newman *et al.* 2000; Schaafsma *et al.* 2000; Mantovi *et al.* 2003). The wetlands have been shown to be efficient at reducing the levels of COD/BOD_5, TSS, P and TKN in the wastewaters. Reductions of up to 95% for COD, 85–90% for BOD_5, 70% for P and 55% for TKN have been obtained. Wetlands have, however, been increasingly used to treat dairy-processing wastewaters in Belgium, where reductions in COD, SS, TN and TP of up to 94, 98, 52 and 70%, respectively, have been reported. Dairy wastewater has been treated in a treatment process including wetlands. Karpiscak *et al.* (1999) investigated a system comprising a solids separator, anaerobic lagoons, aerobic ponds and a wetland system. The wetland was planted with cattail, soft-stem bulrush and reed. BOD_5 was not significantly reduced, but TSS and nitrogen were reduced by 30 and 23%, respectively. Greary and Moore (1999), however, achieved better results from a constructed wetland as part of a treatment system treating dairy parlour wastewater. The wetland had a retention time of 10–14 days, the average hydraulic load was 25 mm day^{-1}, and the average mass loadings were 5.6 g m^{-2} day^{-1} for BOD, 2.6 g m^{-2} day^{-1} for organic nitrogen, 3.2 g m^{-2} day^{-1} for ammonia and 1.5 g m^{-2} day^{-1} for total phosphorus. Calculated mean monthly pollutant reductions due to the treatment wetland were 61% for BOD, 43% for organic nitrogen, 26% for ammonia and 28% for total phosphorus. Dairy wastewater has also been treated by using water hyacinths, which grew exceptionally well in the waste (840 mg BOD l^{-1}) . The BOD was lowered from 840 to 121 mg l^{-1}, COD from 1160 to 164 mg l^{-1}, TSS from 359 to 245 mg l^{-1}, total dissolved solids (TDS) from 848 to 352 mg l^{-1} and total nitrogen from 26.6 to 8.9 mg l^{-1} within 4 days (Trivedy and Pattanshetty 2002).

7.3.4 Anaerobic biological systems

Anaerobic digestion (AD) is a biological biodegradation process characterized by the production of biogas, mainly methane (70–90%) and carbon dioxide (10–30%). The biodegradation is performed by an active microbial consortium in the absence of exogenous electron acceptors. One major economic advantage of the AD process is that up to 95% of the organic load can be converted to biogas, and the remainder is utilized for cell growth and maintenance, resulting in a very low surplus sludge production (typically 1–3% of the COD load) (Lettinga *et al.* 1980). If this sludge is present in a granular form, it can be easily disposed of economically. The anaerobic degradation rate is dependent on the nature of the wastewater composition, the presence of anaerobic sludge, its concentration and activity, the mixing and contact time of the organic material and the biomass, and the impact of other operational parameters. These include substrate pH, process temperature and temperature variations, alkalinity levels, necessary nutrients, presence of toxic or inhibitory substances, etc.

In general, anaerobic systems are seen as more economical for the treatment of dairy wastes, as they have lower energy requirements in comparison with that associated with aerobic systems. Anaerobic digestion also yields biogas which can be used as a heat or even a power source. Nitrogen and phosphate requirements are lower than those needed for aerobic systems, and the final biomass has a high soil conditioning value if the concentration of heavy metals is low. One important aspect for the dairy industry is that pathogens do not survive the process. The possibility of treating high COD dairy wastes without dilution, as required by aerobic systems, leads to a smaller 'foot-print' requirement and lower costs associated with compact installations. Bad odours are generally absent if the system is operated efficiently.

There are several disadvantages associated with anaerobic systems. These include higher installation capital costs, extended start-up periods and stricter operational control. It is also important that the substrate pH be kept at around 7, as a result of the pH sensitivity of the methanogens. It is also important that the temperature must be maintained in the 33–37°C range for efficient kinetics. As ammonia nitrogen is not removed in an anaerobic system, it is consequently discharged with the digester wastewater, creating an oxygen demand in the receiving water. Complementary treatment to achieve acceptable discharge standards may be required.

7.3.4.1 Conventional systems

7.3.4.1.1 Anaerobic ponds. The anaerobic pond normally consists of a covered design to exclude air and to prevent biogas loss. The main drawbacks of this design are the large surface-area requirements, low OLR and the necessity of a final polishing step. Ponds have been successfully employed for the treatment of dairy wastewaters (IDF 1990) at OLRs of 1.5 kg COD m^{-3} day^{-1} and HRTs of 1–2 days. The ponds wastewater had to be clarified, and the settled biomass was recycled through the substrate feed. The clarified wastewater was then treated in an 18,000 m^3 aerated lagoon. The efficiency of the total system reached a 99% reduction in COD.

7.3.4.1.2 Completely stirred tank reactors. In these CSTR designs, the OLR ranges from 1 to 4 kg dry matter m^{-3} day^{-1} (Sahm 1984), and the digesters usually have capacities between 500 and 700 m^3. The limitation of the CSTR is that there is no biomass retention (Demirel *et al.* 2005), and consequently the HRT and sludge retention time (SRT) are not separated, thus necessitating long retention times that are dependent on the growth rate of the slowest-growing bacteria involved in the digestion process. The CSTRs are normally used for concentrated wastes (Feilden 1983) where the COD values are higher than 30,000 mg l^{-1}, and the polluting matter is present mostly as suspended solids. The CSTR has been used to treat cheese factory wastewater with a COD of 17,000 mg l^{-1} (Lebrato *et al.* 1990), and at an HRT of 9.0 days, a COD removal of 90% was obtained. With ice-cream wastewaters, shorter HRTs were obtained but at lower OLRs (Ramasamy and Abbasi 2000). While the CSTR is useful for lab studies, it is not a practical option for full-scale treatment, mostly due to the HRT limitation and biomass washout.

7.3.4.1.3 Contact digesters. The anaerobic contact process is basically an anaerobic activated sludge process that consists of a CSTR operated at 30–40°C, followed by some form of biomass separation (Schroepfer *et al.* 1955). The biomass is then recycled to reduce the HRT from the conventional 20–30 days to about 1 day. Even though the contact digester is considered to be obsolete, there are still many small dairies all over the world that use the system (Ross 1989). In these systems, the OLR can vary from 1 to 6 kg m^{-3} day^{-1} COD, with COD removal efficiencies of 80–95% (Sahm 1984). The major difficulty encountered with this process is the poor settling properties of the anaerobic biomass from the digester wastewater, which subsequently impacts the treatment efficiency.

7.3.4.2 High-rate anaerobic systems

7.3.4.2.1 Anaerobic filter. The anaerobic filter (AF) reactor (Ross 1991) is filled with some form of inert support material (gravel, coke or plastic media), so there is no need for biomass separation or sludge recycling. The AF can easily be operated as either a downflow or an upflow filter system, with OLRs ranging from 1 to 15 kg m^{-3} day^{-1} COD, HRTs in the order of 0.2–3 days, and COD removals of 75–95%. The main disadvantage of this design is the potential risk of clogging by suspended solids, mineral precipitates or biomass. Several examples of pilot-plant and full-scale AF designs have been used to treat dairy wastewaters (Bonastre and Paris 1989; Demirel *et al.* 2005). These anaerobic filters were operated at HRTs between 12 h and 10 days, while COD removals ranged between 60 and 98%. The OLRs varied between 1.7 and 20.0 kg COD m^{-3} day^{-1}. The AF design has also been successfully evaluated on an industrial scale for the treatment of raw-milk wastewaters. OLRs of 5–6 kg m^{-3} day^{-1} COD were applied over a 2-year period with consistent COD removals of more than 90%. No biomass loss was observed, and most of the wastewater fat was degraded. However, alkalinity control was found to be crucial, and a polishing step was required to obtain a COD below 200 mg l^{-1}.

A novel high-rate 'Buoyant Anaerobic Filter Bioreactor' (BFBR) has also successfully been evaluated for the treatment of lipid-rich dairy wastewaters (Haridas *et al.* 2005).

In the BFBR, the biomass and insoluble COD retention time are decoupled from the HRT by means of a granular filter bed of buoyant polystyrene beads. Filter clogging was prevented by an automatic backwash driven by biogas release. This fluidizes the granular filter bed in a downward direction. The BFBR gave a COD removal of >85% at an OLR of 10 kg COD m^{-3} day^{-1}. The treated wastewater had 120 mg COD l^{-1}.

7.3.4.2.2 Fixed-bed reactors. This design is very similar to the AF but contains an open cylindrical permanent porous material to which the biomass can attach to and still remain in close contact with the passing wastewater. This design offers the advantage of simplicity of construction, elimination of stirring, better stability at higher loading rates and capacity to withstand toxic and environmental shocks (Rajeshwari *et al.* 2000). The fixed beds can be operated in an upflow or downflow configuration (De Haast *et al.* 1986).

Cheese whey at an OLR of 3.8 kg COD m^{-3} day^{-1} was treated in a lab-scale fixed-bed reactor equipped with an inert polyethylene carrier (De Haast *et al.* 1986). The best COD removal efficiency of 85–87% was obtained at an HRT of 3.5 days with biogas yields of 0.42 m^{-3} kg^{-1} COD_{added} and a methane content of 55–60%. Sixty-three per cent of the calorific value of the substrate was conserved in the methane. A similar design was used by Cánovas-Diaz and Howell (1987a) to treat deproteinized cheese whey with an average COD of 59,000 mg l^{-1}. At an OLR of 12.5 kg COD m^{-3} day^{-1} the digester achieved a COD reduction of 90–95% at an HRT of 2.0–2.5 days. The deproteinized cheese whey had an average pH of 2.9, while the digester pH was consistently above pH 7.0 (Cánovas-Diaz and Howell 1987b).

7.3.4.2.3 Fluidized and expanded-bed reactors. These bioreactor configurations are designed so that wastewaters pass upwards through a bed of bacterial attached suspended media (Switzenbaum and Jewell 1980). The carrier medium is constantly kept in suspension by powerful recirculation of the liquid phase. The carrier media include plastic granules, sand particles, glass beads, clay particles and activated charcoal fragments. Factors that contribute to the effectiveness of the process include: maximum contact between the liquid and the fine particles carrying the bacteria; avoidence of problems of channelling, plugging and gas hold-up and optimization of the biological film thickness is by the recirculation flow (Sahm 1984). OLRs of 1–20 kg m^{-3} day^{-1} COD can be achieved with COD removal efficiencies of 80–87% at treatment temperatures of 20–35°C. On the practical side Toldrá *et al.* (1987) used the fluidized process to treat dairy wastewater with a COD of only 200–500 mg l^{-1} at an HRT of 8.0 h and obtained a COD removal of only 80%. Bearing in mind the wide variations found between different dairy wastewaters, it can be deduced that this particular wastewater is at the bottom end of the scale in terms of its COD concentration. The dairy wastewater probably was produced by a dairy with very good product-loss control and rather high water use (Strydom *et al.* 2001).

7.3.4.2.4 Upflow anaerobic sludge blanket. This UASB reactor design has successfully been used on full-scale for dairy wastewater treatment for more than two decades (IDF 1990; Lettinga and Hulshoff Pol 1991). The design of the UASB is based on the settling properties of a granular sludge (Lettinga *et al.* 1980; Britz *et al.* 2002). The presence of granules in the UASB system ultimately serves to separate the HRT from the solids retention time (SRT). Thus, good granulation is essential to achieve a short HRT without inducing biomass washout. The wastewater is fed from below and leaves at the top via an internal baffle system for separation of the gas, granular sludge and liquid phases. With the internal baffle system, the granular sludge and biogas are separated. Under optimal conditions, a COD-loading of 30 kg m^{-3} day^{-1} can be applied with a COD removal efficiency of 85–95%. The methane content of the biogas is 80–90% (v/v). With excellent settling, granular sludge HRTs of as little as 4 h and SRTs of more than 100 days are feasible. The treatment temperature ranges from 7 to 60°C with the optimum at 35°C.

Goodwin *et al.* (1990) evaluated the UASB process to treat a synthetic ice-cream wastewater. The UASB was operated at an HRT of 18.4 hr and obtained a total organic carbon (TOC) removal of 86% at an OLR of 3.06 kg TOC m^{-3} day^{-1}. Cheese wastewater has also been treated in the UASB digester at a cheese factory in Wisconsin, USA (De Man and De Bekker 1986). The UASB was operated at a HRT of 16 h and an OLR of 49.5 kg COD m^{-3} day^{-1} with a plant wastewater COD of 33,000 mg l^{-1}, and a COD removal of 86% was achieved. The UASB digester was, however, only a part of a complete full-scale treatment plant. The wastewater from the UASB was recycled to a mixing tank, which also received the incoming wastewater. Although the system is described as an UASB system, it could also pass as a separated or two-phase system, since some degree of pre-acidification is presumably attained in the mixing tank. Furthermore, the pH in the mixing tank was controlled by means of lime dosing, when necessary. The wastewater emerging from the mixing tank was treated in an aerobic system, serving as a final polishing step, to provide an overall COD removal of 99%.

One Finnish full-scale UASB treatment plant (IDF 1990) at the Mikkeli Cooperative Dairy produces Edam type cheese, butter, pasteurized and sterilized milk, and has a wastewater volume of 165 million litres per year. The digester has an operational volume of 650 m^3, which includes a balancing tank of 300 m^3 (Ikonen *et al.* 1985; Carballo-Caabeira 1990). The COD value was reduced by 70–90%, and 400 m^3 biogas is produced daily, with a methane content of 70%, which is used to heat process water in the plant.

Over the last few years, a new generation of more advanced anaerobic reactor systems based on the UASB concept have been developed. A successful version is the internal circulation (IC) reactor (Demirel *et al.* 2005). The IC reactor system is able to handle high upflow liquid and gas velocities, which enables treatment of low-strength wastes at very short HRTs as well as treatment of high-strength wastewaters at high volumetric loading rates feasible. Recently, Ramasamy *et al.* (2004) evaluated the feasibility of using UASB reactors for dairy wastewater treatment at an HRT of 3–12 h and an OLR of 2.4–13.5 kg COD m^{-3} day^{-1}). At an HRT of 3 h, the COD reduction obtained was 95–96%.

One of the most successful full-scale 2000 m^3 UASBs described in the literature was in the UK at South Caernarvon Creameries, to treat whey and other wastewaters (Anon. 1984). The whey alone reached volumes of up to 110,000 l per day. In the system, which included a combined UASB and aerobic denitrification system, COD was reduced by 95%, and sufficient biogas was produced to meet the total energy need of the whole plant. The final wastewater passed to a sedimentation tank, which removed suspended matter. From there, it flowed to aerobic tanks where the BOD was reduced to 20.0 mg l^{-1} and the NH_3-nitrogen reduced to 10.0 mg l^{-1}. The wastewater was finally disposed of into a nearby river. The whey disposal costs, which originally amounted to £30,000 per year, were reduced to zero; the biogas also replaced heavy fuel oil costs. On full output, the biogas had a value of up to £109,000 per year as an oil replacement and a value of about £60,000 as an electricity replacement. These values were, however, calculated in terms of the oil and electricity prices of 1984, but this illustrates the economic potential of the anaerobic digestion process.

A more recent high-rate digester design was that of the EGSB (Lettinga 1995). This is a modified UASB design where a higher liquid velocity is applied (5 ± 10 m h^{-1} as compared with 3 m h^{-1} for soluble wastewater and 1 ± 1.25 m h^{-1} for partially soluble wastewater in an UASB). As a result of the higher upflow velocities, mainly the granular sludge will be retained in an EGSB system, whereas a large portion of the granular sludge bed will be in an expanded or possibly even in a fluidized state in the higher regions of the bed. The contact between the wastewater and sludge is thus enhanced. A further advantage is that the movement of the substrate into the granules is faster when compared with the normal UASB where the mixing intensity is lower. The maximum achievable loading rate in EGSB is higher than that of a conventional UASB design, especially for a low-strength VFA containing wastewater and at lower ambient temperatures.

7.3.4.2.5 Membrane digesters. In a membrane anaerobic reactor system (MARS), the digester wastewater is filtrated by means of a filtration membrane. The use of membranes enhances biomass retention and immediately separates the HRT from the SRT (Strydom *et al.* 2001).

Li and Corrado (1985) evaluated a completely mixed MARS digester with an operating volume of 37,850 l (combined with a micro-filtration membrane system) on cheese whey with a COD of up to 62,000 mg l^{-1}. The digester wastewater was filtered through the membrane and the permeate discharged, while the retentate, containing biomass and suspended solids, was returned to the digester. The COD removal was 99.5% at an HRT of 7.5 days. The most important conclusion the authors made was that the process control parameters obtained in the pilot plant could be effectively applied to their full-scale demonstration plant.

A similar membrane system, the anaerobic digestion-ultra-filtration system (ADUF) has successfully been used in bench and pilot-scale studies on dairy wastewaters (Ross *et al.* 1989). The ADUF system does not use micro-filtration but rather uses an ultra-filtration membrane; and so a far greater biomass retention efficiency is possible.

7.3.4.3 Separated phase digesters

The design of the separated phase anaerobic digesters is specifically to spatially separate acid-forming bacteria from the acid-consuming bacteria. The performance of an acidogenic reactor is especially important during two-phase anaerobic stabilization of wastes, since the acid reactor provides the most appropriate substrate for the subsequent methanogenic digester (Strydom *et al.* 1995). Guidelines for the design of preacidification digesters have previously been published (Demirel *et al.* 2005). These separated phase digesters are especially useful for the treatment of wastewaters where unbalanced carbon-to-nitrogen (C:N) ratios are found. These include wastewaters with high protein levels, or wastes such as dairy wastewaters that acidify quickly (IDF 1990; Strydom *et al.* 2001). High OLRs and short HRTs are the major advantages of the separated phase digester design.

Numerous studies using separated phase digesters are found in the literature (Strydom *et al.* 2001; Britz *et al.* 2004; Demirel *et al.* 2005). In one of these studies, anaerobic digestion of three different dairy wastewaters from cheese, fresh milk and milk powder/butter factories was evaluated using a laboratory-scale mesophilic two-phase system (Strydom *et al.* 2001). For the cheese-factory wastewater, 97% COD removal was achieved at an OLR of 2.82 kg COD m^{-3} day^{-1}, while at an OLR of 2.44 kg COD m^{-3} day^{-1}, 94% COD removal was obtained for the fresh-milk wastewater. For the powder milk/butter factory wastewater, 91% COD removal was achieved at an OLR of 0.97 kg COD m^{-3} day^{-1}. It was the opinion of Kisaalita *et al.* (1987) that two-phase digesters are especially successful in the treatment of lactose containing dairy wastes. The researchers studied the acidogenic fermentation of lactose, determined the kinetics of the process and also found that the presence of whey protein had little influence on the kinetics of lactose acidogenesis. Venkataraman *et al.* (1992) also used a two-phase packed-bed anaerobic filter system to treat dairy wastewater. Their main goals were to determine the kinetic constants for biomass and biogas production rates and substrate utilization rates in this configuration.

According to Burgess (1985), the digester design where the acid-forming and acid-consuming bacteria were separated has successfully been used on full-scale to treat dairy wastewaters. In one case, a dairy had a wastewater with a COD of 50,000 mg l^{-1} and a pH of 4.5. Both digester phases were operated at 35°C, while the acidogenic reactor was operated at an HRT of 24 h and the methanogenic reactor at a HRT of 3.3 days. In the acidification tank, 50% of the COD was converted to organic acids, while only 12% of the COD was removed. The OLR for the acidification reactor was 50.0 kg COD m^{-3} day^{-1} and, for the methane reactor, 9.0 kg COD m^{-3} day^{-1}. An overall COD reduction of 72% was achieved. The biogas had a methane content of 62%, and from the data it was calculated that a methane yield ($Y_{CH4}/COD_{removed}$) of 0.327 m^3 kg $COD_{removed}$ was obtained.

In another case, Lo and Liao (1986, 1988) used anaerobic rotating biological contact reactors (AnRBC) as the separate digesters to treat cheese whey. The COD of the whey was 6720 mg l^{-1} and was diluted approximately 10-fold before being used as substrate. The digesters were of a tubular fixed-film design and orientated horizontally, with internally rotating baffles. The acidogenic phase reactor was mixed by the recirculation of the biogas. However, it achieved a COD reduction of only 4%. More importantly, the total volatile fatty acids concentration was increased from 168 to 1892 mg l^{-1}. This was then used as a substrate for the second digester where a COD reduction of up to 87% was achieved.

7.3.5 Chemical systems

7.3.5.1 Chemical treatments

Chemical-treatment methods rely on chemical reactions to bring about changes in water quality (Tchobanoglous and Schroeder 1985). This is one of the inherent disadvantages associated with chemical treatment processes (activated carbon adsorption is an exception). In most cases, something is added to the wastewater to achieve the removal of something else (Tchobanoglous and Burton 1991). The most important chemical-treatment methods are those used for disinfection, precipitation of dissolved materials, coagulation of colloids, oxidation and activated-carbon adsorption.

Chemical additions to attain disinfection are used in the treatment of both domestic and industrial wastewaters. Industrially, disinfection is mainly used to control biological slime buildup in piping and to lower bacterial loads during dairy processing. In contrast, precipitation is used domestically and industrially for water softening and iron removal and for the removal of organics and soluble ions such as phosphates from wastewaters. Coagulation is used almost entirely for the destabilization of colloids found in surface waters (Tchobanoglous and Burton 1991).

Activated carbon, in either granular or powder form, is also used to treat wastewater, often in combination with other processes (i.e. activated sludge process). Activated carbon, due to the large surface area, can absorb large amounts of compounds, mainly refractory organics and inorganic compounds such as nitrogen, sulfides and heavy metals (Tchobanoglous and Burton 1991). Carbon adsorption led to significant COD removal efficiencies with final levels in the range of 10–20 mg l^{-1}. However, high influent SS concentrations (more than 20 mg l^{-1}) will lead to a loss in adsorption due to deposits forming on the carbon. This leads to pressure loss, flow channelling or blockages. Lack of consistency in pH, temperature and flow rate may also affect the performance of carbon adsorption (Tchobanoglous and Burton 1991). Thus, the main disadvantage of chemical treatment remains the fact that costs are incurred when adding chemicals to wastewaters on a continuous basis, and in many cases, additional costs are incurred with the disposal of the final precipitate.

Chemical oxidation, which includes chlorination and ozonation, has also been used in the removal or breakdown of ions such as iron, manganese and cyanide (Tchobanoglous and Burton 1991). Iron and manganese are much more soluble in the +2 state than in the +3 state and are easily oxidized with oxygen. Thus, their removal is a combination oxidation–precipitation process. Cyanide removal involves oxidation to the innocuous end products CO_2 and N_2 (Tchobanoglous and Schroeder 1985). The uses of ozone (O_3), hydrogen peroxide (H_2O_2), ultraviolet light (UV) and certain catalysts in wastewater treatment are generally classified as advanced oxidation processes.

7.3.5.2 Oxidation technology

Two of the strongest chemical oxidants are ozone and hydroxyl radicals (Mourand *et al.* 1995; Droste 1997; Hoigné 1997). Ozone can react directly with a compound, or it can produce hydroxyl radicals, which then react with a compound. Owing to the low reactivity of ozone towards the majority of organic compounds, degradation by ozone is rather selective. When a more general oxidant is required, the OH radical is the reagent of choice. It is the most reactive oxidant that can be applied in water treatment (Beschkov *et al.* 1997; Hoigné 1997). It reacts with most organic compounds at rates which are close to diffusion-controlled.

In water remediation, a number of OH-radical generating systems are currently in use, or under study for potential future use: ozone/hydrogen peroxide (O_3/H_2O_2), ultraviolet light/hydrogen peroxide (UV/H_2O_2), iron (II)/hydrogen peroxide (Fe^{2+}/H_2O_2), ultraviolet-light/ozone (UV/O_3), ultraviolet light/titanium dioxide (UV/TiO_2) and ionizing radiation (electron beam). Hydroxyl radicals can also be produced by the photolysis of aqueous chlorine, nitrate, nitrite, dissolved aqueous iron (III) and in Fenton reactions. Advanced oxidation processes are alternative techniques for catalysing the production of these radicals. When OH radicals react with the organic substrate, radicals are formed which in the majority of cases react with the dissolved O_2 yielding the corresponding peroxyl radical. Much of the oxidative degradation of the organic matter is affected through the ensuing reactions of the peroxy radicals (Mourand *et al.* 1995; Hoigné 1997; Von Sonntag *et al.* 1997; Gottschalk *et al.* 2000).

7.4 Conclusions

Growth in the dairy and other agro-industries has exacerbated the pressure which the industry exerts on natural resources. This growth has occurred at a time when international legislation (EIPPCB 2005) and local and foreign markets are becoming increasingly stringent in their demands that all factors which have the potential to affect the environment be controlled. Traditionally, in many European countries, the dairy sector has not been overly regulated by environmental legislation, but there is now a trend towards proactive implementation of environmental management systems and the implementation of waste minimization techniques (IDF 2003). In the agro-industries, both primary production and processing are critically dependent on a reliable water source and water quality in conformity with statutory requirements. Now, water consumption and wastewater production have become one of the key environmental issues for the dairy sector as most of the water not used as an ingredient ultimately appears as highly polluted wastewater. Sadly, as the dairy industry is a 'market-driven' industry, many dairy-processing plants still perceive waste management to be a necessary 'evil' that ties up valuable capital which could be better utilized for core business activity. More stringent environmental legislation and escalating costs for the purchase of freshwater as well as the economic impact of setting up and operating an efficient treatment plant have increased the impetus on improving waste control. As environmental management of dairy wastes becomes an ever-increasing concern, treatment strategies will need to be based on state and local regulations. It is thus essential that environmental management becomes a common activity of all dairy-processing plants. More positively for the dairy industry, as the IDF (2003) has recently pointed out, there are rewards for implementing a waste-management strategy, which may include easier compliance with legislation, economic gains from more efficient use of resources, lower freshwater purchase and lower compliance costs. A further incentive identified by the IDF (2003) is the good publicity, which is important for a consumer product such as milk.

The specific treatment level is normally dictated by environmental regulations applicable to the specific area. As the dairy industry is a major user and generator of water, it is a prime candidate for efficient waste management and water reuse. Even if the purified wastewater is initially not reused, the dairy industry will still benefit from in-house waste management, since reducing waste at the source can only help in reducing costs or improving the performance of any downstream treatment facility.

All wastewater-treatment systems are unique. Before selecting any treatment method, a complete process evaluation should be undertaken along with economical analysis. This should include the wastewater composition, concentrations, volumes generated and treatment susceptibility as well as the environmental impact of the solution to be adopted (Britz *et al.* 2004). All options are expensive, but an economical analysis may indicate that slightly higher maintenance costs may be lower than the increased operating costs. What is appropriate for one site may be unsuitable for another. The most useful processes are those that can be operated with a minimum of supervision and are inexpensive to construct or even mobile enough to be moved from site to site. The varying quantity and quality of dairy wastewater must also be included in the design and operational procedures. Biological methods appear to be the most cost-effective organic removers, with aerobic methods easier to control, but anaerobic methods having lower energy requirements, smaller footprints and lower sludge production rates. In the anaerobic case, the produced sludge may lead to an excellent economic incentive. Since no single treatment process is capable of producing final water that complies with minimum effluent discharge requirements, it is necessary to consider combining processes to solve specific dairy wastewater problems.

References

Al-Khateeb, B.M.A. and Tebbutt, T.H.Y. (1992) The effect of physical configuration on the performance of laboratory-scale oxidation ponds. *Water Research* **26**, 1507–1513.

Anon. (1984) South Caernarvon Creameries converts whey into energy. *Dairy Industries International* **49**, 16–17.

Azad, H.S. (1976) *Industrial Wastewater Management Handbook*. McGraw-Hill, New York. pp. 4.1–5.49.

Bakka, R.L. (1992) Wastewater issues associated with cleaning and sanitizing chemicals. *Dairy, Food and Environmental Sanitation* **12**, 274–276.

Barnett, J.W., Kerridge, G.J. and Russell, J.M. (1994) Effluent treatment systems for the dairy industry. *Australian Biotechnology* **4**, 26–30.

Beschkov, V., Bardarska, G., Gulyas, H. and Sekoulov, I. (1997) Degradation of triethylene glycol dimethyl ether by ozonation combined with UV irradiation or hydrogen peroxide addition. *Water Science and Technology* **36**, 131–138.

Bitton, G. (1999) *Wastewater Microbiology*, pp. 155–426. John Wiley, New York.

Bonastre, N., Paris, J.M. (1989) Survey of laboratory, pilot and industrial anaerobic filter installations. *Process Biochemistry* **24**, 15–20.

Britz, T.J., Van Sckalkwyk, C. and Hung, Y-T. (2004) Treatment of dairy processing wastewaters. In: Wang, L. K., Hung, Y-T., Lo, H.H. and Yapijakis, C. (eds) *Handbook of Industrial and Hazardous Wastes Treatment,* 2nd edn, pp. 619–646. Marcel Dekker, New York.
Britz, T.J., Van Schalkwyk, C. and Roos, P. (2002) Method for the enhancement of granule formation in batch systems. *Water SA,* **28**, 49–54.
Burgess, S. (1985) Anaerobic treatment of Irish creamery effluents. *Process Biochemistry* **20**, 6–7.
Burton, C. (1997) FOG clearance. *Dairy Industries International* **62**, 41–42.
Cammarota, M.C., Teixeira, G.A. and Freire, D.M.G. (2001) Enzymatic pre-hydrolysis and anaerobic degradation of wastewaters with high fat contents. *Biotechnology Letters* **23**, 1591–1595.
Cánovas-Diaz, M. and Howell, J.A. (1987a) Down-flow anaerobic filter stability studies with different reactor working volumes. *Process Biochemistry* **22**, 181–184.
Cánovas-Diaz, M. and Howell, J.A. (1987b) Stratified ecology techniques in the start-up of an anaerobic down-flow fixed film percolating reactor. *Biotech. Bioeng.* **10**, 289–296.
Carballo-Caabeira, J. (1990) Depuracion de aguas residuales de centrales lecheras. *Rev. Española de Lech.* **13**, 13–16.
Danalewich, J.R., Papagiannis, T.G., Belyea, R.L., Tumbleson, M.E. and Raskin, L. (1998) Characterization of dairy waste streams, current treatment practices, and potential for biological nutrient removal. *Water Research* **32**, 3555–3568.
De Haast, J., Britz, T.J. and Novello, J.C. (1986) Effect of different neutralizing treatments on the efficiency of an anaerobic digester fed with deproteinated cheese whey. *Journal of Dairy Research* **53**, 467–476.
De Man, G. and De Bekker, P.H.A.M.J. (1986) New technology in dairy wastewater treatment. *Dairy Industries International* **51**, 21–25.
Demirel, B., Yenigun, O. and Onay, T.T. (2005) Anaerobic treatment of dairy wastewaters: a review. *Process Biochemistry* **40**, 2583–2595.
Droste, R.L. (1997) *Theory and Practice of Water and Wastewater Treatment.* John Wiley, New York. pp. 670–708.
Eckenfelder, W.W. (ed.) (1989) *Industrial Water Pollution Control.* McGraw-Hill, New York.
EIPPCB. (2005) *Draft Reference Document on Best Available Techniques in the Food, Drink and Milk Industries.* Final Draft June 2005. European Commission, European Integrated Pollution Prevention and Control Bureau. Edificio Expo, Seville, Spain.
Eroglu, V., Ozturk, I., Demir, I. and Akca, A. (1992) Sequencing batch and hybrid reactor treatment of dairy wastes. In: *Proceedings of the 40th Industrial Waste Conference, Purdue,* West Lafayette, IN, pp. 413–420.
Feilden, N.E.H. (1983) The theory and practice of anaerobic reactor design. *Process Biochemistry* **18**, 34–37.
Garrido, J.M., Omil, F., Arrojo, B., Méndez, R. and Lema, J.M. (2001) Carbon and nitrogen removal from a wastewater of an industrial dairy laboratory with a coupled anaerobic filter-sequencing batch reactor system. *Water Science and Technology* **43**, 249–256.
Gilde, L.C. and Aly, O.M. (1976) Water pollution control in the food industry. In: Azad, H.S. (ed.). *Industrial Wastewater Management Handbook*, pp. 5–51. McGraw-Hill, New York.
Goodwin, J.A.S., Wase, D.A.J. and Forster, C.F. (1990) Anaerobic digestion ice-cream wastewaters using the UASB process. *Biological Wastes* **32**, 125–144.
Gottschalk, C., Libra, J.A. and Saupe, A. (2000) *Ozonation of Water and Waste Water*, pp. 5–36. Wiley-VCH, Weinheim, Germany.
Gough, R.H. and McGrew, P. (1993) Preliminary treatment of dairy plant waste water. *Journal of Environmental Science and Health* **A28**, 11–19.
Graz, C.J.M. and McComb, D.G. (1999) Dairy CIP – A South African review. *Dairy, Food and Environmental Sanitation* **19**, 470–476.
Greary, P.M. and Moore, J.A. (1999) Suitability of a treatment wetland for dairy wastewaters. *Water Science and Technology* **40**, 179–185.

Green, J.H. and Kramer, A. (1979). *Food Processing Waste Management*, pp. 339–498. Avi, Westport, CT.

Grismer, M.E., Ross, C.C., Edward Valentine, G., Smith, B.M. and Walsh, J.L. (2002) Food processing wastes. *Water Environment Research* **74**, 377–378.

Hale, N., Bertsch, R., Barnett, J. and Duddleston, W.L. (2003) Sources of wastage in the dairy industry. In: *Guide for Dairy Managers on Wastage Prevention in Dairy Plants,* pp. 7–30. IDF Bulletin Document 382, International Dairy Federation, Brussels.

Hammer, M.J. (1986) *Water and Waste-Water Technology*. John Wiley, New York. pp. 345–437.

Haridas, A., Suresh, S., Chitra, K.R. and Manilal, V.B. (2005) The buoyant filter bioreactor: a high-rate anaerobic reactor for complex wastewater process dynamics with dairy effluent. *Water Research* **39**, 993–1004.

Hayward, D.J., Lorenzen, L., Bezuidenhout, S., Barnardt, N., Prozesky, V. and van Schoor, L. (2000) Environmental compliance or complacency – can you afford it? Modern trends in environmental management for the wine industry. *WineLand* **69**, 99–102.

Hemming, M.L. (1981) The treatment of dairy wastes. In: Herzka, A. and Booth, R.G. (eds), *Food Industry Wastes: Disposal and Recovery,* pp. 109–121. Applied Science, London.

Herzka, A. and Booth, R.G. (1981) *Food Industry Wastes: Disposal and Recovery*. Applied Science, London.

Hoigné, J. (1997) Inter-calibration of OH radical sources and water quality parameters. *Water Science and Technology* **35**, 1–8.

Holdsworth, S.D. (1970) Effluents from fruit and vegetable processing. *Effluents and Water Treatment Journal* **3**, 131–268.

IDF (1984) Balance tanks for dairy effluent treatment plants. *Bulletin of the International Dairy Federation Document No. 174.*

IDF (1990) Anaerobic treatment of dairy effluents – The present stage of development. *Bulletin of the International Dairy Federation Document No.* 252.

IDF (1997) Removal of fats, oils and grease in the pre-treatment of dairy wastewaters. *Bulletin of the International Dairy Federation Document No. 327.*

IDF (2003) Guide for dairy managers on wastage prevention in dairy plants. *Bulletin of the International Dairy Federation Document No. 382.*

Ikonen, M., Latola, P., Pankakoski, M. and Pelkonen, J. (1985) Anaerobic treatment of waste water in a Finnish dairy. *Norddisk Mejeriindustri* **12 (8)**, 81–82.

Jameson, G.J. (1999). Hydrophobicity and floc density in induced-air flotation for water treatment. *Colloids and Surfaces A: Physicochemical and Engineering Aspects* **151**, 269–281.

Karpiscak, M.M., Freitas, R.J., Gerba, C.P., Sanchez, L.R. and Shamir, E. (1999) Management of dairy waste in the Sonoran Desert using constructed wetland technology. *Water Science and Technology* **40**, 57–65.

Kessler, H.G. (1981) *Food Engineering and Dairy Technology*. Verlag A. Kessler, Freisburg, Germany.

Kilani, J.S. (1992) Studies on the treatment of dairy wastes in an algal pond. *Water SA* **18**, 57–62.

Kim, Y., Pipes, W.O. and Chung, P-G. (1998) Control of activated sludge bulking by operating clarifiers in a series. *Water Science and Technology* **38**, 1–8.

Kisaalita, W.S., Pinder, K.L. and Lo, K.V. (1987) Acidogenic fermentation of lactose. *Biotechnology and Bioengineering* **30**, 88–95.

Leal, M.C.M.R., Cammarota, M.C., Freire, D.M.G. and Sant'Anna, G.L. (2002) Hydrolytic enzymes as coadjuvants in the anaerobic treatment of dairy wastewaters. *Brazilian Journal of Chemical Engineering* **19**, 175–180.

Lebrato, J., Perez-Rodriguez, J.L., Maqueda, C. and Morillo, E. (1990) Cheese factory wastewater treatment by anaerobic semicontinuous digestion. *Resources, Conservation and Recycling* **3**, 193–199.

Lettinga, G. (1995) Anaerobic reactor technology. Lecture notes by Prof. G. Lettinga. In: *International Course on Anaerobic Wastewater Treatment*. Wageningen Agriculture University, Delft, Netherlands.

Lettinga, G. and Hulshoff Pol, L.W. (1991) UASB-process design for various types of wastewaters. *Water Science and Technology* **24**, 87–107.

Lettinga, G., Van Velsen, A.F.M., Hobma, S.W., De Zeeuw, W. and Klapwijk, A. (1980) Use of the upflow sludge blanket (USB) reactor concept for biological wastewater treatment especially for anaerobic treatment. *Biotechnology and Bioengineering* **22**, 699–734.

Li, A.Y. and Corrado J.J. (1985) Scale up of the membrane anaerobic reactor system. In: *Proceedings of the 40th Industrial Waste Conference, Purdue,* West Lafayette, IN, pp. 399–404.

Li, X. and Zhang, R. (2002) Aerobic treatment of dairy wastewater with sequencing batch reactor systems. *Bioprocess and Biosystems Engineering* **25**, 103–109.

Lo, K.V. and Liao, P.H. (1986) Digestion of cheese whey with anaerobic rotating biological contact reactor. *Biomass* **10**, 243–252.

Lo, K.V. and Liao, P.H. (1988) Laboratory scale studies on the mesophilic anaerobic digestion of cheese whey in different digester configurations. *Journal of Agricultural Engineering Research* **39**, 99–105.

Mantovi, P., Marmiroli, M., Maestri, E., Tagliavini, S., Piccinini, S. and Marmiroli, N. (2003) Application of a horizontal subsurface flow constructed wetland on treatment of dairy parlour wastewater. *Bioresource Technology* **88**, 85–94.

Maris, P.J., Harrington, D.W., Biol, A.I. and Chismon, G.L. (1984) Leachate treatment with particular reference to aerated lagoons. *Water Pollution Control* **83**, 521–531.

Mourand, J.T., Crittenden, J.C., Hand, D.W., Perram, D.L. and Notthakun, S. (1995) Regeneration of spent adsorbents using homogeneous advanced oxidation. *Water Environmental Research* **67**, 355–363.

Nadais, H., Capela, I., Arroja, L. and Duarte, A. (2001) Effects of organic, hydraulic and fat shocks on the performance of UASB reactors with intermittent operation. *Water Science and Technology* **44**, 49–56.

Nameche, T.H. and Vasel, J.L. (1998) Hydrodynamic studies and modelization for aerated lagoons and waste stabilization ponds. *Water Research* **32**, 3039–3045.

Nazaroff, W.W. and Alvarez-Cohen, L. (2001) *Environmental Engineering Science*, pp. 306–364, 555–558. John Wiley, New York.

Newman, J.M., Clausen, J.C. and Neafsey, J.A. (2000) Seasonal performance of a wetland constructed to process dairy milkhouse wastewater in Connecticut. *Ecological Engineering* **14**, 181–198.

Odlum, C.A. (1990) Reducing the BOD level from a dairy processing plant. In: *Proceedings of the 23rd International Dairy* Congress, pp. 835–851, October, Montreal, Canada.

Pearson, H.W. (1996) Expanding the horizons of pond technology and application in an environmentally conscious world. *Water Science and Technology* **33**, 1–9.

Rajeshwari, K.V., Balakrishnan, M., Kansal, A., Lata, K. and Kishore, V.V.N. (2000) State-of-the-art of anaerobic digestion technology for industrial wastewater treatment. *Renewable and Sustainable Energy Reviews* **4,** 135–156.

Ramasamy, E.V. and Abbasi, S.A. (2000) Energy recovery from dairy waste waters: impacts of biofilm support on anaerobic CST reactors. *Applied Energy* **65**, 91–98.

Ramasamy, E.V., Gajalakshmi, S., Sanjeevi, R., Jithesh, M.N. and Abbasi, S.A. (2004) Feasibility studies on the treatment of dairy wastewaters with upflow anaerobic sludge blanket reactors. *Bioresource Technology* **93,** 209–212.

Robinson, T. (1994) How to be affluent with effluent. *Milk Industry* **96**, 20–21.

Ross, W.R. (1989) Anaerobic treatment of industrial effluents in South Africa. *Water SA* **15**, 231–246.

Ross, W.R. (1991) Anaerobic digestion of industrial effluents with emphasis on solids–liquid separation and biomass retention. Ph.D. thesis, University of the Orange Free State Press, Bloemfontein, South Africa.

Ross, W.R., Barnard, J.P. and De Villiers, H.A. (1989) The current status of ADUF technology in South Africa. In: *Proceedings of the 2nd Anaerobic Digestion Symp*, pp. 65–69. University of the Orange Free State Press, Bloemfontein, South Africa.

Rusten, B., Eikebrokk, B. and Thorvaldsen, G. (1990) Coagulation as pre-treatment of food industry wastewater. *Water Science and Technology* **22**, 1–8.

Rusten, B., Lundar, A., Eide, O. and Ødegaard, H. (1993) Chemical pre-treatment of dairy wastewater. *Water Science and Technology* **28**, 67–76.

Rusten, B., Odegaard, H. and Lundar, A. (1992) Treatment of dairy wastewater in a novel moving-bed biofilm reactor. *Water Science and Technology* **26**, 703–709.

Sahm, H. (1984) Anaerobic wastewater treatment. *Advances in Biochemical Engineering and Biotechnology* **29**, 83–115.

Samkutty, P.J., Gough, R.H. and McGrew, P. (1996) Biological treatment of dairy plant wastewater. *Journal of Environmental Science and Health* **A31**, 2143–2153.

Schaafsma, J.A., Baldwin, A.H. and Streb, C.A. (2000) An evaluation of a constructed wetland to treat wastewater from a dairy farm in Maryland, USA. *Ecological Engineering* **14**, 199–206.

Schroepfer, G.J., Fuller, W.J., Johnson, A.S., Ziemke, N.R. and Anderson, J.J. (1955) The anaerobic contact process as applied to packinghouse wastes. *Sewage and Industrial Wastes* **27**, 460–486.

Steffen, Robertson and Kirsten Inc. (1989) *Water and Wastewater Management in the Dairy Industry. WRC Project No. 145.* Water Research Commission, Pretoria, South Africa.

Sterrit, R.M. and Lester, J.N. (1982) Biological treatment of fruit and vegetable processing waste water. *Science and Technology Letters* **3**, 63–68.

Strydom, J.P., Mostert, J.F. and Britz, T.J. (1995) Anaerobic treatment of a synthetic dairy effluent using a hybrid digester. *Water SA* **21**, 125–130.

Strydom, J.P., Mostert, J.F. and Britz, T.J. (2001) *Anaerobic Digestion of Dairy Factory Effluents. WRC Report No. 455/1/01.* Water Research Commission, Pretoria, South Africa.

Switzenbaum, M.S. and Jewell, W.J. (1980) Anaerobic attached-film expanded-bed reactor treatment. *Journal of the Water Pollution Control Federation* **52**, 1953–1965.

Tamime, A.Y. and Robinson, R.K. (eds.) (1999) *Yoghurt Science and Technology*. Woodhead, Cambridge.

Tanner, C.C., Clayton, J.S. and Upsdell, M.P. (1994) Effect of loading rate and planting on treatment of dairy farm wastewaters in constructed wetlands – I. Removal of oxygen demand, suspended solids and faecal coliforms. *Water Research* **29**, 17–26.

Tchobanoglous, G. and Burton, F.L. (1991) *Wastewater Engineering: Treatment, Disposal, and Reuse,* 3rd edn. McGraw-Hill, New York.

Tchobanoglous, G. and Schroeder, E.D. (1985) *Water Quality*. Addison-Wesley, Reading, MA.

Tetra Pak (2003) *Dairy Processing Handbook*. Tetra Pak Processing Systems, Lund, Sweden.

Toldrá, F., Flors, A., Lequerica, J.L. and Vall, S.S. (1987) Fluidized bed anaerobic biodegradation of food industry wastewaters. *Biological Wastes* **21**, 55–61.

Torrijos, M., Vuitton, V. and Moletta, R. (2001) The SBR process: an efficient and economic solution for the treatment of wastewater at small cheese making dairies in the Jura Mountains. *Water Science and Technology* **43**, 373–380.

Trivedy, R.K. and Pattanshetty, S.M. (2002) Treatment of dairy waste by using water hyacinth. *Water Science and Technology* **45,** 329–334.

Venkataraman, J., Kaul, S.N. and Satyanarayan, S. (1992) Determination of kinetic constants for a two-stage anaerobic up-flow packed bed reactor for dairy wastewater. *Bioresource Technology* **40**, 253–261.

Vidal, G., Carvalho, A., Méndez, R. and Lema, J.M. (2000) Influence of the content in fats and proteins on the anaerobic biodegradability of dairy wastewaters. *Bioresource Technology* **74**, 231–239.
Von Sonntag, C., Dowideit, P., Fang, X., Mertens, R., Pan, X., Schuchmann, M.N. and Schuchmann, H-P. (1997) The fate of peroxyl radicals in aqueous solution. *Water Science and Technology* **35**, 9–15.
Wayman, M.J.V. (1996) Water supplies, effluent disposal and other environmental considerations. In: Arthey, D. and Ashurst, P.R. (eds) *Fruit Processing*, pp. 221–243. Blackie Academic & Professional, London.
Wendorff, W.L. (2001). Treatment of dairy wastes. In: Marth, E.H. and Steele, J.L. (eds.) *Applied Dairy Microbiology*, 2nd edn, pp. 681–704. Marcel Dekker, New York.
Wendorff, W.L., Westphal, S.J. and Yau, J.C.Y. (1997) Phosphorus reduction in dairy wastes by conservation of burst rinses. *Dairy, Food and Environmental Sanitation* **17**, 72–75.

Index